Aviation and International Cooperation

Ruwantissa Abeyratne

Aviation and International Cooperation

Human and Public Policy Issues

Springer

Ruwantissa Abeyratne
Global Aviation Consultancies Inc.
Cote Saint Luc, Québec
Canada

ISBN 978-3-319-17021-3 ISBN 978-3-319-17022-0 (eBook)
DOI 10.1007/978-3-319-17022-0

Library of Congress Control Number: 2015939050

Springer Cham Heidelberg New York Dordrecht London

Printed on acid-free paper

Springer International Publishing AG Switzerland is part of Springer Science+Business Media (www.springer.com)

Preface

This book was commenced in December 2014 when the world was celebrating 100 years of commercial aviation and 70 years of the existence of the Convention on Civil Aviation signed at Chicago on 7 December 1944. Now, popularly known as the Chicago Convention, this multilateral treaty created the International Civil Aviation Organization (ICAO) which has, as one of its aims and objectives, to meet the needs of the people of the world for safe, regular, efficient and economical air transport. The Chicago Convention, in its *Preamble* says *inter alia* that the future development of international civil aviation can greatly help to create and preserve friendship and understanding among the nations and peoples of the world. These are the only two references in the entirety of the treaty to people. All other references are to States, aircraft, airports and other component elements of the aviation world. Although passengers are mentioned in other provisions of the treaty, such references are always in association with aircraft or States or other instrumentalities. There is a bewildering and startling lack of focus on or attention to the person affected by aviation, and in particular passenger, let alone his or her rights.

A perceived anomaly in aviation and human rights is the obvious deleterious effect aviation has had on people, despite the exhortation in the Chicago Convention on international civil aviation having a benevolent effect on friendship and understanding among people. On numerous occasions, aviation has been used to attack and bombard whole societies and people. This book will address public policy issues on this subject.

Notably, in early December of 2014, there were numerous ceremonies to mark the 70th Anniversary of the Chicago Convention. Books were written where authors representing industry wrote in the Prefaces that air transport generates 580 million jobs and $4.1 trillion dollars in economic activity. One commentator noted that for every $100 earned in aviation an income of $350 was triggered and that every 100 direct jobs in aviation created over 600 indirect jobs. Some wrote of tremendous leaps of humankind in aviation in the 100 years past. These are incontrovertible facts that we can be proud of. However, the fact that they are highlighted at the expense of ignoring the consumer of the air transport product is indeed lamentable.

The Chicago Convention is not about people nor airline passengers. Nor is it about the rights of the passengers. It is about regulating States, airlines, airports and other service providers. Another disturbing fact is that there are no major organizations representing the passenger, as there are for airlines, airports and air navigation service providers.

2014 was also the year when Malaysian Airlines Flight MH 17, operated by a Boeing 777-200ER aircraft flying from Amsterdam to Kuala Lumpur on 17 July 2014, and carrying 283 passengers and 15 crew, was shot down by a BUK surface to air missile over Donetsk Oblast in Eastern Ukraine, while at an altitude of 10,000 m. All those on board perished. It was also the year when, earlier, many of us woke up on a Saturday morning to the disturbing news that a Malaysian Airlines plane, which took off from Kuala Lumpur for Beijing at 12.41 a.m. on Saturday, had lost contact with air traffic control two hours into the flight. The Boeing 777-200 carrying 239 people including 12 crew members carried 14 nationalities. At the time of writing, 9 months later, neither the aircraft or parts thereof, nor any of the passengers or their remains had been found.

If those were not enough, on 28 December of 2014, Air Asia flight QZ 8501 crashed into the Java sea on its way to Singapore, killing all 162 passengers and crew on board. It was later reported that the flight did not have authorization to be operated on the route that day.

At the outset of this book, in Chap. 1, this book discusses Flights MH 370 and MH 17 that resulted in the deaths of hundreds of passengers on board as well as touching hundreds of relatives and several States who lost their nationals, purely because the fundamental human right to life, liberty and security of person was adversely affected through aviation. The book also discusses the manner in which aviation has been used to attack whole communities in States; how the airline passenger's rights are identified and demarcated, as well as the rights of the disabled passenger.

No one would deny the fact that there has been little focus on the main protagonist of the air transport product—the passenger. This book, although bearing reference to consumer protection in general in its title, focuses solely on the rights of the passenger from a human rights perspective. In an earlier work,[1] I have addressed the passenger's rights from the perspectives of providing him or her with connectivity, availability of services, reasonable pricing and value for money, which neither the Chicago Convention nor ICAO addresses adequately. In this book, I draw on the international conventions on human rights as the fundamental postulates that lead to specific air passenger rights; steps that have been taken so far in the air transport context; and measures that could be taken to ensure the

[1] *Regulation of Air Transport – The Slumbering Sentinels*, Springer: Heidelberg, 2013. See also, Ruwantissa Abeyratne, *Aeronomics and Law: Fixing Anomalies*, Springer: Heidelberg, 2012.

implementation of passenger rights appropriately, and in a just and equitable manner.

Therefore, this book also discusses public policy affecting aviation that would bear upon humanity as a whole as well as consumer protection focused on the rights of the airline passenger.

Montreal, Canada
January 2015

Ruwantissa Abeyratne

Contents

Table of Cases

1962 case of *Thornburg v. Port of Portland*, P. 133

Aaron v. City of Los Angeles, 40 Cal.App. 3d. 471, 115 Cal. Rptr. 62 (1974), P. 133
Abdulrahman Al-Zamil v. British Airways Inc. 770 F2d. 3 (2nd Circ. 1985), P. 233
Abnet v. British Airways PLC Lloyd's Aviation Law, Vol. 3, No. 2, 15 January 1992 at 4–5, P. 208
Abramson v. Japan Airlines Company Ltd. 739 F. 2d. 130 (3rd Circ. 1984), P. 235
Air France v. Saks 105 S. Ct. 1338 (1985), P. 233–234
Airlines v. Floyd 499 U.S.530, 111 S.Ct.1489 (1991), P. 249
Aldridge v. Clough, Unreported S.C. Decisions of Western Australia No. 1246 of 1979, P. 136, 139
Archibald v. Pan American World Airways Inc., 460 F.2d 14, 16 (9th Cir. 1972), P. 209
Arkin v. Trans International Airlines Inc. 19 Avi Cas 18, 311 (EDNY 1985), P. 233
Atlantic Mutual Insurance Co. v. R [1919] 1 K.B. 307, P. 28
Azubuko v. Varig Airline Lloyd's Aviation Law, Vol. 15, No. 9 May 11 1995 at 2, P. 207

Baba v. Compagnie Nationale Air France 866 F. Supp. 588 (D.D.C. 1994), P. 230
Bankstown Municipal Council v. Berzins [1962] N.S.W.R. 641, P. 131
Barber v. Penley [1893] 2 Ch. 447, P. 140
Barboni v. Cie Air-France (1982) 36 RFDA 355, P. 219
Barras v. Aberdeen Steam Trawling Co. Ltd. [1933] AC 402, P. 177
Bart v. British West India Airways Ltd. (1967) 1 Lloyds Rep. 239 (Guyana Ct. App. 1966), P. 205
Batten v. United States 306 F.2d. 580 (10th Cir. 1962), P. 133
Bell v. Swiss Air Transport Co. Ltd 25 Avi. Cas (CCH) 17, 259 (Sup. Ct. App. Tm. N.Y. 1st Dep't. 1996) 83 920, P. 231
Bone v. Seale [1975] 1 All.E.R. 787, P. 142–143
Boomer v. Atlantic Cement Company 257 N.E. 2d. 870, P. 138, P. 143

Chapter 1
Public Policy and Human Rights

1.1 Human Rights Covenants Relevant to Aviation

The human being at international law is collectively referred to as "humankind" encompassing all members of the human species as a whole,[1] and it is in this context that the aviation perspective should look at the people affected by it both as a whole and in the singular context of the passenger carried by air. *The Universal Declaration of Human Rights* of the United Nations recognizes the inherent dignity and of the equal and inalienable rights of all members of the human family as the foundation of freedom, justice and peace in the world, and provides that everyone has the right to freedom of movement and residence within the borders of each state and everyone has the right to leave any country, including his own, and to return to his country.[2] The Declaration also provides that everyone has the right to life, liberty and security of person,[3] implying that such rights should not be arbitrarily taken away by any means.

The protection of human rights is the most significant and important task for the modern world, particularly since multi ethnic States are the norm in today's world. The traditional nation State in which a district national group rules over a territorial unit is fast receding to history. Globalization and increased migration across borders is gradually putting an end to the concept of the nation State, although resistance to reality can be still seen in instances where majority or dominant cultures impose their identity and interests on groups with whom they share a territory. In such instances, minorities frequently intensify their efforts to preserve and protect their identity, in order to avoid marginalization.

In the above context, and in the perspective of the misuse of aviation in particular, Article 1 of the United Nations Charter provides that the aim of the

---

[1] Cancado Trinidade (2010), p. 281.

[2] *Universal Declaration of Human Rights*, UN: New York, 10 December 1948, Article 13.

[3] *Id.*, Article 3.

R. Abeyratne, *Aviation and International Cooperation*,
DOI 10.1007/978-3-319-17022-0_1

Charter is to maintain international peace and security, and to that end: to take effective collective measures for the prevention and removal of threats to the peace, and for the suppression of acts of aggression or other breaches of the peace, and to bring about by peaceful means, and in conformity with the principles of justice and international law, adjustment or settlement of international disputes or situations which might lead to a breach of the peace. Another aim of the Charter is to develop friendly relations among nations based on respect for the principle of equal rights and self-determination of peoples, and to take other appropriate measures to strengthen universal peace and to achieve international co-operation in solving international problems of an economic, social, cultural, or humanitarian character, and in promoting and encouraging respect for human rights and for fundamental freedoms for all without distinction as to race, sex, language, or religion, while ensuring that the United Nations will be a centre for harmonizing the actions of nations in the attainment of the aforesaid common ends.

Article 2.4 of the Charter explicitly prohibits intervention (which includes aerial intervention) when it provides that all Members of the United Nations are required to refrain in their international relations from the threat or use of force against the territorial integrity or political independence of any state, or in any other manner inconsistent with the purposes of the United Nations.

The Geneva Convention relative to the Protection of Civilian Persons in Time of War adopted on 12 August 1949[4] stipulates that In the case of armed conflict not of an international character occurring in the territory of one of the High Contracting Parties, each Party to the conflict shall in all circumstances treat humanely persons taking no active part in the hostilities, including members of armed forces who have laid down their arms and those placed hors de combat by sickness, wounds, detention, or any other cause without any adverse distinction founded on race, colour, religion or faith, sex, birth or wealth, or any other similar criteria. To this end, the Convention prohibits inter alia violence to life and person, in particular murder of all kinds, mutilation, cruel treatment and torture; the taking of hostages; outrages upon personal dignity, in particular humiliating and degrading treatment.

The 1984 Declaration on the Right of Peoples to Peace, approved by General Assembly Resolution 39/11 of 12 November 1984 reaffirmed that the principal aim of the United Nations is the maintenance of international peace and security, bearing in mind the fundamental principles of international law set forth in the Charter of the United Nations, expressing the will and the aspirations of all peoples to eradicate war from the life of mankind and, above all, to avert a world-wide nuclear catastrophe. The Resolution also reflected the belief of the United Nations that life without war serves as the primary international prerequisite for the material well-being, development and progress of countries, and for the full implementation of the rights and fundamental human freedoms proclaimed by the United Nations.

---

[4] Diplomatic Conference for the Establishment of International Conventions for the Protection of Victims of War, held in Geneva from 21 April to 12 August 1949 The Convention entered into force on 21 October 1950.

The Resolution solemnly proclaimed that the peoples of planet Earth have a sacred right to peace and declared that the preservation of the right of peoples to peace and the promotion of its implementation constitute a fundamental obligation of each State. The resolution appealed to all States and international organizations to do their utmost to assist in implementing the right of peoples to peace through the adoption of appropriate measures at both the national and the international level.

The World Conference on Human Rights held in Vienna in 1993 recognized and affirmed that all human rights derive from the dignity and worth inherent in the human person, and that the human person is the central subject of human rights and fundamental freedoms, and consequently should be the principal beneficiary and should participate actively in the realization of these rights and freedoms. The Conference also reaffirmed the solemn commitment of all States to fulfil their obligations to promote universal respect for, and observance and protection of, all human rights and fundamental freedoms for all in accordance with the Charter of the United Nations, other instruments relating to human rights, and international law, stating that the universal nature of these rights and freedoms is beyond question.

The United Nations Millennium Declaration, contained in General Assembly Resolution 55/2 of 8 September 2000, recognizes that, in addition to separate responsibilities of States to their individual societies, they have a collective responsibility to uphold the principles of human dignity, equality and equity at the global level. States leaders recognized that as leaders, they had a duty therefore to all the world's people, especially the most vulnerable and, in particular, the children of the world, to whom the future belongs. States reaffirmed their commitment to the purposes and principles of the Charter of the United Nations, which have proved timeless and universal, concluding that their relevance and capacity to inspire have increased, as nations and peoples have become increasingly interconnected and interdependent.

In the Declaration on the Right and Responsibility of Individuals, Groups and Organs of Society to Promote and Protect Universally Recognized Human Rights and Fundamental Freedoms adopted by General Assembly Resolution 53/144 of 9 December 1998, participating States recognized in Article 1 that everyone has the right, individually and in association with others, to promote and to strive for the protection and realization of human rights and fundamental freedoms at the national and international levels. Article 2 of the Declaration states that each State has a prime responsibility and duty to protect, promote and implement all human rights and fundamental freedoms, inter alia, by adopting such steps as may be necessary to create all conditions necessary in the social, economic, political and other fields, as well as the legal guarantees required to ensure that all persons under its jurisdiction, individually and in association with others, are able to enjoy all those rights and freedoms in practice and to that extent each State is bound to adopt such legislative, administrative and other steps as may be necessary to ensure that the rights and freedoms referred to in the Declaration are effectively guaranteed.

Against the backdrop of the aforementioned international approach aimed at ensuring human rights and eschewing violence against nations and people, one of the worst incursions of human rights in the perspective of aviation has been unlawful interference with civil aviation. These acts have affected human rights and entitlement to the peaceful enjoyment of life and good health.

During its 24th Session on December 1969, the United Nations General Assembly discussed the problem of "forcible diversion of civil aircraft" and adopted Resolution 2551 (XXIV),[5] in which the General Assembly stated its deep concern over acts of unlawful interference with international civil aviation. The General Assembly also called upon States to take every appropriate measure to see that their respective national legislation provides an adequate framework for effective legal measures against all kinds of acts of unlawful seizure of civil aircraft. It furthermore called upon States to ensure that persons on board who perpetrate such acts are prosecuted. The General Assembly urged that States give their fullest support to the International Civil Aviation Organisation in its endeavours towards the speedy preparation and adoption of a convention which would provide for appropriate measures which would make the offence of unlawful seizure of aircraft punishable. The commission of the offence would lead to the prosecution of persons who commit it. By this resolution, the General Assembly also invited States to ratify and accede to the Convention on Offenses and Certain Other Acts Committed on Board Aircraft, signed in Tokyo on 14 September 1963.

On 25 November 1970 the General Assembly adopted Resolution 2645 (XXV)[6] which condemned without exception whatsoever all aerial hijacking or other interference with civil air travel caused through the threat or use of force. The Resolution also condemned all acts of violence which may be directed against passengers, crew and aircraft engaged in, and air navigation facilities and aeronautical communication used by civil air transport. The Assembly called upon States to take all appropriate measures to deter, prevent or suppress such acts within their jurisdiction, at every stage of the execution of those acts, and to provide for the prosecution and punishment of persons who perpetrate such acts, in a measure commensurate with the gravity of those crimes, or extradite such persons for the purpose of their prosecution and punishment. Furthermore, the Assembly condemned the exploitation of unlawful seizure of aircraft for the purpose of taking of hostages, calling upon States to take joint and separate action, in accordance with the United Nations Charter and in co-operation with the United Nations and International Civil Aviation Organisation so that passengers, crew and aircraft engaged in civil aviation are not used for purposes of extortion.

The international community thus condemned terrorism against air transport by giving official recognition to such condemnation and called upon all States to contribute to the eradication of the offence by taking effective, preventive and deterrent measures. Notwithstanding the weight of these resolutions the General Assembly has seemingly deprived itself of the opportunity of declaring the offence

---

[5] Resolution 2551 (XXIV). The Resolution was adopted by a vote of 77 in favor, 2 against with 17 abstentions.

[6] Resolution 2645 was adopted by 105 in favor, non against and 8 abstentions.

of hijacking an international crime under international law. The world condemnation of the offence has left the question open to States as to whether the international community would collectively respond in the face of a crisis related to unlawful interference with civil aviation. Another blatant weakness of the Resolution is that the provisions of the resolution regarding extradition are ambivalent. The Resolution has also remained silent as to whether political motive would be a valid ground against extradition or not. It is submitted that the General Assembly should have considered adopting the principle that political motive will not be a factor affecting the extradition of hijackers.

The Resolution, with all its lapses, has many advantages, such as its condemnation of the offence of unlawful interference and call for international action against the offence. The persuasive nature of Resolutions will facilitate nations in interacting with each other and assisting each other.

The United Nations has, over the past two decades extended an invitation to nations, to co-operate with each other in eradicating or controlling international terrorism. For instance Resolution 2645 (XXV) recognized that international civil aviation is a vital link in the promotion and preservation of friendly relations among States, and that the Assembly was gravely concerned over acts of aerial hijacking or other wrongful interference with civil air travel. The resolution condemned without exception, all acts of aerial hijacking or other interference with civil air travel and called upon States to take all appropriate measures to deter, prevent or suppress such acts within their jurisdiction.[7] Earlier, the Security Council had adopted Resolution 286 (1970) which expressed the Council's grave concern at the threat to innocent civilian lives from the hijacking of aircraft and any other interference in international travel. The Security Council appealed to all parties concerned for the immediate release of all passengers and crews without exceptions, held as a result of hijackings and other interference in international travel, and called on States to take all possible legal steps to prevent further hijackings or any other interference with international civil air travel.[8]

On 18 December 1972, the United Nations General Assembly, at its 27th Session adopted a resolution[9] expressing the deep concern of the Assembly over acts of international terrorism which are occurring with increasing frequency and recalled the declaration on principles of international law which called for friendly relations and co-operation among States in accordance with the Charter of the United Nations. The resolution urged States to devote their immediate attention to finding quick and peaceful solutions to the underlying causes which give rise to such acts of violence.[10]

---

[7] A/RES/2645 (XXV), 30 November 1970. The Resolution was approved by the United Nations General Assembly on November 25, 1970 by a vote of 105 in favor, none against, and 8 abstentions.

[8] S/RES/286 (1976), 9 September 1970.

[9] A/RES/3034 (XXVII), 18 December 1972.

[10] *Ibid.*

One of the salutary effects of this resolution was the sense of urgency it reflected in reaffirming the inalienable right to self-determination and independence of all people and the condemnation it issued on the continuation of repressive acts by colonial, racist and alien regimes in denying peoples their legitimate right to the enforcement of their human rights. The resolution followed up with the invitation to States to become parties to the existing international Conventions which relate to various aspects of the problem of international terrorism.[11]

On 21 January 1977, the General Assembly commenced drafting an international convention against the taking of hostages, which was authorized by Resolution A/RES/31/103 which broadly invoked the Universal Declaration of Human Rights; and the International Convention on Civil and Political Rights which provides that everyone has the right to life, liberty and security. The resolution established an *ad hoc* Committee on the drafting of an international convention against the taking of hostages. The Committee was mandated to draft, as early as possible, an international convention. The President of the General Assembly was requested by the Assembly to appoint the members of the *ad hoc* Committee on the basis of equitable geographical distribution and representing the principal legal systems of the world.[12] The resolution was adopted on 15 December 1976.

Three years later in December 1979, the General Assembly adopted a resolution[13] which revised the work of the *ad hoc* Committee and called for international co-operation dealing with acts of international terrorism. The resolution, while welcoming the results achieved by the Committee, called upon States to fulfil their obligations under international law to refrain from organizing, instigating, assisting or participating in civil strife or terrorist acts in another State, or acquiescing in organized activities within their territory directed towards consensus of such acts.[14]

A major contribution of this Resolution was its recognition that in order to contribute to the elimination of the causes and the problem of international terrorism, both the General Assembly and the Security Council should pay special attention to all situations, including, *inter alia*, colonialism, racism and situations involving alien occupation, that may give rise to international terrorism and may endanger international peace and security. The application, when feasible and necessary, of the relevant provisions of the Chapter of the United Nations, was also recommended. The resolution also requested the Secretary General of the United Nations to prepare a compilation on the basis of material provided by Mentor States of relevant provisions of material legislation dealing with the combating of international terrorism.

In December 1985 the United Nations General Assembly adopted Resolution 40/61 which unequivocally condemned as criminal, all acts, methods and practices

[11] *Ibid.*

[12] A/RES/31/103, 21 January 1977.

[13] A/RES/34/145, 22 January 1980.

[14] *Ibid.*

of terrorism, whenever committed, including those which jeopardise international peace and security which affect States or their property.[15] The Resolution referred to the international conventions that had been adopted in relation to unlawful interference with civil aviation and called upon States to fulfil their obligations under international law to refrain from organizing, instigating, assisting or participating in any terrorist acts against other States, their people or property.

The Resolution, while citing the relevant conventions relating to unlawful interference with international civil aviation (a discussion of which will follow), once again appealed through the General Assembly to States that had not done so, to become parties to such conventions, including others which related to the suppression of international terrorism. While encouraging ICAO to continue its efforts aimed at promoting universal acceptance of and strict compliance with the international air services conventions, the Resolution also called upon all States to adhere to the ICAO conventions that provide for the suppression of terrorist attacks against civil aviation transport and other forms of public transport.[16]

Simultaneously, the Security Council, in December 1985 adopted Resolution *S/RES/579* which expressed deep concern at the prevalence of incidents of hostage taking and abduction following terrorist acts. The Resolution appealed to all States to become parties *inter alia* to the ICAO Conventions. This resolution further urged the development of international co-operation among States according to international law, in the facilitation of prevention, prosecution, and punishment of all acts of hostage taking and abduction which were identified as manifestations of international terrorism.[17]

The General Assembly, in December 1987, adopted another Resolution[18] which referred to the recommendations of the *ad hoc* Committee which had called for stringent measures of international co-operation in curbing international terrorism, which repeated the appeal of the previous resolutions for more participation by States in controlling the problem and welcomed the efforts of ICAO and IMO (International Maritime Organization) to curb unlawful interference with civil aviation and shipping respectively. The Resolution also called upon other specialized agencies and inter-governmental organizations, in particular, the Universal Postal Union, the World Tourism Organization and the International Atomic Energy Agency, within their respective spheres of competence, to consider what further measures could usefully be taken to combat and eliminate terrorism.[19] This resolution was followed by another, in December 1989, which called for a universal policy of firmness and effective measures to be taken in accordance with international law in order that all acts, methods and practices of international terrorism

---

[15] A/RES/40/61, 14 January 1986. United Nations Resolutions (Dusan. J. Djonovich ed. 7 Series), Volume XXIV, 1985–1986, p. 507.

[16] *Ibid.*

[17] S/RES/579 (1985), 18 December 1985.

[18] A/RES/42/159, 7 December 1987.

[19] *Ibid.*

may be brought to an end.[20] The Resolution also expressed the grave concern of the United Nations Mentor States at the growing and dangerous link between terrorist groups, condemned traffickers of drugs and paramilitary gangs which had been known to perpetrate all types of violence, and thereby endanger the constitutional order of States and violating basic human rights.[21]

In 1991, the United Nations General Assembly once again unanimously condemned as criminal and unjustifiable all acts, methods and practices of terrorism; called firmly for the immediate and safe release of all hostages and abducted persons; and called upon all States to use their political influence in accordance with the Charter of the United Nations and the principles of international law to secure the safe release of all hostages and abducted persons and do their utmost to prevent commission of acts of hostage-taking and abduction.[22] The plea for international co-operation was reviewed by the General Assembly in December 1993 where the Assembly urged the international community to enhance co-operation in the fight against the threat of terrorism at national, regional and international levels.[23]

In Resolution 2178 (2014) adopted by the Security Council at its 7,272nd meeting, on 24 September 2014 The Security Council reaffirmed that all States were required to prevent the movement of terrorists or terrorist groups by effective border controls and controls on issuance of identity papers and travel documents, and through measures for preventing counterfeiting, forgery or fraudulent use of identity papers and travel documents, *underscores*, in this regard, the importance of addressing, in accordance with their relevant international obligations, the threat posed by foreign terrorist fighters, and encouraged Member States to employ evidence-based traveller risk assessment and screening procedures including collection and analysis of travel data, without resorting to profiling based on stereotypes founded on grounds of discrimination prohibited by international law.

In the aftermath of the shooting down by unidentified individuals of Malaysian Airlines Flight MH 17, The UN Security Council adopted Resolution 2175 (2014) at its 7,256th meeting, on 29 August 2014, reiterating that the primary responsibility of the United Nations was to maintain international peace and security and, in this context, the need to promote and ensure respect for the principles and rules of international humanitarian law.

1.1.1 Flight MH 17

On Tuesday 29 July 2014 the International Civil Aviation Organization (ICAO), the International Air Transport Association (IATA), Airports Council International

---

[20] A/RES/44/29, 4 December 1989.

[21] *Id.* Clause 9.

[22] A/RES/46/51, 9 December 1991, Clauses 1 and 8.

[23] A/RES/48/122, 20 December 1993, Clause 2.

(ACI) and the Civil Air Navigation Services Organization (CANSO) met at ICAO's headquarters to address risks to civil aviation arising from conflict zones. The meeting was a direct response to the shooting down of a Malaysian Airlines Boeing 777 aircraft operating flight MH 17 from Amsterdam to Kuala Lumpur on 17 July 2014, and carrying 283 passengers and 15 crew. All on board perished. The four Organizations, in a joint statement issued at the end of the meeting said: "we have met at ICAO today with collective resolve to urgently review the issues and potential responses to be pursued". The statement recognized that in the aftermath of the MH 17 disaster and as a response, ICAO had reminded its member States of their responsibilities to address any potential risks to civil aviation in their airspace.

The four parties to the meeting agreed upon the importance of ICAO's work, in the context of the subject, in urgently working with its member States, in coordination with the aviation industry and other bodies within the United Nations, to ensure "the right information reaches the right people at the right time". The following discussion examines the legal principles applicable to the shooting down of the aircraft and the regulatory background relating to the possibilities of the objective of ensuring the timely dissemination of threat information.

The shooting down in July 2014 of Flight MH 17 in the territory of Ukraine is a recent example of the vulnerability of human rights from an aviation perspective. It amply demonstrated the erosion of public security that could affect aviation and how aviation could be used to threaten public security both within and beyond national borders. As the discussion to follow reflects, much work needs to be done by the regulators concerned in obviating the possibility of persons with ill intent using aviation to attack human lives and destabilize societies.

Malaysian Airlines Flight MH 17, operated by a Boeing 777-200ER aircraft flying from Amsterdam to Kuala Lumpur on 17 July 2014, and carrying 283 passengers and 15 crew, was shot down by a BUK surface to air missile over Donest Oblast in Eastern Ukraine, while at an altitude of 10,000 m. Two thirds of the passengers on board were of Dutch origin. All those on board perished.

A similar event had occurred in September 1983 when a Russian SU-15 Interceptor plane shot down a Korean Airlines Boeing 747 aircraft operating flight KE 007 bound from New York City to Seoul via Anchorage. The plane was destroyed over Sakhalin Island while navigating over prohibited Russian airspace. All 269 passengers and crew on board died.

Consequent upon the 1983 shooting down of KL 007, and amidst a vociferous international outcry, the ICAO convened a special Assembly of ICAO member States which adopted article 3 *bis* to the Convention on International Civil Aviation (Chicago Convention)[24] which now provides that ICAO member States undertake to refrain from using force against civil aircraft. Arguably, this provision *ex facie* does not apply to the destruction of the aircraft which operated flight MH 17. At the time of writing, there was no formal pronouncement, unlike in the KE 007 disaster,

[24] Convention on International Civil Aviation signed at Chicago on 7 December 1944. See ICAO Doc 7300/9, 9th Edition: 2006.

that a State was involved in bringing down flight MH 17. Some alleged that it had been brought down by pro-Russian rebels, which left both Ukraine and Russia as presumed innocent.

Although the destruction of aircraft at high altitude was addressed in Article 3 *bis*[25] of the Chicago Convention, the ICAO Assembly, in addition, addressed the analogous issue of Man Portable Air Defence Systems (MANPADS) at its 36th Session of the Assembly held from 18 to 28 September 2007, where member States of the Organization adopted Resolution A36-19 [threat to civil aviation posed by man-portable air defence systems (MANPADS)]. In this Resolution, States expressed their deep concern regarding the global threat posed to civil aviation by terrorist acts, in particular the threat posed by MANPADS, other surface to air missiles systems light weapons and rocket propelled grenades. The Assembly urged all member States to take the necessary measures to exercise strict and effective controls on the import, export, transfer or retransfer and stockpile management of MANPADS and associated training and technologies, as well as limiting the transfer of MANPADS production capabilities; It also called upon all Contracting States to cooperate at the international, regional and sub-regional levels with a view to enhancing and coordinating international efforts aimed at implementing countermeasures carefully chosen with regard to their effectiveness and cost, and combating the threat posed by MANPADS.

Relevant to the adoption of Resolution A36-19 was the fact that the United Nations General Assembly, on 8 September 2006, had adopted its Counter-Terrorism Strategy, which is a unique global instrument that was calculated to

---

[25] Article 3 *bis* provides as follows:

a) The contracting States recognize that every State must refrain from resorting to the use of weapons against civil aircraft in flight and that, in case of interception, the lives of persons on board and the safety of aircraft must not be endangered. This provision shall not be interpreted as modifying in any way the rights and obligations of States set forth in the Charter of the United Nations.

b) The contracting States recognize that every State, in the exercise of its sovereignty, is entitled to require the landing at some designated airport of a civil aircraft flying above its territory without authority or if there are reasonable grounds to conclude that it is being used for any purpose inconsistent with the aims of this Convention; it may also give such aircraft any other instructions to put an end to such violations. For this purpose, the contracting States may resort to any appropriate means consistent with relevant rules of international law, including the relevant provisions of this Convention, specifically paragraph a) of this Article. Each contracting State agrees to publish its regulations in force regarding the interception of civil aircraft.

c) Every civil aircraft shall comply with an order given in conformity with paragraph *b)* of this Article. To this end each contracting State shall establish all necessary provisions in its national laws or regulations to make such compliance mandatory for any civil aircraft registered in that State or operated by an operator who has his principal place of business or permanent residence in that State. Each contracting State shall make any violation of such applicable laws or regulations punishable by severe penalties and shall submit the case to its competent authorities in accordance with its laws or regulations.

enhance national, regional and international efforts to counter terrorism. The Strategy emphasizes the need to combat the illicit arms trade, in particular small arms and light weapons, including MANPADS. Member States have agreed to a common strategic approach to fight terrorism, not only by sending a clear message that terrorism is unacceptable but also resolving to take practical steps individually and collectively to prevent and combat it. These steps include a wide range of measures ranging from strengthening State capacity to counter terrorist threats, to better coordinating United Nations System's counter-terrorism activities.

The use of surface to air missiles and anti-tank rockets by terrorists goes back to 1973. On 5 September 1973 Italian police arrested five Middle-Eastern terrorists armed with SA-7 s. The terrorists had rented an apartment under the flight path to Rome Fumicino Airport and were planning to shoot down an El Al airliner coming in to land at the airport. This arrest proved a considerable embarrassment to Egypt because the SA-7 s were later traced back to a batch supplied to it by the Soviet Union. It was alleged that the Egyptian government was supplying some of the missiles to the Libyan army but inexplicably, the SA-7 s had been directly rerouted to the terrorists. This incident also placed the Soviet Union in an awkward position because of the possibility that its new missile and its policy of the proxy use of surrogate warfare against democratic states were revealed to the West.

Another significant incident occurred on 13 January 1975 when an attempt by terrorists to shoot down an El Al plane with a missile was believed to have brought civil aviation to the brink of disaster. Two terrorists drove their car onto the apron at Orly airport, where they set up a rocket launcher and fired at an El Al airliner which was about to take off for New York with 136 passengers. The first round missed the target thanks to the pilot's evasive action and hit the fuselage of a Yugoslav DC-9 aeroplane waiting nearby to embark passengers for Zagreb. The rocket failed to explode and no serious casualties were reported. After firing again and hitting an administration building, which caused some damage, the terrorists escaped by car. A phone call from an individual claiming responsibility for the attack was received at Reuters. The caller clearly implied that there would be another such operation, saying 'Next time we will hit the target'.

Missile attacks are common.[26] There was a marked increase in missile attacks since 1984. On 21 September 1984 Afghan counter-revolutionaries fired a surface-

[26] On 21 February 1973 a sandstorm and mechanical problems combined to force a Boeing 727 (Libyan Arab Airlines Flight 114) flying from Tripoli to Cairo off-course, drifting east across the Suez Canal and over the Sinai Peninsula (then Israeli territory). Israeli interceptors were launched, and their pilots signalled to the Libyan plane to land; Flight 114's crew instead turned west. The Israelis' cannon fire damaged the plane, forcing it to make an emergency landing in the desert. Out of 113 people on board, 108 died. On 20 April 1978 a Boeing 707 operating Korean Airlines Flight 902 was ordered shot down after violating Soviet airspace. The damaged plane made an emergency landing on a frozen lake near the border of Finland. Only two of the 107 people on board died. On 3 September 1978, during the Rhodesian Bush War, guerrilla fighters used a surface-to-air missile to shoot down an Air Rhodesia plane (Flight 825) shortly after the Russian-made Strela 2 airliner took off from Kariba. After the plane crashed and broke up into pieces, the guerrillas located 10 people who were still alive and executed them. Only eight people on board survived.

to-air missile and hit a DC-10 Ariana Airliner carrying 308 passengers. The explosion tore through the aircraft's left engine, damaging its hydraulic system and a wing containing a fuel tank. The captain of the aircraft, however, managed to land the aircraft safely at Kabul International Airport. Another significant incident took place on 4 April 1985, when a member of Abu Nidal group fired an RPG rocket at an Alia airliner as it took off from Athens Airport. Although the rocket did not explode, it left a hole in the fuselage.

Advanced missiles and rockets have been found in many terrorist and insurgent armouries. It is suspected that some terrorist organizations, including Iranian militia in Lebanon, the Provisional Irish Republican Army and various African and Latin American insurgents, possess the sophisticated Russian-made RPG-7 portable rocket launcher, but it is disturbing to note that some terrorist organizations, most notably Palestinian groups, have their own RPG-7-manufacturing facilities. In addition, more than a dozen other terrorist and insurgent groups were known to possess portable surface-to-air missiles. These groups included various Cuban surrogates, Colombian drug dealers, and a number of African, European and Palestinian terrorist organizations.

1.1.1.1 Some Counter Measures

The gathering of reliable intelligence remains the first line of defence. Although modern technologies clearly aid terrorists in terms of weapons and targets, technology can also be used against terrorists. Governments which are endowed with the necessary technology can keep track of terrorist organizations and their movements with the aid of computers. At the same time, electronic collection methods and signals intelligence afford the possibility of eavesdropping on and intercepting terrorist communications, leading to better predictions of their operations. One of the instances where intelligence gathering worked well to prevent terrorism occurred in September 1984, when the Provisional IRA spent an estimated £1.5 million in the United States on a massive shipment of seven tons of arms. With the help of an informer about a forthcoming shipment of weapons, including rockets, to the Provisional IRA from the United States, the FBI informed British intelligence, who in turn contacted the Irish, and the ship carrying the arms was tracked by a US satellite orbiting 300 km above the earth. The satellite photographed the transfer of the arms to a trawler. Finally, two Irish Navy vessels intercepted the trawler and British security forces arrested the crew. This incident shows that intelligence gathering with the help of high technology can cut off the transfer of missiles and other weapons to terrorists.

The installation of a sophisticated antimissile system similar to that employed on military aircraft to divert surface-to-air missiles is an effective deterrent. One good example is the measure taken by the British government which, immediately after the discovery of 20 SA-7 s in the coaster *Eksund*, which was intercepted by French authorities off the coast of Brittany in November 1987 when bound for the IRA, fitted all British Army helicopters flying in Northern Ireland with electronic and other decoy systems to confuse the missile's heat-seeking guidance system. These included the US-made Saunders, AN/ALG 144. This system, when linked to the Tracor AN/ALE 40 chaff dispenser, works by jamming the missile's homing radar and sending infra-red flares and chaff to act as a decoy for the heat-seeking device. The system is used by both the US and the Israeli Armies, which have been well-pleased with its performance. Until the British realised that the IRA might be in possession of SAMs, the Ministry of Defense hesitated to install such a system because of the high cost involved, and its decision to do so shows the seriousness of the threat. Another example of a good counter-measure is the response of El Al airlines to the threat of such an attack which included the installation of electronic countermeasure equipment similar to that employed on military aircraft to divert surface-to-air missiles.

1.1.1.2 The Legal Fallout

The aircraft was shot down in the air space over Ukraine. Therefore technically, according to the Chicago Convention, Article 3 *bis*, Ukraine was obligated to make every effort to refrain from using force against the Malaysian aircraft. One could argue that, in the exercise of its sovereignty, Ukraine should have required the landing of Flight MH 17 at some designated airport If Ukraine believed that the aircraft was flying above its territory without authority or if there were reasonable grounds to conclude that it was being used for any purpose inconsistent with the aims of the Convention; it was also entitled to give such aircraft any other instructions to put an end to such violations. For this purpose, Ukraine could have resorted to any appropriate means consistent with relevant rules of international law, including the relevant provisions of the Chicago Convention. Also, Ukraine was required to specifically publish its regulations in force regarding the interception of civil aircraft.

According to Article 28 of the Chicago Convention, Ukraine was required to provide in its territory air navigation facilities *inter alia* to facilitate international air navigation. This provision imputes to Ukraine the obligation to provide air traffic services.

It has been reported that Malaysia accused Ukraine of being responsible for the loss of all on board the aircraft and the destruction of the aircraft, on the ground that the air traffic controllers of Ukraine had directed the flight towards the flight path it was on and that Ukraine had a duty to warn the aircraft not to traverse the particular flight path.[27] From the perspective of State responsibility, separatists or no, the aircraft was shot down over Ukraine and it is relevant to examine its legal

[27] http://www.wantchinatimes.com/news-subclass-cnt.aspx?cid=1101&MainCatID=11&id=20140729000007.

responsibility towards the aircraft in its territory.[28] Responsibility would devolve upon a State in whose territory an act of unlawful interference against civil aviation might occur, to other States that are threatened by such acts. The International Court of Justice (ICJ) recognised in the famous *Corfu Channel* Case:

> Every State's obligation not to allow knowingly its territory to be used for acts contrary to the rights of other States.[29]

Ukraine may anchor itself on the judicial dictum in the *Corfu Channel* case, where the International Court of Justice applied the subjective test and applied the fault theory. The Court was of the view that:

> It cannot be concluded from the mere fact of the control exercised by a State over its territory and waters that the State necessarily knew, or ought to have known, of any unlawful act perpetrated therein, nor yet that it necessarily knew, or should have known the authors. This fact, by itself and apart from other circumstances, neither involves prima facie responsibility nor shifts the burden of proof.[30]

Malcolm Shaw is of the view:

> The rights accorded to states under international law imply responsibilities. States are liable for breaches of their obligations, provided that the breach is attributable to the state itself. A state is responsible for direct violations of international law—e.g., the breach of a treaty or the violation of another state's territory. A state also is liable for breaches committed by its internal institutions, however they are defined by its domestic law; by entities and persons exercising governmental authority; and by persons acting under the direction or control of the state. These responsibilities exist even if the organ or entity exceeded its authority. Further, the state is internationally responsible for the private activities of persons to the extent that they are subsequently adopted by the state.[31]

The Malaysian accusation brings to mind the shooting down on 3 July 1988 of Iran Air Flight 655 which was operated by an Airbus A-300B aircraft, and the killing of its 290 passengers and crew by two surface-to-air missiles launched from the *U.S.S. Vincennes*, a guided-missile cruiser on duty with the United States Persian Gulf/Middle East Force in the Iranian airspace over the Islamic Republic's territorial waters in the Persian Gulf. The incident occurred in the midst of an armed engagement between U.S. and Iranian forces, in the context of a long series of attacks on U.S. and other vessels in the Gulf. While the United States sought the intervention of the ICAO Council in considering the attack, Iran sought the United Nations Security Council's condemnation of the United States. The latter did not materialize. In July 1988, the ICAO Council convened an extraordinary session of the Council at which it resolved *inter alia* to conduct a fact-finding investigation to

---

[28] For a detailed discussion on State responsibility see Abeyratne (2010a), pp. 177–211 at 195–209.

[29] The *Corfu Channel* Case, ICJ Reports, 1949, *I.C.J.R.* 1, p. 22.

[30] *Id.*, p. 4.

[31] http://www.britannica.com/EBchecked/topic/291011/international-law/233508/The-responsibility-of-states.

determine all relevant facts and technical aspects of the incident. The direction and purpose of the investigation was based on the theme of ensuring the safeguarding of civil aviation, particularly with a view to determining the feasibility of introducing possible revisions to ICAO standards and recommended practices, as necessary. Accordingly, a team of experts was assembled which later submitted its Report to the Council. Upon discussion of the Report by the Council, in December 1988, the Government of Iran sought to have the Report examined to identify any violations of the Chicago Convention and drew the Council's attention to Article 54[32] of the Convention. However, Iran did not request that the Council undertake dispute-resolution procedures under Article 84[33] of the Chicago Convention, nor did Iran seek to apply the comprehensive and exclusive ICAO Rules for the Settlement of Differences promulgated to address disputes arising under that Article 84 of the Chicago Convention.

If, as the Malaysian authorities claimed, the Ukranian air traffic controllers were negligent in not warning the aircraft against flying over separatist held territory, one could argue that the principle of objective responsibility could apply to Ukraine. Brownlie states:

> Technically, objective responsibility rests on the doctrine of the voluntary act: provided that agency and causal connection are established, there is a breach of duty by result alone. Defences such as act of third party are available, but the defendant has to exculpate himself.[34]

[32] Article 54 n) provides that the Council of ICAO has, as its mandatory duty, to consider any matter relating to the Convention which any contracting State refers to it.

[33] Article 84 stipulates: "if any disagreement between two or more contracting States relating to the interpretation or application of the Chicago Convention and its Annexes cannot be settled by negotiation, it shall, on the application of any State concerned in the disagreement, be decided by the Council. No member of the Council shall vote in the consideration by the Council of any dispute to which it is a party. Any contracting State may, subject to Article 85, appeal from the decision of the Council to an ad hoc arbitral tribunal agreed upon with the other parties to the dispute or to the Permanent Court of International Justice. Any such appeal shall be notified to the Council within sixty days of receipt of notification of the decision of the Council. Article 85 provides that If any contracting State party to a dispute in which the decision of the Council is under appeal has not accepted the Statute of the Permanent Court of International Justice and the contracting States parties to the dispute cannot agree on the choice of the arbitral tribunal, each of the contracting States parties to the dispute shall name a single arbitrator who shall name an umpire. If either contracting State party to the dispute fails to name an arbitrator within a period of three months from the date of the appeal, an arbitrator shall be named on behalf of that State by the President of the Council from a list of qualified and available persons maintained by the Council. If, within thirty days, the arbitrators cannot agree on an umpire, the President of the Council shall designate an umpire from the list previously referred to. The arbitrators and the umpire shall then jointly constitute an arbitral tribunal. Any arbitral tribunal established under Article 84 or 85 shall settle its own procedure and give its decisions by majority vote provided that the Council may determine procedural questions in the event of any delay which in the opinion of the Council is excessive".

[34] Brownlie (1990), p. 437.

In the famous *Caire* claim of 1929,[35] The President of the Claims Commission stated that according to the objective responsibility principle, a State has to accept responsibility for the actions of their officials whether such officials were at fault or not, on delictual (tortious) principles under international law, regardless of whether such officials exceeded their parameters of competency or not.[36]

However, over Donetsk Oblast in Eastern Ukraine where the aircraft was shot down, armed separatist groups had taken over the territory and Ukraine was therefore not in control. In other words, the area concerned was a "conflict zone". These armed separatist groups were in full control of the crash site, even preventing international investigators from entering the site which prompted the the United Nations Security Council to unanimously adopt Resolution 2166 (2014) calling on those controlling the MH 17 crash site to allow unfettered access to international investigators. It must be noted in this context that the Chicago Convention provides that in case of war,[37] the provisions of the Convention do not affect the freedom of action of any of the contracting States affected, whether as belligerents or as neutrals. The same applies in the case of any contracting State which declares a state of national emergency and notifies the fact to the Council.[38] It is therefore arguable that Article 3 *bis* of the Chicago Convention would not apply to Ukraine in the circumstances of Flight MH 17.

In the *Home Missionary Society* case[39] of 1920 the tribunal stated that it was a well established principle of international law that no State could be held accountable or responsible for the acts of rebellious bodies of men committed in violation of its authority where the State is not guilty of a breach of good faith or of no negligent in suppressing insurrection.[40]

1.1.1.3 The Regulatory Fallout

ICAO identifies air routes but has no mandate or competence to issue warnings based on political and war situations. This is entirely dependent on the State concerned which has the responsibility to issue what is called a NOTAM (Notice to Airmen) which advises aircraft which airspace to traverse and which not to. If a pilot traverses many airspaces he/she has to consider all relevant NOTAMs.

The Chicago Convention prescribes in Articles 9 and 12 the principle of the right of States to prescribe where aircraft should go and should not, in accordance with

---

[35] (1929), RIAA v. 516.

[36] Id., pp. 529–531.

[37] War is a state of armed conflict between different nations or states or different groups within a nation or State. See https://www.google.ca/#q=definition+of+war.

[38] Article 89 of the Chicago Convention. Ukraine had not advised the Council of ICAO of it situation.

[39] (1920) *RIAA* vi. 42.

[40] *Id.* 44.

requirements set by such States. Article 12 on *Rules of the Air* provides that where flights over the high seas are concerned the rules prescribed under the Chicago Convention shall prevail. Those rules are contained in Annex 2[41] to the Convention. Technically, ICAO cannot take on the task of publishing advisories on navigation in airspace. That is not ICAO's job, which is clearly enshrined in Article 44 of the Chicago Convention. According to this provision, ICAO is only required to develop principles and techniques of air navigation and foster the development of air transport. Advisories are purely matters of State which have sovereignty over the airspace above their territories.

According to the Joint Statement issued by ICAO, IATA, ACI and CANSO in July 2014, ICAO has agreed to immediately establish, together with industry partners, a senior level Task Force composed of State and industry experts to address the civil aviation and national security aspects of this challenge, in particular how information can be effectively collected and disseminated. The findings of the task force are expected to be submitted with the greatest urgency to a special meeting of the ICAO Council for action.

ICAO has also been requested (by the industry) to address fail-safe channels for essential threat information to be made available to civil aviation authorities and industry as well as the need to incorporate into international law, through appropriate UN frameworks, measures to govern the design, manufacture and deployment of modern anti-aircraft weaponry.

The entire joint statement boils down to one objective: avoidance of risk through information sharing. A noteworthy feature of the statement is the recognition that ICAO has an important role to play to "ensure" the right threat information reaches the right people at the right time.

The Joint Statement, although reflecting a sound approach to the issue, is deficient in not requiring the task force to address the individual role to be played by each international and industry partner and the legal justification and legitimacy of those roles. Without this the Council would not get the purpose and direction of the thrust of the task force's findings.

Another consideration is that if, as the joint statement seemingly requires, ICAO were to ensure disseminating threat information, it would make the Organization go way beyond the basic expectations of its aims and objectives, which is to develop principles and techniques of air navigation. There is also the issue of what is required in the entire process of developing an efficient and effective threat information sharing system. The key players responsible for ensuring the success of this process would have to be given sufficient resources, be able to identify problems early, engage in detailed training and continuous practice, maintain excellent flows of communication, and encourage trust and shared value amongst all concerned.

Drazenovich and *Skovira*[42] state that "open communications and the free flow of shared information should lead to the development of shared mental models of the

---

[41] Annex 2 (Rules of the Air).

[42] Drazenovich and Skovira (2010), pp. 673–688.

system, its hazards, and its warning systems. A strong organizational culture and shared values are important...[43]" In order to achieve these objectives, the political will of States is essential and obtaining it collectively will be ICAO's first challenge. The second challenge would be to obtain continuous commitment of all States concerned to treat the issue of safety of flight over conflict areas as their absolute priority.

As for the Industry request contained in the Joint Statement: that ICAO address the need to incorporate into international law, through appropriate UN frameworks, measures to govern the design, manufacture and deployment of modern anti-aircraft weaponry, one wonders how this request fits within ICAO's aim and objective to develop principles and techniques of air navigation. Perhaps ICAO, IATA, ACI and CANSO should have given this more serious thought, as to how ICAO could dabble in the design, manufacture and deployment of modern anti-aircraft weaponry.

The way forward in responding to the unfortunate and sad fate of the lives on board Flight MH 17 and the total destruction of the aircraft seems to rest on the critical need for sharing threat information in a timely and efficient manner. If Malaysian Airlines Flight MH 370 is anything to go by, where both ICAO and the International Criminal Police Organization (INTERPOL) failed to advise both States and airlines of the existence of a database at INTERPOL on forged or fraudulent passports,[44] ICAO and other key players concerned would have to adopt a more serious approach to the problem. Information sharing is a central process through which team members collectively utilize their available informational resources. Information retrieval becomes a key element in information sharing. The ICAO appointed Task Force would have to perform their task on the basis that, as digitally mediated communication and information sharing increase, collaborative information retrieval merits greater attention and support.

The question therefore is: how does ICAO ensure the timely dissemination of threat information and to whom should this information be relayed? Since threat information must essentially and exclusively come from States, does the joint statement mean that ICAO should be a mere conduit in transmitting information to its member States?

Risk avoidance has been defined to involve the risk assessment technique that entails eliminating hazards, activities and exposures that place valuable assets at risk. In the case of civil aviation within the context of conflict zones this would mean eliminating hazards by avoiding the airspace over that zone entirely. Unlike risk management, which is calculated to control dangers and risks, risk avoidance totally bypasses a risk. The information to States on threats posed to their civil aviation over conflict zones would therefore have to be disseminated through policy and procedure, training and education and technology implementations.

---

[43] *Id.*, p. 681.

[44] See Abeyratne (2014a), pp. 238–249.

Therefore firstly, ICAO would have to, with the involvement of industry partners and other relevant players, establish the policy and procedure involved in information sharing. It would also have to consider the best way in which adequate training and knowledge can be passed on.

The important issue though is not the logistics of establishing a senior level task force. Nor is it to have the task force come up with assessments and recommendations. The critical factor is who in the end "ensures" that the information disseminated is timely and reaches the right people at the right time. In the ultimate analysis, it should be the States themselves and this is already being accomplished through NOTAMs (notices to airmen) issued by the States in accordance with guidelines of ICAO in Annex 15 to the Chicago Convention (Aeronautical Information Services). One has to take it that the Joint Statement envisions an extension to this process through a more efficient system.

In this context, it is incontrovertible that the most critical factor in achieving the objective of ensuring "the right information reaches the right people at the right time" would be the political will and commitment of every one of the 191 member States of ICAO. These States must act along the bottom line which is that those who perished in Flight MH were not involved in the situation between the Russian Federation and Ukraine. One of the cardinal principles enunciated by the International Committee of the Red Cross (ICRC) is that persons *hors de combat* and those who do not take a direct part in hostilities are entitled to respect for their lives and physical and moral integrity.[45] They should, in all circumstances, be protected and treated humanely without any adverse distinctions. It is obvious that the "protection" referred to must come from the State and no one else.

1.1.2 Risk Management

On 28 October 2014, the ICAO Council adopted a resolution in response to the shooting down of Flight MH 17 in July 2014,[46] stating *inter alia*, that the Council welcomed the leading role and efforts of ICAO, supported by States, organizations and industry players, for further improvement of international standards and sharing of best practices in relation to the safety of civil aircraft at or near conflict zones to prevent recurrence of such tragic events in future. The Council urged States to take all necessary measures to safeguard the safety of air navigation, including the

[45] The fundamental principle of the International Red Cross is humanity. The Red Cross, born of a desire to bring assistance without discrimination to the wounded on the battlefield, endeavours—in its international and national capacity—to prevent and alleviate human suffering wherever it may be found. Its purpose is to protect life and health and to ensure respect for the human being. It promotes mutual understanding, friendship, co-operation and lasting peace amongst all peoples. See *Proclamation of the Fundamental Principles of the Red Cross*, http://www.icrc.org/eng/resources/documents/misc/fundamental-principles-commentary-010179.htm.

[46] For a detailed discussion on Flight MH 17 See Abeyratne (2014b), pp. 329–342.

establishment of robust arrangements to identify, assess and share information and respond to risks to civil aircraft from activities in conflict zones, through, *inter alia*, effective coordination of civil and military activities within conflict zones and, if considered necessary when the safety of civil aircraft is deemed to be compromised, take appropriate airspace management measures within their jurisdictions such as access restrictions or the closure of airspace, or the issuance of advisories to airspace users.

What follows from the gist of this resolution is that while States are ultimately responsible for measures to safeguard the safety of air navigation ICAO has the responsibility of taking a leadership role in enunciating the best practices and standards that would direct and guide the States in preventing future MH 17 type disasters. Therefore ICAO has the daunting task of being a beacon to States in such areas as coordination of civil and military activities within conflict zones and air space management and the issuance of advisories.

Surprisingly, the Council Resolution mentioned above does not make reference in its Preamble to Article 9 of the Chicago Convention[47] which provides inter alia that each contracting State may, for reasons of military necessity or public safety, restrict or prohibit uniformly the aircraft of other States from flying over certain areas of its territory, which explicitly recognizes the overall responsibility of States to be vigilant of military danger to civil aviation and divert civil aircraft accordingly.[48] In the context of the daunting role ICAO has ascribed to itself, the Council earlier examined a report of an ICAO appointed *Task Force on Risks to Civil Aviation Arising from Conflict Zones*[49] which intriguingly said that one of the objectives of the Task Force was to ensure that arrangernents apply and relevant information is available to assure the safety of passengers and crew on civil aircraft irrespective of which airline they are travelling with or which cities they are travelling between. In other words, ICAO has taken it upon itself to ensure that arrangements are in place and information is available that would enable States to invoke Article 9 of the Chicago Convention.

The report waffles between pragmatism and impracticality, often bordering on startling incongruence and careless disregard to the roles to be played by States and

[47] Convention on International Civil Aviation, signed at Chicago on 7 December 1944. See ICAO Doc 7300/9, Ninth Edition 2006. Article 9 also states that each contracting State reserves the right, in exceptional circumstances or during a period of emergency, or in the interest of public safety, and with immediate effect, temporarily to restrict or prohibit flying over the whole or any part of its territory, on condition that such restriction or prohibition shall be applicable without distinction of nationality to aircraft of all other States. A State may also require any aircraft entering its territory to effect a landing as soon as practicable at some designated airport within its territory and if necessary to effect a landing as soon as practicable thereafter at some designated airport within its territory.

[48] The Resolution irrelevantly refers to Article 3 *bis* of the Chicago Convention which imposes an obligation on States to refrain from resorting to the use of weapons against civil aircraft in flight. There is no evidence that either the Russian Federation or Ukraine shot down the aircraft operating Flight MH 17.

[49] C-WP/14220, Appendix B.

industry. While on the one hand the report supports proactively the necessity to augment risk management so as to provide guidance and to mentor States on exchange of risk assessments and supporting information,[50] it somewhat frivolously advocates enhancing transparency by developing best practices on the provision of information to passengers and flight crew regarding the use of airspace over or near conflict zones.[51]

It boggles the mind as to why ICAO, any State or other organization or any airline would develop best practices to advise an airline passenger of the flight plan of his flight or of conflict zones that he would possibly fly over. Although several jurisdictions have adopted passenger rights charters and legislation pertaining to the rights of the passenger to be advised of flight delays, weather delays, lost luggage and to be provided food and drink, none has alluded to the need to inform the travelling public on risks regarding the use of airspace over conflict zones. If the intention is to assist passengers in choosing a safe airline that does not fly over conflict zones, airlines should be required to assure passengers that they do not fly over conflict zones, without providing advise on the risk of flying over conflict zones.

Bolivia and the Russian federation submitted a working paper to the ICAO Council at its 203rd Session[52] which sensibly focused on risk assessment and risk management, calling for tighter responsibility to be devolved upon States, which, in the ultimate analysis should be accountable for the safety of air navigation within their territories. The two States suggested that:

> Each State in order to eliminate the threat to lives of persons on board civil aircrafts (shall) make all possible efforts for timely and proper restriction of flights of civil aircraft in the national or delegated airspace over zones where military activity, potentially hazardous to civil aircraft operations take place, at a minimum, within the radius of the possible destruction by used types of weaponry. In this way, Contracting States would agree to inform ICAO immediately and other Contracting States on the existence of threats or on the probability of the emergence of such threats, while understanding their full responsibility for the possible consequences in case of concealment of such information or delays in its distribution.[53]

The two States suggested that the matter of enhanced State responsibility be remanded to the ICAO Working Group on Governance and Efficiency (WGGE) in the course of the current session of the ICAO Council, where the WGGE could undertake a preliminary discussion on the feasibility of developing an amendment to the Chicago Convention, and that the matter be brought to the attention of the High Level Safety Conference to be held in February 2015.[54]

[50] *Id.*, Paragraph 4.

[51] *Id.*, Paragraph 7.

[52] Risk Assessment of Operations over Airspace Affected by Armed Conflict—Responsibility of States for Ensuring the Flight Safety of Civil Aircraft Within Their National and Delegated Airspace Over Armed Conflict Zones or Zones of Military Exercises, *C-WP/14227*, 20/10/14.

[53] *Id.*, p. 4, Para 3.5.

[54] *Id.*, Paragraphs 3.6 and 4.1. c).

Although the adoption of an amendment may be a cumbersome process[55] the overall thrust of the proposal of insistence on enhanced State responsibility is both practical and consistent with the attendant duties of States to ensure safety of air navigation over their territories.

1.1.2.1 Principles of Military and Civil Aviation

1.1.2.1.1 ICAO Guidance

ICAO's Air Traffic Management Operational Concept is grounded on the principle that airspace management is the process by which airspace options are selected and applied to meet the needs of the air traffic management (ATM) community. Key conceptual principles include: (a) all airspace will be the concern of ATM and will be a usable resource; (b) airspace management will be dynamic and flexible; (c) any restriction on the use of any particular volume of airspace will be considered transitory; and (d) all airspace will be managed flexibly. Airspace boundaries are required to be adjusted to particular traffic flows and should not be constrained by national or facility boundaries.[56] The Concept recognizes that the accommodation of mixed capabilities and worldwide implementation needs should be addressed to enhance safety and efficiency; relevant ATM data will be fused for an airspace user's general, tactical and strategic situational awareness and conflict management; and (c) relevant airspace user operational information will be made available to the ATM system.[57]

Part of the ICAO Council Resolution alluded to[58] refers to vigilance over military and civil aviation activities in conflict zones. ICAO's *Circular 330-AN/189* focuses on the global interoperability of the ATM system on the basis that air space management should follow the principles and strategies that admit of all available airspace being managed flexibly; air space management processes accommodate dynamic flight trajectories and provide optimum operational solutions; different types of traffic be segregated as required by airspace organization;

[55] Article 94 of the Chicago Convention prescribes that any proposed amendment to the Convention must be approved by a two-thirds vote of the Assembly and shall then come into force in respect of States which have ratified such amendment when ratified by the number of contracting States specified by the Assembly. The number so specified shall not be less than two thirds of the total number of contracting States. If in its opinion the amendment is of such a nature as to justify this course, the Assembly in its resolution recommending adoption may provide that any State which has not ratified within a specified period after the amendment has come into force shall thereupon cease to be a member of the Organization and a party to *the Convention.*

[56] Global Air Traffic Management Operational Concept, *Doc 9854 An/458*, First Edition: 2005 at 2.1.2.

[57] *Id.* 2.1.6.

[58] *Supra*, note 46.

airspace use be coordinated and monitored; and airspace reservations be planned in advance.[59]

Another ICAO principle provides that States are required to ensure that the level of air traffic services (ATS) and communications, navigation and surveillance, as well as the ATS procedures applicable to the airspace or aerodrome concerned, are appropriate and adequate for maintaining an acceptable level of safety in the provision of ATS.[60] Where this is deemed necessary by the appropriate ATS authority or authorities, co-ordination between ATS units providing flight information service in adjacent flight information regions (FIRs) shall be effected in respect of IFR and VFR flights, in order to ensure continued flight information service to such aircraft in specified areas or along specified routes. Such co-ordination shall be effected in accordance with an agreement between the ATS units concerned.[61] Air traffic services authorities are required to establish and maintain close cooperation with military authorities responsible for activities that may affect flights of civil aircraft.[62] The arrangements for activities potentially hazardous to civil aircraft, whether over the territory of a State or over the high seas, are required to be co-ordinated with the appropriate air traffic services authorities.[63] The Global Air Navigation Capacity and Efficiency Plan (GANP)[64] of ICAO recommends the achievement of service improvement through integration of all digital air traffic management information that would result in enhanced global interoperability and greater and timelier access to up-to-date information by a wider set of users.[65]

---

[59] Civil/Military Cooperation in Air Traffic Management, *Cir* 330 AN/189 at 3.1.2.

[60] Procedures for Air Navigation Services (PANS), Air Traffic Management, *Doc 4444 ATM 501* at 2.1. Confusing air traffic signals and information may lead to the endangering of the safety of air navigation. The Federal Aviation Administration of the United States (FAA) had safety and national security concerns regarding flight operations in a portion of the Simferopol (UKFV) FIR concerning Ukraine and the Russian Federation. On March 28, 2014, the Russian Federation issued a Notice-to-Airmen (NOTAM) purporting to establish unilaterally a new FIR, effective April 3, 2014, in a significant portion of the Simferopol (UKFV) FIR. The affected airspace includes sovereign Ukrainian airspace over the Crimean Peninsula and the associated Ukrainian territorial sea, as well as international airspace managed by Ukraine over the Black Sea and the Sea of Azov under a regional air navigation agreement approved by the Council of ICAO. In response to the Russian Federation's actions, Ukraine established a prohibited area over the Crimean Peninsula for flight operations below flight level 290 by means of a NOTAM and closed various ATS route segments. The Russian Federation further responded by the issuance of a NOTAM that rejected and directly conflicts with Ukrainian NOTAMs concerning the establishment of the prohibited area and the route segment closures. The FAA claimed that this action by the Russian Federation contradicted international law, including provisions of the Chicago Convention.

[61] Manual Concerning Interception of Civil Aircraft, *Doc 9433-AN926* (Second Edition, 1990) at 3.1.4.1.

[62] Manual Concerning Safety Measures Relating to Military Activities Potentially Hazardous to Civil Aircraft Operations, *Doc 9554-AN/932* First Edition: 1990 at 2.1.5.1.

[63] *Id.* 2.1.6.1.

[64] 2013–2028 Global Air Navigation Capacity and Efficiency Plan, *Doc 9750.*

[65] *Id.* B1-30 Service Improvement through Integration of all Digital ATM Information.

Some examples of military activities that may pose a threat to civil aviation are: (a) practice firing or testing of any weapons air-to-air, air-to-surface, surface-to-air or surface-to-surface in an area or in a manner that could affect civil air traffic; (b) certain military aircraft operations such as air displays, training exercises, and the intentional dropping of objects or of paratroopers; (c) launch and recovery of space vehicles; and (d) operations in areas of conflict, or the potential for armed conflict, when such operations include a potential threat to civil air traffic.[66] ATS authorities should be alert to military operations in areas of conflict, or the potential for armed conflict, when such operations include a potential for hazardous activity, and react accordingly. If the potentially hazardous activities are planned to take place on a regular or continuing basis, a co-ordinating group should be given the task of ensuring that the operational needs of all parties concerned are adequately co-ordinated. This group should consist of representatives of the military organization(s) concerned, the appropriate ATS authorities and the operators of civil aircraft.[67]

Although the Chicago Convention does not constrain States or require them to apply provisions of the Convention in instances of armed conflict,[68] nonetheless close coordination between civil and military authorities become critical. The responsibility for initiating the co-ordination process rests with the States whose military forces are engaged in the conflict. The responsibility for instituting special measures to ensure the safety of international civil aircraft operations remains with the States responsible for providing air traffic services in the airspace affected by the conflict, even in cases where co-ordination is not initiated or completed. Based on the information which is available, the State responsible for providing air traffic services should identify the geographical area of the conflict, assess the hazards or potential hazards to international civil aircraft operations, and determine whether such operations in or through the area of conflict should be avoided or may be continued under specified conditions. An international Notice to Airmen (NOTAM) containing the necessary information, advice and safety measures to be taken should then be issued and subsequently updated in the light of developments. All those concerned with initiating and issuing of NOTAM should be aware of the provisions governing the duration of the published NOTAM. Annex 15, 5.3.1.2 states that a NOTAM given Class I distribution shall be superseded by a NOTAM given Class II distribution when the duration of the circumstances notified is likely to exceed 3 months or the NOTAM has remained in force for three months. A copy of the NOTAM should be forwarded to the appropriate regional office of ICAO.[69]

---

[66] *Id.* 3.2.

[67] *Id.* 3.2.1 and 3.3.

[68] Article 89 states that In case of war, the provisions of this Convention shall not affect the freedom of action of any of the contracting States affected, whether as belligerents or as neutrals. The same principle shall apply in the case of any contracting State which declares a state of national emergency and notifies the fact to the Council.

[69] Doc 9554, *supra*, note 46 at 10.2 and 10.3.

1.1.2.1.2 Resolutions and Annex Provisions

Airspace is a resource common to both civil and military aviation, and given that many air navigation facilities and services are provided and used by both civil and military aviation, the ICAO Assembly, at its 38th Session (September/October 2013) adopted Resolution A38-12[70] which laid down that the common use by civil and military aviation of airspace and of certain facilities and services shall be arranged so as to ensure the safety, regularity and efficiency of civil aviation as well as to ensure the requirements of military air traffic are met. It was also resolved that the regulations and procedures established by Member States to govern the operation of their state aircraft over the high seas are required to ensure that these operations do not compromise the safety, regularity and efficiency of international civil air traffic and that, to the extent practicable, these operations comply with the rules in Annex 2 to the Chicago Convention (Rules of the Air). The State concerned should coordinate the matter with all States responsible for the provision of air traffic services over the high seas in the area in question. The Secretary General of ICAO would provide guidance on best practices for civil/military coordination and cooperation.

Annex 11 to the Chicago Convention, which addresses the subject of air traffic services, lays down requirements for coordination of activities that are potentially hazardous to civil aircraft. The International Standards and Recommended Practices in the Annex, Chapter 2 (2.17 and 2.18 in particular), contain provisions for co-ordination between military authorities and air traffic services and co-ordination of activities potentially hazardous to civil aircraft. These provisions specify that air traffic services authorities are required to establish and maintain close co-operation with military authorities responsible for activities that may affect flights of civil aircraft. The provisions also prescribe that the arrangements for activities potentially hazardous to civil aircraft shall be co-ordinated with the appropriate air traffic services authorities and that the objective of this co-ordination shall be to achieve the best arrangements which will avoid hazards to civil aircraft and minimize interference with the normal operations of such aircraft. Standard 2.17.1 stipulates that arrangements for activities potentially hazardous to civil aircraft, whether over the territory of a State or over the high seas, shall be coordinated with the appropriate air traffic services authorities, such coordination to be effected early enough to permit timely promulgation of information regarding the activities in accordance with the provisions of Annex 15 to the Chicago Convention. Standard 2.17.2 of Annex 11 explains that the objective of the coordination referred to in the earlier provision is to achieve the best arrangements that are calculated to avoid hazards to civil aircraft and minimize interference with the normal operations of aircraft. One must of course hasten to add that Article 89 of the Convention stipulates that in case of war, the provisions of the Convention (and, by implication

[70] Resolution A38-12: Consolidated statement of continuing ICAO policies and associated practices related specifically to air navigation, Appendix I.

its Annexes) shall not affect the freedom of action of any of ICAO's member States affected, whether as belligerents or as neutrals. The same principle would apply in the case of any member State which declares a state of national emergency and notifies the fact to the ICAO Council.

Another challenge would lie in the testing by States of missiles. If the response of State authorities who fire missiles into the air without paying heed to applicable regulations and guidelines were to be that, since the State concerned has sovereignty over its airspace (as recognized by Article 1 of the Chicago Convention) and that whatever it does over its airspace is its concern, it must be pointed out that air routes are used by airlines of various nationalities carrying persons of various nationalities and that there must be recognition that the concept of sovereignty, in its pristine purity and simplistic interpretation cannot be sustained in this instance. The role of the State in the modern world is a complex one. According to legal theory, each State is sovereign and equal. In reality, with the phenomenal growth in communications and consciousness, and with the constant reminder of global rivalries, not even the most powerful of States can be entirely sovereign. Interdependence and the close knit character of contemporary international commercial and political society ensures that virtually any action of a State could well have profound repercussions upon the system as a whole and the decisions under consideration by other States.

1.1.2.2 Managing Risk

1.1.2.2.1 ICAO's Role

One of the key tasks that the ICAO Task Force has identified in its report is the identification of types of information required for making reliable risk assessments, including reputable open source information which could augment existing risk assessments.[71] The preceding discussion reflects the many creditable initiatives and efforts of ICAO with regard to issuing guidance material on the safety of air navigation. However, ICAO should start by assessing its own risk of plunging into an abyss of uncertainty as to its role in leading the aviation community with regard to conflict zones and aviation. A perceived inadequacy of the Report of the Task Force is that it ascribes and allocates every function in regard to aviation over conflict zones primarily to the ICAO Secretariat, which the report claims would be supported by States and the industry. By doing so the Task Force has upended the legal situation. Instead of placing the main focus on States and their responsibility, the Task Force has devolved the entire responsibility on the ICAO Secretariat.

ICAO's aims and objectives in this regard are clear under the Chicago Convention: it has to develop principles and techniques of air navigation[72] and, in this

[71] *Supra*, note 49 at paragraph 1.

[72] Article 44.

regard inter *alia meet* the needs of the people of the world for safe, regular, efficient and economical air transport.[73] Risk management and providing advice to the traveller on how to avoid risk does not come anywhere near these objectives. Safety, which is one of ICAO's aims, relates to the safety of the act of aeronautical navigation of the aircraft according to technical requirements specified and not in accordance with principles of avoidance of political risks that may be posed as a result of conflict. Besides, Article 9 of the Convention (which has already been discussed) clearly lays down the principle of State responsibility in terms of diverting aircraft in times of military necessity. Article 89 effectively precludes the application of the Chicago Convention in war situations in the context of the Convention prescribing the manner in which States should act, thus obviating any involvement of ICAO (which was established by the Convention and derives its authority therefrom). Therefore it is the States that should take the initiative and agree on strengthening risk management and ensuring the safety of aircraft over their territories and of course, by all means, they could call upon the expertise of the ICAO Secretariat to support them in this endeavour.

ICAO is a club of States and its activities and output are determined by the extent to which the member States empower ICAO. That does not mean that a matter in which the sole responsibility devolves upon the States should be thrown back at the ICAO Secretariat when the Council—a representative body of States—is resident at ICAO. It is also somewhat surprising and indeed confusing that, in this instance, the Council did not establish a subordinate body of experts representing the ICAO member States to take on the task of developing risk management principles for the States to follow. The Committee on Aviation Environmental Protection (CAEP)—a group of experts that develops recommendations for States—and the ICAO Legal Committee (representing ICAO member States) that by itself develops and drafts multilateral treaties with only the support of the ICAO Secretariat, are examples.

1.1.2.2.2 Understanding Risk Management

The ICAO Task Force Report speaks of "risk assessments". Current societal thinking is more geared toward considering risk as danger. Whoever takes on the task of developing best practices based on risk management of air transport over conflict zones has first to have a clear concept of what risk means. One philosopher has argued that the western world is moving towards creating a society that is risk based which hovers between security and destruction, where one's overriding perception of threatening risks drives the thinking process that eventually leads to action.[74] Today's risk society is the result of globalization where boundaries have been lifted and interaction between the State and private actors are both seamless and prolific. The air transport industry is no different. On the one hand there is the

---

[73] Article 44 d).

[74] Ulrich Beck, *World Risk Society*, 135 cited in Leloudas (2009), p. 11.

State with its legally entrenched claim to sovereignty and resultant political calluses. On the other hand is the airline, often a private economic entity trying to ply its trade through sovereign State territory. The risk assessment should therefore be founded on the fundamental postulate of State responsibility to the airline in ensuring that persons and property on board are free of risk of danger while its aircraft traverses over State territory.

Here again responsibility devolves squarely on the State and the risk is often attributable to the State. One commentator observes:

> In the global risk society, no one any longer knows with certainty the extent of the risks we face through our collective technologies and innovations. Science now fails us, with conflicting reports, contradictory assessments and wide variance in risk calculations. Faith in the risk technocrats evaporates, the hegemony of experts dissolves and risk assessment becomes no more than a political game that advances sectional interests.[75]

In the air transport context the responsibility of the State becomes even more prominent since it is the air carrier that has to make amends to the injured or the deceased's family when an accident on board occurs.[76] Conflict zones often portend danger to civil aviation as has been demonstrated in recent events where the airport of Israel was threatened by rockets fired by HAMAS and man portable air defence systems (MANPADS) have been used against aircraft elsewhere. The evaluators of risk assessment and risk management have therefore to measure risk to civil air transport in terms of geo political reality as well as the true meaning of war and conflict which appear to be blurred at law. War is conventionally defined as a behaviour pattern of organized violent conflict typified by extreme aggression, societal disruption, and high mortality. This behaviour pattern involves two or more organized groups War has also been defined as a course of hostility engaged in by entities that have at least significant attributes of sovereignty.[77] Courts have recognized that there could be war between quasi sovereign entities[78] and that undeclared *de facto* war could exist between sovereign States.

The term *hostilities*, as in the case of the situation in Donetsk when Flight MH 17 was shot down, has been defined as acts or operations of war committed by belligerents.[79] Justice Bailhache, in handing down his decision in the 1919 case of *Atlantic Mutual Insurance Co.* v. *R*[80] said:

---

[75] Jarvis (2007), p. 32.

[76] Article 17 of the *Montreal Convention of 1999* prescribes *inter alia* that the carrier is liable for damage sustained in case of death or bodily injury of a passenger upon condition only that the accident which caused the death or injury took place on board the aircraft or in the course of any of the operations of embarking or disembarking.

[77] *British Steamship Co.* v. *R* [1921] 1 AC 99.

[78] *Pan America World Airways Inc* v. *Aetna Casualty and Surety Co*, [1974] 1 Lloyds Rep. 207 at 211.

[79] See Margo (2000), p. 338.

[80] [1919] 1 K.B. 307.

> The word "hostilities" means hostile acts by persons acting as the agents of sovereign powers, or of such organized and considerable forces as are entitled to the dignified name of rebels as contrasted with mobs or rioters and does not cover the act of a mere private individual acting entirely on his own initiative, however hostile his action may be.[81]

Therefore the terms "war", "conflict" and "hostilities" bring to bear the essential link with the State as the protagonist which has to take primary responsibility for promulgating best practices in risk assessment and management.[82] One commentator, who has proposed a treaty that would impose State liability for acts of unlawful interference against aircraft, has commented that the distinctive feature of his proposed treaty is that it would make States, as opposed to operators, answerable for damage caused on the surface of the earth by aircraft as a result of hijacking or other unlawful interference. He goes on to say that holding governments fiscally responsible and accountable and therefore liable for aircraft hijackings and terrorism is not in-and-of-itself a novel idea.[83] One such precedent cited is the Libyan Government's agreement to settle lawsuits concerning the destruction of PANAM flight 103 over Lockerbie, Scotland in 1988.[84]

The overriding and fundamental theme of risk avoidance should involve the realization that current political and diplomatic problems mostly emerge as a result of the inability of the world to veer from its self-serving concentration on individual perspectives to collective societal focus. This distorted approach gives rise to undue emphasis being placed on rights rather than duties; on short-term benefits rather than long-term progress and advantage and on purely mercantile perspectives and values rather than higher human values. Another sensitivity is the thin line which exists between international law and international politics, which, when applied to aviation becomes even thinner and the responsibility of the State becomes even stronger.

Two fundamental issues emerge from the foregoing discussion: whether, in the face of Articles 9 and 89 of the Chicago Convention, ICAO stands stultified in taking the "lead" as so far claimed, in prescribing best practices in risk management and providing advice to travellers on how to avoid conflict zones; and whether the approach that so far stands on its head—that primary responsibility which devolves on the ICAO Secretariat—should be shifted to the States themselves so that they could come up with an appropriate scheme of prevention and pre-emption of attacks against aircraft in conflict zones through geo political risk assessment and management.

---

[81] *Id.* at 310.

[82] In March 2011, the United Nations Security Council adopted Resolution 1973 which *inter alia* decided to establish a ban on all flights in the airspace of the Libyan Arab Jamahiriya in order to help protect civilians. The Resolution also authorized Member States to take all necessary measures, to protect civilians and civilian populated areas under threat of attack in the Libyan Arab Jamahiriya, including Benghazi.

[83] Petras (2004).

[84] *Smith* v. *Socialist Peoples' Libyan Arab Jamahiriya*, 866 F. Supp 306 (1995).

As the discussion in this article reflects, the vast number of initiatives of ICAO in providing manuals and guidance to States on the safety of aircraft over their territories and the harmonized use of a common airspace by both civil and military aviation are indeed worthy of credit of an Organization which has amply fulfilled its role in the field of air navigation. However, in some instances, ICAO has shown a feckless tendency to jump in too soon without preliminary assessment of the legality of its ventures, whether it be on the subject of commercial space transport or risk management over conflict zones. This should be rectified as any ICAO venture should start primarily with an examination of *locus standi* and legal justification.

1.2 Cyber Threats to Human Rights in Aviation

The disappearance of Malaysia Airlines Flight 370 in March 2014 brought to bear the fearful possibility that the event could be the start of cyber attacks on aircraft. Philip Ross writing in the *International Business Times* of 16 March 2014 said: "As the search for the missing Malaysia Airlines Flight MH 370 continues, investigators have come across some startling evidence that the plane could have been hijacked using a mobile phone or even a USB stick. The theory comes from a British anti-terrorism expert who says cyber terrorists could have used a series of 'codes' to hack the plane's in-flight entertainment system and infiltrate the security software. According to Sally Leivesley, a former scientific adviser to the UK's Home Office, the Boeing 777's speed, direction and altitude could have been changed using radio signals sent from a small device. The theory comes after investigators determined that someone with knowledge of the plane's system intentionally flew the jet off course".

Author Michael Hanlon envisions the consequences of a cyber attack: "at first, it would be no more than a nuisance. No burning skyscrapers, no underground explosions, just a million electronic irritations up and down the land. Thousands of government web pages suddenly vanish... the disruption continues: thousands of popular websites, from eBay to YouTube, start malfunctioning or are replaced by malicious parodies. Tens of millions of pounds are wiped off the share price of companies like Amazon as fears grow that the whole Internet credit card payment network is now vulnerable and insecure... eventually, reports start to flood in that hundreds of thousands of personal bank accounts have been raided overnight".[85]

James D. Zirin, writing to the *Washington Times* said: "It is an irony of the digital age that technology has aided the security forces in detecting and thwarting terrorist operations and has helped terrorists do their evil" (http://bit.ly/d41gsV).

In taking action against cyber crimes, then US President Bill Clinton, in a 1999 speech to the National Academy of Sciences said: "open borders and revolutions in

---

[85] See Hanlon (2007).

technology have spread the message and the gifts of freedom, but have also given new opportunities to freedom's enemies... we must be ready... ready if our adversaries try to use computers to disable power grids, banking, communications and transportation networks, police, fire, and health services—or military assets".

Jack Phillips writing to *EpochTimes* of 17 March 2014 recounts the instance in 2013, at a conference entitled "the Hack in the Box Conference", that: "security researcher Hugo Teso went on stage and took out his phone. He accessed an app, *Planesploit*, that he coded himself, which he said could affect a plane's navigation systems. Teso, who is a researcher, said that he could theoretically change a plane's route and make it crash with the app. He reportedly did a demonstration on stage to show that systems on board planes are vulnerable".

This was reportedly discounted by regulators of the United States as impossible on the ground that such a hacking technique would not work on certified flight hardware incorporated into the avionics system of a certified commercial jetliner. It was reported in the article that the authorities had further said: "the described technique cannot engage or control the aircraft's autopilot system using the FMS or prevent a pilot from overriding the autopilot. Therefore, a hacker cannot obtain 'full control of an aircraft' as the technology consultant has claimed". This view had been corroborated by the European Aviation Safety Agency (EASA) which had downplayed Teso's findings, on the basis that embedded software has a "robustness that is not present on ground-based simulation software".

Be that as it may, and since anything is up for grabs in the search for the missing aircraft, with theories galore, the cyber attack theory deserves at least a cursory discussion.

Cyber crimes and cyber terrorism are becoming increasingly menacing and the latter has been identified internationally as a distinct threat requiring attention. At the 21st Aviation Security Panel Meeting of The International Civil Aviation Organization (ICAO) (AVSECP/21, 22–26 March 2010) a new Recommended Practice related to cyber threats was proposed for adoption by the Council as part of amendment 12 to Annex 17 (Security) to the Convention on International Civil Aviation (Chicago Convention). It was adopted on 17 November 2010, became effective on 26 March 2011 and applicable on 1 July 2011. This Recommended Practice suggests that each Contracting State develop measures in order to protect information and communication technology systems used for civil aviation purposes from interference that may jeopardize the safety of civil aviation. At the 22nd Meeting of the Panel, conducted by ICAO from 21 to 25 March 2011, the Panel noted the value of vulnerability assessments pertaining to cyber security in aviation whose objectives are to evaluate the efficiency of existing mitigation measures and identify any vulnerabilities from a threat-based perspective and further noted that better understanding of residual risks will support a State's efforts to refine its risk response.

Member States of ICAO, at ICAO's 38th Session of the Assembly (September/October 2013) adopted Resolution A38-15 (Consolidated statement of continuing ICAO policies related to aviation security) which *inter alia* strongly condemned all acts of unlawful interference against civil aviation wherever and by whomsoever and for whatever reason they are perpetrated. The Resolution noted with abhorrence acts and attempted acts of unlawful interference aimed at the destruction in flight of civil aircraft including any misuse of civil aircraft as a weapon of destruction and the death of persons on board and on the ground and reaffirmed that aviation security must continue to be treated as a matter of highest priority and appropriate resources should be made available by ICAO and its Member States.

The first international convention which addressed the subject of interference with aviation by cyber hacking was the 2010 *Convention on the Suppression of Unlawful Acts Relating to International Civil Aviation* adopted in Beijing (which unhappily is not in force, still begging for 22 ratifications) which China has signed (but not ratified) and Malaysia is yet to sign. The Convention states in Article 1d) that an offence is committed when a person destroys or damages air navigation facilities or interferes with their operation, if any such act is likely to endanger the safety of aircraft in flight. This clearly refers, *inter alia*, to cyber terrorism, yet anomalously links the offence exclusively to the safety of aircraft in flight. Article 2a) of the Convention provides that an aircraft is considered to be in flight at any time from the moment when all its external doors are closed following embarkation until the moment when any such door is opened for disembarkation. In the event of a forced landing, the flight would be deemed to continue until the competent authorities take over responsibility for the aircraft and for persons and property on board.

Perhaps the events unfolding from MH 370 would spur more States to ratify the Beijing Convention?

More generally, yet with relevance to the field of aviation, are the activities conducted since the 1980s by international organizations such as the United Nations, Council of Europe, INTERPOL, and the Organization for Economic Co-operation and Development in response to the challenges posed by cyber crime. A significant result of such collective efforts was the publication of the *United Nations Manual on Cybercrime* and 2001 United Nations Resolution (*United Nations Resolution on Combating the Criminal Misuse of Information Technologies* General Assembly Resolution 55/63) exhorting States, in the context of an earlier United Nations Resolution on Millennium Goals, which recognized that the benefits of new technologies, especially information and communication-related technologies, are available to all, to ensure that their laws and practices eliminate safe havens for those who criminally misuse information technology. The Resolution also urged States to ensure the cooperation of law enforcement authorities in the investigation and prosecution of international cases of the criminal misuse of information technology, and that this should be coordinated among all concerned States. The Resolution further required information to be exchanged between States regarding the challenges faced in combating such criminal misuse and stated that law enforcement personnel should be trained and equipped to address any criminal misuse of information technology.

A particular feature of cyber terrorism is that the threat is enhanced by globalization and the ubiquity of the Internet. Given such a global problem, requiring a global solution, the one forum that can provide a global framework against cyber terrorism is ICAO. A sustained global process of security risk assessment is the first necessary step. One definition of security risk assessment considered by the ICAO Aviation Security Panel at its Twenty-second Meeting was: "an outcome based process, coordinated by the Appropriate Authority utilising all appropriate resources, consisting of an analysis of prevailing threat factors compared against current mitigation measures, with a view to determining levels of risk that result in the application of appropriate mitigation measures".

In pursuance of these objectives, ICAO, in collaboration with its Member States, could undertake a study to identify critical aviation information systems; review the effectiveness of existing mitigation measures established for such systems; identify any vulnerabilities in current security arrangements; analyse best practices on how to address these vulnerabilities; and determine how to better manage identified residual risks.

1.3 MH 370 and Other Threats to Human Rights

Every now and then, the world of aviation gets a rude awakening which makes the aviation community spring into action to take steps in closing loopholes in aviation security. One such example was the attempt by shoe bomber Richard Reid who in December 2001 tried to blow up an American Airlines jet flying from Paris to Miami. He was foiled as he fumbled with matches and attempted to detonate a bomb hidden in his shoe. Reid was spotted by a stewardess and overpowered just in time. The ICAO Council sprang into action immediately after this event and that is how the prohibition of liquids, aerosols and gels on aircraft came into being as an international requirement. Malaysia Airlines Flight MH 370 is another disturbing wake-up call which brings to bear the need for cohesive and coherent international requirements that mandate States to check the genuineness and integrity of travel documents of those who board aircraft in their territories. ICAO's attempts to address this problem in its Annexes and Security Manual are non-existent, except perhaps for a jumbled and convoluted approach of confining travel documents purely to the realm of facilitation, thereby relegating them to the background.

A Malaysian Airlines aircraft which took off from Kuala Lumpur for Beijing at 12.41 a.m. on Saturday, 8 March 2014 lost contact with air traffic control 2 h into the flight. The Boeing 777-200 carrying 239 people including 12 crew members carried 14 nationalities (mostly Chinese but Malaysian, Indonesian and Australians as well) on board. There were infants and toddlers on board. It was reported that the plane had lost contact with Vietnamese air traffic control in Ho Chi Minh City at 1.21 a.m. and that it had deviated course from the North to the West and down towards the Southern Indian Ocean. At the time of writing—more than 2 weeks later—the Prime Minister of Malaysia officially announced (on 24 March 2014) that it was assumed that Flight MH 370 had terminated in the Southern Indian Ocean and that there were no survivors. The search however, went on.

1.3.1 Malaysia Airlines

Malaysia Airlines has a good record of operational safety and efficiency since its first commercial flight in 1947 as Malayan Airways and many iterations thereafter.[86] Malaysia was formed in 1963 causing the airline to change its name to Malaysian Airlines Limited (MAL), which, almost immediately thereafter incorporated Borneo Airways into MAL.

In 1965, with the separation of Singapore from Malaysia, MAL became a bi-national airline with a tie up between Malaysia and Singapore and was renamed Malaysia–Singapore Airlines (MSA). A new logo was introduced and the airline grew exponentially with new services to Perth, Taipei, Rome and London. However, in 1972, the partners went separate ways. Malaysia introduced Malaysian Airlines Limited, with its establishment as a standalone airline. Malaysia Airlines which was subsequently renamed Malaysian Airline System and took to the skies on 1 October 1972. It was later branded Malaysia Airlines and continues to be the country's national flag carrier.

In 2012, Malaysia Airlines operated 105,077 flights, averaging 292 flights per day system-wide. With a fleet of 109 aircraft, the airline carried 13.4 million passengers to 60 destinations in 30 countries across Asia, Australasia, Europe, and North America. On Time Performance (OTP) improved considerably to 87.22 % in 2012, the best in 5 years.[87] On 1 February 2013, Malaysia Airlines became a full-fledged member of *Oneworld* alliance and is now connected to some 850 destinations in 150 countries across the *Oneworld* alliance network. The airline holds a lengthy record of service and best practices excellence, having received more than 100 awards in the last 10 years.

1.3.2 Flight MH 370

At the time of writing, it was assumed by all concerned (for lack of any specific information or data that might lead to credible scenarios) that the ill-fated aircraft operating Flight MH 370 had plunged into the sea. No one ventured to even hazard a conjecture as to what could have happened, except that aviation experts who were interviewed by the media confirmed that both Malaysia Airlines and the B 777 aircraft had splendid safety records. There were two troubling enigmas that surfaced: there were no distress calls at all from the flight deck in an aircraft flying at an altitude of 35,000 ft; and two of the passengers were travelling with stolen passports of Austrian and Italian nationality (both Austria and Italy have reportedly confirmed that none of their nationals were on board). Both had bought their tickets simultaneously. Although this led to a possible terrorist angle, this theory was

[86] The airline had two previous accidents. The first was in 1977 which occurred as a result of a hijacking and the other took place in 1995 where a plane crash landed.

[87] Malaysian Airline System Berhad, *Annual Report 2012*, at 012.

abandoned on the ground that at least one of the two (both being Iranian nationals) may have been travelling to Germany for purposes of immigration and that none of the two was connected to any terrorist group. However, the fact that the plane had veered course to the left instead of proceeding north to Beijing brought in various possible scenarios and theories: that it may have been a crew problem; a passenger illness that necessitated landing at the nearest airport; or even a hijacking attempt and all these left speculation open. The stark and incontrovertible fact behind all this intrigue was the portentous danger of stolen passports being used prolifically in international air travel that may be a serious threat to aviation security.

Of course neither the Malaysian civil aviation authorities nor Malaysia Airlines stands alone in the omission of not properly checking the passports against available databases, as claimed by several experts. A formal statement issued by the International Criminal Police Organization (INTERPOL)[88] INTERPOL confirmed that at least two passports—Austrian and Italian—recorded in its Stolen and Lost Travel Documents (SLTD) database[89] were used by passengers on board missing Malaysia Airlines flight MH 370. The Austrian and Italian passports were added to the database of SLTD database after their theft in Thailand in 2012 and 2013 respectively. After the disappearance of flight MH 370 INTERPOL conducted checks on all other passports used to board flight MH 370 which may have been reported stolen. INTERPOL further stated that there had been no checks of the stolen Austrian and Italian passports carried out by any country between the time they were entered into INTERPOL's database and the departure of flight MH 370. Therefore, INTERPOL was unable to determine as to how many other occasions these passports had been used to board flights or cross borders.[90]

1.3.3 Issues Involved

1.3.3.1 The Stolen Passports

Two persons travelled on Flight MH 370 on stolen passports[91] belonging to an Austrian national and an Italian national who had reported the thefts almost

[88] INTERPOL today is based in Lyon, France, and counts 188 member countries. The core tenet of the organization has not wavered since its inception—INTERPOL is still in the business of fighting transnational crime.

[89] In June 2002, with the support of the United Nations Security Council, INTERPOL created the Stolen or Lost Travel Document database (SLTD) to register all reported stolen and lost documents in order to prevent their misuse. ICAO has endorsed the use of this database without explicitly requiring its member States to ensure that they use the database in the embarkation of passengers on commercial aircraft in the territories.

[90] http://www.thestandard.com.hk/breaking_news_detail.asp?id=47179&icid=a&d_str.

[91] It is presumed that these two persons altered a genuinely issued document to make it fit the bearer who is not the person to whom the passport was issued See Machine Readable Passports—*ICAO Doc 9303*, Part I, Informative Appendix 3 to Section III at 2.2.

immediately after they were stolen a considerable time before the flight took off. As already mentioned, details of these passports were registered in the database of stolen passports maintained by INTERPOL. However, at the time of check-in, these passports had not been checked against the INTERPOL database. The Secretary General of INTERPOL stated:

> Whilst it is too soon to speculate about any connection between these stolen passports and the missing plane, it is clearly of great concern that any passenger was able to board an international flight using a stolen passport listed in INTERPOL's databases...This is a situation we had hoped never to see. For years INTERPOL has asked why should countries wait for a tragedy to put prudent security measures in place at borders and boarding gates...Now, we have a real case where the world is speculating whether the stolen passport holders were terrorists, while INTERPOL is asking why only a handful of countries worldwide are taking care to make sure that persons possessing stolen passports are not boarding international flights.[92]

The stolen passports issue, whether or not it had a link to the disappearance of the aircraft, is nonetheless worrying, as eloquently explained by the chief of INTERPOL. He clearly equivocated INTERPOL's concern over the years as to why countries should wait for a tragedy to happen to take security measures at borders and gates.

First of all, for States to take such security measures, they have to have the necessary guidance from whoever gives that guidance and direction. The United Nations specialized agency that has the responsibility to provide that guidance and direction is the International Civil Aviation Organization (ICAO).[93] At the 38th Session of the ICAO Assembly Member States endorsed five Strategic Objectives for ICAO among which is one concerning Security and Facilitation where ICAO "enhances global civil aviation security and facilitation. This Strategic Objective reflects the need for ICAO's leadership in aviation security, facilitation and related border security matters".[94]

In this context ICAO states that:

> With the collaboration of Member States, ICAO plays an essential leadership role in the field of aviation security with the ultimate goal of enhancing civil aviation security worldwide. To this end, its efforts are focused primarily on developing and coordinating

[92] Supra, note 91, *Ibid.*

[93] The International Civil Aviation Organization is the United Nations specialized agency dealing with international civil aviation. ICAO was established by the Convention on International Civil Aviation (Chicago Convention), signed at Chicago on 7 December 1944. Fifty-two States signed the Chicago Convention on 7 December 1944. The Convention came into force on 4 April 1947, on the thirtieth day after deposit with the Government of the United States. Article 43 of the Convention states that an Organization to be named the International Civil Aviation Organization is formed by the Convention. ICAO is made up of an Assembly, which is the sovereign body of the Organization composed of the entirety of ICAO member (Contracting) States, and a Council which elects its own president. The Assembly, which meets at least once every 3 years, is convened by the Council. The Council is a permanent organ responsible to the Assembly, composed of 36 Contracting States. ICAO has 191 Member States.

[94] http://www.icao.int/about-icao/Pages/Strategic-Objectives.aspx.

an effective global policy and legal framework in response to the evolving threat to civil aviation, conducting audits that identify aviation security shortcomings, and assisting States in implementing security Standards and resolving deficiencies.[95]

1.3.3.2 The ICAO Annexes

In the field of facilitation, ICAO's Standards and Recommended Practices (SARPs)[96] are contained in Annex 9 to the Chicago Convention.[97] Annex 9, in Standard 3.7 requires ICAO member States to regularly update security features in new versions of their travel documents, to guard against their misuse and to facilitate detection of cases where such documents have been unlawfully altered, replicated or issued. Recommended Practice 3.9 suggests that member States incorporate biometric data in their machine readable passports, visas and other official travel documents, using one or more optional data storage technologies to supplement the machine readable zone, as specified in Doc 9303, Machine Readable Travel Documents. The required data stored on the integrated circuit chip is the same as that printed on the data page, that is, the data contained in the machine-readable zone plus the digitized photographic image. Fingerprint image(s) and/or iris image(s) are optional biometrics for member States wishing to supplement the facial image with another biometric in the passport. Member States incorporating biometric data in their Machine Readable Passports are to store the data in a contactless integrated circuit chip complying with ISO/IEC 14443 and programmed according to the Logical Data Structure as specified by ICAO.

Recommended Practice 6.9.1 of the Annex (which is not a mandatory requirement but a recommendation) provides that each Contracting State, in consultation with airport and aircraft operators, should notify travellers, via signage, leaflets, video, audio, internet websites or other media, of the penalties for breaching regulations with regard to entry and departure and attempting to import or export any banned or restricted item. There is also Recommended Practice 6.17 which

---

[95] http://www.icao.int/Security/Pages/default.aspx.

[96] A *Standard* is any specification for physical characteristics, configuration, materiel, performance, personnel or procedure, the uniform application of which is recognized as necessary for the safety or regularity of international air navigation and to which Contracting States will conform in accordance with the Chicago Convention. In the event if impossibility of compliance, notification to the Council is compulsory under Article 38 of the Convention. A *Recommended Practice* is any specification for physical characteristics, configuration, materiel, performance, personnel or procedure, the uniform application of which is recognized as desirable in the interests of safety, regularity or efficiency of international air navigation, and to which Contracting States will endeavour to conform in accordance with the Convention. These definitions are taken from Annex 17 (Security) to the Chicago Convention. See Annex 17 to the Convention on International Civil Aviation (Security)—Safeguarding International Civil Aviation Against Acts of Unlawful Interference Eighth Edition: April 2006 at (vi).

[97] *Op.cit.*

suggests that each Contracting State should ensure that airport operators and the State's security or border control personnel use efficient screening and examination techniques, in security and border control examinations of passengers and their baggage, in order to facilitate aircraft departure. As to why this was a mere recommendation for States to choose or not to implement is a mystery.

There are of course no SARPs in the Annex requiring that States check the passports of checking-in passengers with any database maintained by a criminal police organization or institution. The only provision that might be applicable is Standard 3.46 which requires the public authorities of each Contracting State to seize fraudulent, falsified or counterfeit travel documents. This includes seizing the travel document of a person impersonating the rightful owner of the travel document. As to how the authorities would determine that a travel document is counterfeit or that a person is impersonating the rightful owner of the travel document is not specified or even hinted at. However, worthy of note is Recommended Practice 3.47 (again, just a recommendation) which suggests that wherever appropriate, Contracting States should introduce a system of advance passenger information which involves the capture of certain passport or visa details prior to departure and their transmission of electronic details to the destination prior to departure.

1.3.3.3 Inconsistencies and Inadequacies

There are glaring disconnects in these ambivalent provisions that detract from what was expected of ICAO as the central regulatory forum in the MH 370 context. The first is that these provisions are contained in the Facilitation Annex (Annex 9) and not in the Security Annex (Annex 17) where ICAO clearly veers from the security relevance of the provisions. They are merely given as an Attachment to Annex 17 as extracts of Annex 9. The second is that, apart from Recommended Practice 3.47 which is on advance passenger information that is relayed to the destination—explicitly as the recommendation says "to expedite clearance" at the destination—there is no provision for information to be determined prior to departure.

The jumble of provisions in two separate Annexes where provisions are irrelevantly placed in one Annex instead of the other, while the Strategic Objective of ICAO pertains to Security and facilitation, where common sense would require one consolidated Annex, seriously calls to question ICAO's claim of its "leadership" role in "responding to the evolving threat to civil aviation". Annex 17 to the Chicago Convention contains SARPs on aviation security. ICAO knows, or ought to know that in 2013 stolen passports were used over a billion times and that INTERPOL's SLTD records over a million stolen passports, as confirmed by INTERPOL in the aftermath of the disappearance of Flight MH 370.[98] In such a

[98] http://www.thestar.com/news/world/2014/03/09/stolen_passports_used_on_missing_malaysia_flight_werent_checked_in_INTERPOL_database.html.

scenario, should Annex 17 have specific Standards requiring States to check and screen embarking passengers with measures that are specified and certain?

Ironically, Standard 3.4.1 of Annex 17 requires each State to ensure that the persons implementing security controls are subject to background checks. The Annex also requires such persons to possess all competencies required to perform their duties[99] and be certified according to a national civil aviation programme.[100] There is also Recommended Practice 4.2.7 (again, not a Standard requiring compliance) pertaining only to airline crew which recommends that each Contracting State should ensure that identity documents issued to aircraft crew members provide a harmonized and reliable international basis for recognition and validation of documentation to permit authorized access to airside and security restricted areas.

The only provision in Annex 17 which is even remotely connected to the screening of passports is found in Standard 4.4.1 which requires Contracting States to establish measures to ensure that originating passengers of commercial air transport operations and their cabin baggage are screened prior to boarding an aircraft departing from a security restricted area.[101] Ironically, the word "screening" is defined in the Annex as "the application of technical or other means which are intended to identify/and or detect weapons, explosives or other dangerous devices, articles or substances which may be used to commit an act of unlawful interference",[102] effectively ignoring the screening of passports. Without a specific requirement in Annex 17 that Contracting States should link their software to the INTERPOL SLTD, how could one expect any State to have a mandatory directive to screen passports of checking in passengers?

1.3.3.4 The ICAO Security Manual

If the provisions of Annex 9 and Annex 17 are too vague and inadequate to sufficiently direct States to ensure that stolen passports are not used, at least the ICAO Security Manual[103] should, as it claims in its introduction that it "provides details of how States can comply with the various SARPs of Annex 17 of the Chicago Convention" should have filled this lacuna. Regrettably, there is no mention in the Manual of stolen, fraudulent or altered passports being a threat to

---

[99] Standard 3.4.2.

[100] Standard 3.4.3.

[101] Annex 17 defines a security restricted area as an area of the airside of an airport which are identified as priority risk areas where in addition to access control, other security controls are applied. A security area includes all commercial passenger departure areas between the screening checkpoint and the aircraft. See Annex 17 to the Convention on International Civil Aviation, Eighth Edition: April 2006 at Chapter 1—Definitions.

[102] *Ibid.*

[103] *Security Manual for Safeguarding Civil Aviation Against Acts of Unlawful Interference*, Doc 8973, Sixth Edition: 2002.

security or their proper identification and screening against available databases or lists. Chapter 4 addresses preventive security measures which involve passenger terminal security; surveillance; briefing of public; terminal personnel; VIP facilities; extra security measures (which do not address passengers) and a host of other security resources with no mention of passengers or their travel documents. The same chapter has a sub paragraph on screening of passengers and cabin baggage. Here "screening" is defined by the Manual the same as in Annex 17 as:

> The application of technical or other means which are intended to identify and/or detect weapons, explosives or other dangerous devices which may be used to commit an act of unlawful interference.[104]

The Manual states that the screening of all passengers and their cabin baggage is recognized as an essential element in the fabric of safeguarding measures advocated by ICAO, the basic rule being that all passengers and all their cabin baggage must undergo screening before being allowed access to an aircraft Such screening may be carried out by the use of metal detectors, X-ray equipment or other explosive detection devices, by manual search or a combination of equipment and manual search.[105] There is no mention in the Manual of a tie up between States and INTERPOL that could strengthen aviation security and minimize threats to unlawful interference with civil aviation.

The 38th Session of the ICAO Assembly, held in September/October of 2013 in one of its Resolutions urges ICAO Member States to intensify their efforts for implementation of existing SARPs and procedures relating to aviation security, to monitor such implementation, to take all necessary steps to prevent acts of unlawful interference against international civil aviation and to give appropriate attention to the guidance material contained in the Aviation Security Manual.[106] This does not mean anything at all in terms of the threat posed by stolen passports as the Manual deals with physical screening of the person of the passenger and his baggage.

Interestingly, the same Resolution recognizes that machine readable travel documents strengthen security by improving the integrity of the documents which verify the identity of the travellers and air crew, and that they enable high level cooperation among States to strengthen resistance to passport fraud, including, *inter alia* the use of valid passports by imposters and the use of fraudulently obtained passports. There is no mention in the Resolution of cooperation between States and INTERPOL although, on the subject of cargo, the Resolution makes mention of such organizations as the World Customs Organization, Universal Postal Union and the International Maritime Organization.[107]

Consistent with its practice of confining travel documents solely to the area of facilitation, as the preceding discussion showed, and totally ignoring the relevance

---

[104] *Id.* Chapter 1—Definitions at 1-1-4.

[105] *Id.* Chapter 4.2 at 1-4-7.

[106] Resolution A38-15 *Consolidated statement of continuing ICAO policies related to aviation security*, Appendix C.

[107] *Ibid.*

of the need to check passports against international databases, another Resolution of the 38th Assembly—on facilitation—waxes eloquent on national and international action in ensuring the security and integrity of traveller identification and border controls. Ironically, this Resolution[108] although calling for national and international cooperation, focuses only on cooperation between States on machine readable travel documents (e-Passports) and the Public Key Directory (PKD)[109] as tools that strengthen the security and integrity of border control. The problem with the PKD is that it is limited in usage among ICAO's Member States where, by summer 2013 only 37 e-Passport issuers have already decided to join the PKD and take advantage of its benefits.[110]

1.3.3.5 Interaction Between States

During the first few days of searching for debris in the ocean by several countries including China, Malaysia and Vietnam, the Chinese Foreign Ministry in a press conference made the statement that China considered its responsibility to request Malaysia to step up its search and rescue operations. By this statement the Chinese authorities were implying that Malaysia and its national carrier had a responsibility towards ensuring that the Chinese citizens (and passengers of other nationalities of course) are found. Flight MH 370 was presumably operated on the legitimacy it was awarded under a bilateral air services agreement (BASA) between Malaysia and China. Typically the BASA does not specify what nationalities are to be carried between Kuala Lumpur and Beijing in a flight such as MH 370. However such agreements fall under the Vienna Convention on the Law of Treaties,[111] Article 2 of which defines a treaty as an international agreement concluded between States in written form and governed by international law, whether embodied in a single instrument or in two or more related instruments and whatever its particular designation.

The Vienna Convention while recognizing treaties as a source of law, accepts free consent, good faith and the *pacta sunt servanda* as universally recognized elements of a treaty.[112] Article 11 of the Vienna Convention provides that the

---

[108] Resolution A38-16 *Consolidated statement of continuing policies and practices related to facilitation*, Appendices B and C.

[109] For a discussion on the Public Key Directory see Abeyratne (2005), pp. 255–268.

[110] *ICAO MRTD Report*, Vol. 8, No. 2—Summer 2013 at 6.

[111] Vienna Convention on the Law of Treaties, which provides that such treaties shall be binding upon the Parties and be performed by them in good faith. The Vienna Convention further states that a treaty shall be interpreted in good faith in accordance with the ordinary meaning to be given to the terms of the treaty. Article 31.1 of the *Vienna Convention on the Law of Treaties* provides that "a treaty shall be interpreted in good faith in accordance with the ordinary meaning to be given to the terms of the treaty in their context and in the light of its object and purpose". See *Vienna Convention on the Law of Treaties 1969*, done at Vienna on 23 May 1969. The Convention entered into force on 27 January 1980. United Nations, *Treaty Series*, vol. 1155, p. 331.

[112] *Vienna Convention*, Preamble.

consent of a State to be bound by a treaty may be expressed by signature, exchange of instruments constituting a treaty, ratification, acceptance, approval or accession, or by any other means agreed upon. A State demonstrates its adherence to a treaty by means of the *pacta sunt servanda*, whereby Article 26 of the Vienna Convention reflects the fact that every treaty in force is binding upon the parties and must be performed by them in good faith.

Based on the above, one could argue that the principles of State responsibility would be critical in the operation of air services from one country to another and that a given State is obligated to take all necessary measures to ensure the safety of persons and property carried by its national carrier. State responsibility is a recognized principle of international law. The law of international responsibility involves the incidence and consequence of acts which are irregular at international law, leading to the payment of compensation for the loss caused. The sustainability of aviation security measures and arrangements is an important strategic issue for all entities with aviation security-related responsibilities and risk-based security measures, outcomes-focused security measures, rationalization of security measures, optimization of technology, mutual recognition of equivalence and one-stop security, harmonization, and preparedness for crisis events are policy principles and practices whose implementation can contribute significantly to the sustainability of aviation security measures and arrangements.

The need for each State to carry out continuous risk assessments as a preliminary measure is paramount for the sustainability of security. Aviation security has to be sustained in a balanced manner so that, on the one hand, the application of security measures to mitigate identified threats, and on the other hand, the essential task of facilitating operations, passengers' experience and trade could be ensured. Security should not accumulate layer upon layer of controls and associated costs, but should rather ensure the sustainability of the system from the perspectives of cost, efficiency, and acceptability by passengers and air transport operators, which should be a central consideration when designing security processes. In the case of a BASA a proactive measure to be adopted by both State Parties concerned would be to achieve and retain sustainability by following the practice of reciprocal acceptance of equivalent security measures across the board, with due regard to the principle of host State responsibility. In that respect, the need for any one State to require extra security measures of another State can be avoided by their working together to align international requirements to the global threat environment. It is this kind of approach that could sit well with Annex 17 to the Chicago Convention.

There is a breach of international obligation when an act or omission of a State is not in conformity with what is required of it by that obligation, regardless of its origin or character.[113] In the case of Flight MH 370, in the absence of an international obligation imposed by explicit requirements either by ICAO or INTERPOL as already discussed, it is highly questionable whether any country involved in the case would be guilty of a derogation of their international responsibility.

[113] *Gabcikove-Nagymaros Project case*, ICJ Reports 1997, pp. 7 and 38.

The above discussions both with regard to Flight MH 17 as well as to MH 370 demonstrate that there could be numerous lapses in aviation regulatory structures that could result in breaches of human rights both to individuals as well as whole societies. Both unfortunate events touched a wider range of persons than the passengers themselves who were involved. For instance, as the flaw in the MH 370 situation demonstrated, the passport is a basic document in the transport by air of persons. Its use therefore is of fundamental importance as a travel document, not only because it reflects the importance of the sovereignty of a State and the nationality of its citizens but also because it stands for the inviolability of relations between States that are linked through air transport. A passport asserts that the person holding the passport is a citizen of the issuing State while a visa confirms that the State issuing the visa has granted the visa holder the non-citizen privilege of entering and remaining in the territory of the issuing State for a specified time and purpose. An ePassport is a type of Machine Readable Passport[114] with an embedded microchip that contains data printed on the data page of the passport, including biographic and biometric information of the holder, and passport data. The chip also contains security features for preventing passport fraud and forgery and misuse of data stored on the chip. ePassports are easily recognised by the international ePassport symbol on the front cover.[115]

Stolen or forged passports portend a serious threat to aviation security, whether they are used by individuals or States. The assassination of a leader of Hamas on 19 January 2010 by a group of individuals in Dubai who used forged passports belonging to various nations, raised a diplomatic outcry and brought to bear an important facet of air transport that is vulnerable to abuse and contention among States.[116]

Although ICAO's disjointed references in its Annexes and Security Manual and the glaring lack of mention of the threat to the integrity of passports and the need for States to link themselves to the SLTD of INTERPOL is apparent, one must hasten

[114] The machine readable passport (MRP) is a passport that has both a machine readable zone and a visual zone in the page that has descriptive details of the owner. The machine readable zone enables rapid machine clearance, quick verification and instantaneous recording of personal data. Besides these advantages, the MRP also has decided security benefits, such as the possibility of matching very quickly the identity of the MRP owner against the identities of undesirable persons, whilst at the same time offering strong safeguards against alteration, counterfeit or forgery. See Abeyratne (1992), pp 1–31. See Also, Abeyratne (2013), pp. 23–39.

[115] http://www.dhs.gov/xtrvlsec/programs/content_multi_image_0021.shtm.

[116] On 19 January 2010, Mahmoud al-Mabhouh, considered to be a senior commander of Hamas, a radical Palestinian group, was assassinated at a hotel in Dubai in a manner usually employed by professionally trained military and secret service agencies. The assassins had used forged passports that belonged to Ireland, France, Germany and the UK—to coordinate their travel into Dubai from various parts of the world, synchronizing their arrival time from various flights into Dubai International Airport and checking into the hotel of the victim contemporaneously. See Abeyratne (2010b), pp. 32–45. Also published on line.

to add that INTERPOL should also bear some responsibility towards urging States to use its database. INTERPOL should also have a monitoring system to evaluate the use of the database by States. Merely attending meetings of each other as observers simply will not do for both ICAO and INTERPOL. Ideally an audit system should develop between ICAO and INTERPOL with both as equally responsible partners. MH 370 was a good enough wake up call for this purpose.

1.4 Does Global Tracking of Aircraft Ensure a Human Right?

The dangers that could be posed to life and limb on the ground as well as to the passengers on board if an aircraft suddenly vanishes off the skies would be ominous. At the time of writing, 9 months had passed since the disappearance of Malaysia Airlines Flight 370. Australian authorities had stopped searching for debris related to the disappearance of the aircraft in the area where the Australian Defence Vessel Ocean Shield (ADV Ocean Shield) detected acoustic signals in early April. Developments in the search led to a specific focus on the southern corridor, particularly areas of the southern Indian Ocean, about 2,000 km west of the Australian coastal city of Perth. In late May 2014 the Australian Transport Safety Bureau (ATSB) advised that the search in the vicinity of the acoustic detections was considered complete and in its professional judgement, the area was being discounted as the final resting place of MH 370.[117] However, the overall search for the debris will not stop and was said to recommence in August.[118]

Over 2 months had passed after the disappearance of the flight when the ICAO, after a long period of silence, convened on 12–13 May 2014 what was called a Multi Disciplinary Meeting Regarding Global Tracking of Aircraft. The meeting resulted in what ICAO called "Conclusions and Recommendations" which regrettably lacked specific details.

[117] http://www.flightglobal.com/news/articles/search-for-mh370-ceases-until-aug-ust-399793/.

[118] The Australian Government, which lead the search and recovery operations was bearing the cost but had sought contributions from other States. See http://the-conversation.com/who-will-bear-the-60m-cost-of-the-search-for-mh370-26050. It has been reported that the search and recovery operations had cost £25–30 m sterling as at May 2014. Most of the financial burden will be borne by the countries who have contributed their forces. Australia deployed a navy replenishment vessel, HMAS Success in early May. It cost AU$550,000 a day to operate, according to the Australian Department of Defence, amounting to $7.7 m ($7.2 m; £4.3 m) for the 2 weeks. See Missing Malaysian plane: How much will MH 370 search cost? http://www.bbc.com/news/world-asia-26927822.

1.4.1 Search for the Aircraft

The first question would be, for how long would States involved in the search and recover operations be obligated to go on searching?

The *Chicago Convention* requires contracting States of the International Civil Aviation Organization to fulfil their obligations under Article 25 which provides:

> Each Contracting State undertakes to provide such measures of assistance to aircraft in distress in its territory as it may find practicable, and to permit, subject to control by its own authorities, the owners of aircraft or authorities of the State in which the aircraft is registered to provide such measures of assistance as may be necessitated by the circumstances. Each Contracting State, when undertaking search for missing aircraft, will collaborate in coordinated measures which may be recommended from time to time pursuant to this Convention.

Annex 12 to the *Chicago Convention* elaborates on this fundamental requirement by qualifying that Contracting States are required to arrange for the establishment and provision of search and rescue (SAR) services within their territories on a 24 h basis. Contracting States are further requested to delineate the SAR process under the Annex on the basis of regional air navigation agreements and provide such services on a regional basis without overlap. A search and rescue region has been defined in the Annex as "an area of defined dimensions within which SAR service is provided". Annex 12 to the *Chicago Convention* requires Contracting States to coordinate their SAR organizations with those of neighbouring Contracting States with a recommendation that such States should, whenever necessary, coordinate their SAR operations with those of neighbouring States and develop common SAR procedures to facilitate coordination of SAR operations with those of neighbouring States. These provisions collectively call upon all Contracting States to bond together in coordinating both their SAR organizations and operations.

At the 32nd Session of the Assembly, held in 1998, ICAO adopted Resolution A32-14, Appendix O which addresses the provision of SAR services. This Resolution refers to Article 25 of the *Convention* in which each Contracting State undertakes to provide such measures of assistance to aircraft in distress in its territory as it may find practicable and to collaborate in coordinated measures which may be recommended from time to time pursuant to the *Convention*.

The Resolution mentions Annex 12 which contains specifications relating to the establishment and provision of SAR services within the territories of Contracting States as well as within areas over the high seas. The resolution recognizes that Annex 12 specifies that those portions of the high seas where SAR services will be provided shall be determined on the basis of regional air navigation agreements, which are agreements approved by the Council normally on the advice of regional air navigation meetings. Annex 12 also recommends that boundaries of SAR regions should, insofar as practicable, be coincident with the boundaries of corresponding flight information regions.

In the case of the Africa-Indian Ocean Region, the ICAO Regional Air Navigation Plan, in Part V addresses issues of Search and Rescue by pointing to the provisions of the ICAO Search and Rescue Manual (Doc 7333), referring in particular to the need for aircraft to carry specified equipment, carry out paper and communications exercises and, more importantly, for the need for States to pool their resources and provide mutual assistance in the case of SAR operations. The Plan calls for precise agreements between States to implement these measures. The ICAO Regional Plan also calls upon States, in order to ensure compatibility between aeronautical and maritime search and rescue regions (SRRs), and aeronautical search and rescue authorities, to maintain close liaison with their maritime counterparts and the International Maritime Organization (IMO).

Since the RAN Plan specifies such aspects State cooperation in SAR—and in the case of MH 370 it might not be search and rescue anymore but (sadly) search and recover—and also since the RAN Plan for the AFI region specifies a formula for costs of the SAR to be apportioned among the States participating in the operations, it would be for the States to decide as to the time limit of the operations. As a last resort, they can come to the ICAO Council, which is obligated by the *Chicago Convention* to consider "any matter referred to it"[119] by a State/s.

1.4.2 The ICAO Multi Disciplinary Meeting on Global Tracking

1.4.2.1 ICAO and Global Tracking

Technically, ICAO has been working on surveillance of aircraft under the Communications, Navigation, Surveillance and Air Traffic Management (CNS/ATM) system since 1972 when the Air Navigation Commission of ICAO recommended at its Seventh Air Navigation Conference that satellites be used for tracking civilian aircraft. This gave way in the early 1980s to the Future Air Navigations System (FANS). The culmination of this initiative came in 1991 at ICAO's Twelfth Air Navigation Conference which approved and endorsed the implementation of CNS/ATM system. It was inevitable that, after this endorsement, global tracking of aircraft would therefore technically come within the purview of the Global Navigation Satellite System (GNSS) which has three main players: the Global Positioning System (GPS) of the United States; the GLONASS system of the Russian Federation and Galileo of the European Commission. There are certain shortcomings of the GPS system which may affect the accuracy and timing of aircraft by satellites and augmentation systems, both ground based and satellite based, have been used to rectify these lapses.[120]

[119] Article 54 n).

[120] See Oster and Strong (2007), pp. 15–16.

The Global Positioning System (GPS) is founded on a constellation of 21 satellites orbiting the Earth at extremely high altitude. These satellites have been called "man made stars" that replace stars traditionally used for navigation through centuries in time.[121] The satellites in the GPS are known to use technology sufficiently accurate to pinpoint positions anywhere in the world around the clock.[122] This is achieved by using satellites and computers to triangulate positions anywhere on Earth. From an aviation perspective, GPS is considered to be the most expeditious and cost effective way that a fool-proof air collision avoidance system could be designed.

GPS uses a technique called "satellite ranging", which is based on measuring the distance of a target from a group of satellites in space, which become a precise reference point in the process of tracking. The distance to a satellite is determined by measuring the time taken by a radio signal to reach the tracking position on Earth from the satellite contacted.

1.4.2.2 ICAO's Response to Flight MH 370

As stated earlier, the ICAO reaction (one could not justifiably call it an initiative) to the disappearance of Flight MH 370 came after 2 months of the event. ICAO decided to convene a Multidisciplinary Meeting Regarding Global Flight Tracking of Aircraft on 12–13 May 2014. The meeting resulted in what ICAO called "Conclusions and Recommendations" which regrettably were riddled with ambivalent statements that lacked specific tasks and other details. The Conclusions and Recommendations are categorized into three time zones: near term; midterm; and long term. In the near term there is a blanket statement that "global tracking of airline flights will be pursued as a matter of priority to provide early notice of and response to abnormal flight behaviour". There is no explanation as to who will do the tracking and how it will be done. Then there is another statement for the near term that a draft concept of operations on flight tracking will be developed that includes a clear definition of the objectives of flight tracking that ensures that information is provided in a timely fashion to the right people to support search and rescue, recovery and accident investigation activities, as well as, the roles and responsibilities of all stakeholders. One is baffled by the term "draft concept of operations" as to what exactly it means. Who will develop this mysterious document or process?

Another unclear statement is "under the ICAO framework, the contribution by the industry through an Aircraft Tracking Task Force (ATTF) will help address the near-term needs for flight tracking". What is this "ICAO framework"? And what is the "industry" referred to here? Is it the manufacturing industry, the air transport

[121] Hurn (1989), p. 7.

[122] *Id.*, p. 8. These satellites assist delivery vehicles and emergency vehicles in getting to their destinations under accurate and prompt guidance.

industry or the service providers? Or is it the entire aviation industry which comprises all the above?

Perhaps the most confusing statement for the near term is "ICAO will *consider* establishing a short term joint ICAO/IATA advisory group to support the global tracking initiative". What is there to consider? When will ICAO make up its mind?

Some of the other befuddled statements are that ICAO should increase its resources allocated to the Search and Rescue in order to improve the effectiveness across national and regional boundaries; ICAO should, in collaboration with a pool of search and rescue experts, identify and address operational search and rescue challenges with implementation of existing Annex 12 provisions, and provide assistance to States, including aiding in the setting of priorities for the mid and long term; ICAO should facilitate the sharing of experience and lessons learned from States that were recently involved in accidents where flight tracking could have facilitated search and rescue efforts to all other States; and ICAO should strongly encourage States to regularly run practice exercises involving airlines operation centres, air navigation service providers (ANSPs) and rescue coordination centres (RCCs) to test and verify their ability to respond and coordinate together in an integrated manner to abnormal flight behaviour scenarios.

The use of the word "should" seems to give ICAO the choice and one wonders why these statements did not say "ICAO will…". The word "should" occurs throughout the mid-term and long term goals making confusion worse confounded.[123]

Tracking of airline flights essentially involves the ability of an airline to know the location of its aircraft at any given time, anywhere in the world. Tracking of airline flights are totally different from the provision of air traffic services (ATS), although ATS could be used in certain instances to provide necessary data in the tracking process. The ICAO High Level Safety Conference of 2010 examined instances of accidents over the high seas, responding to a request by the Accident Investigation Bureau of France following the Air France AF 447 disaster,[124] for an ICAO study on the possibility of making it mandatory for aircraft engaged in transporting the public to regularly transmit basic flight parameters. As expected, the Conference came up with a Declaration which stated that States have a collective responsibility to ensure aviation safety and that there was an increasing awareness of a compelling need for improvements in communications over oceanic and remote areas.[125]

---

[123] For the full text of the Conclusions and Recommendations see http://www.icao.int/Meetings/GTM/Pages/Documentation.aspx.

[124] Air France Flight 447 (AF447/AFR447) was a scheduled, international, long haul passenger flight, operated by the Air France from Rio de Janeiro to Paris. On 1st June 2009 the flight was being operated by Air France with an Airbus A 330 aircraft and, just after 02:10 UTC, the aircraft crashed into the Atlantic Ocean. All 228 passengers, and crew (both technical and cabin crew), aboard the Airbus perished.

[125] See Airline Flight Tracking: The Need and Available Technologies, Global Tracking 2014-WP/1, 5/5/14 Revised 9/5/14, p. 2.

Based on the premise that States and airlines have responsibility towards tracking aircraft under their realm of responsibility, and that there are some airlines that already have tracking systems in place, the ICAO Special Meeting on Global Flight Tracking of Aircraft concluded that there is a near term need for air operators to track their flights. It must be noted that Annex 6 to the *Chicago Convention* (Operations of Aircraft) obligates an air operator to establish and maintain a method of control and supervision of flight operations, but the Annex does not make it mandatory that an air operator track its flights. At the Special Meeting on Global Flight Tracking of Aircraft ICAO suggested that in developing new Standards for flight tracking a holistic review of other related Standards should be conducted. A performance-based approach to global flight tracking may be conducive to a system that could support other requirements, both current and under development, such as: transmission of basic flight data; implementation of triggered transmission of flight data; new generation of emergency locator transmitters (ELTs); deployable flight data recorders; and identification of the location of an accident site. ICAO also recommended that any future flight tracking Standards should also leverage the emergence of new technologies to support air traffic services (ATS) surveillance systems. The emergence of new technologies may not only provide increased capacity and efficiency in remote/oceanic airspace, but may also provide a future data source for flight tracking purposes. Consequently, the development of any future Standards addressing global flight tracking will require a multidisciplinary approach involving, at a minimum, expertise in such areas as flight operations, airworthiness, flight data and air traffic control.[126] ICAO invited the meeting to consider a multidisciplinary and performance-based approach towards the development and the formulation of any international Standards requiring global flight tracking; and also consider the various options for future Standards such as incorporation into Annex 6 and/or Annex 2 to the *Chicago Convention*.[127]

1.4.2.3 Views of Other Key Players

At the Conference, the International Air Transport Association (IATA)[128] proposed that an expert Task Force be convened to address the issue of global tracking of

[126] International Standards for Global Flight Tracking, Global Tracking 2014-WP/3, 5/5/2014, p. 2.

[127] *Id.*, p. 3.

[128] The International Air Transport Association, an association of air carriers, was formed in 1919 as the International Air Traffic Association. Encapsulated in IATA's overall mission are 7 core objectives: to promote safe, reliable and secure air services; to achieve recognition of the importance of a healthy air transport industry to worldwide social and economic development; to assist the air transport industry in achieving adequate levels of profitability; to provide high quality, value for money, industry-required products and services that meet the needs of the customer; to develop cost effective, environmentally-friendly standards and procedures to facilitate the operation of international air transport; to identify and articulate common industry positions and support the resolution of key industry issues; and to provide a working environment which attracts, retains and develops committed employees.

aircraft, such task force to include but not be limited to personnel from ICAO, airlines, flight safety organizations, manufacturers, air navigation service providers, flight tracking equipment and service providers, pilots and air traffic management experts. IATA will ensure that participation is limited to a manageable number and has the required skill mix to achieve desired outcomes.

IATA suggested that the Task Force examine available options for tracking commercial airplanes, considering implementation, time, complexity and cost efficiency to achieve the desired coverage. In particular, the Task Force will assess the responses to the ICAO vendor survey and examine existing capabilities to determine potential solutions against elements such as reporting parameters and intervals, reliability, accuracy and coverage. The Task Force will develop recommended options based on performance requirements to implement global airplane tracking. It was IATA's recommendation that the Task Force would develop a draft report with recommendations within 120 days. Following required coordination, the Task Force will issue its final report within 150 days of the Multidisciplinary Meeting.[129]

The European Union (EU) highlighted the fact that one obvious vulnerability of flight was brought to bear by Flight MH 370 in that there was possible intentional disconnection of airborne air traffic management (ATM) systems. The EU focused on a number of incidents that led to aircraft losses as a result of deliberate action of the pilot and stated that ground control should be alerted to the emergence of an unusual situation in flight. The EU strongly suggested that an approach be followed that addresses the needs of all relevant stakeholders (search and rescue specialists, accident investigators; air traffic controllers; flight safety/airworthiness/maintenance services) rather than letting them develop separate and incomplete solutions, in particular where their technological needs are rather similar, taking into account the current means already in place for search and rescue (COSPAS-SARSAT[130] infrastructure and current ELT requirements) and for locating the wreckage and the flight recorders under water.[131]

---

[129] See Industry Initiative to Develop Options for Global Flight Tracking, Global Tracking 2014-WP/4, 7/5/14, pp. 1–2.

[130] The International COSPAS-SARSAT Programme, which cooperates with ICAO, the International Maritime Organization (IMO) and the International Telecommunications Union (ITU), is a satellite-based search and rescue (SAR) distress alert detection and information distribution system, established by Canada, France, the United States, and the former Soviet Union in 1979. It is best known as the system that detects and locates emergency beacons activated by aircraft, ships and backcountry hikers in distress. COSPAS-SARSAT is implementing a new MEOSAR system based on the use of search and rescue transponders on new global positioning systems (GPS), global navigation satellite systems (GLONASS) and GALILEO satellites and accompanied new ground segment. This new MEOSAR system will significantly improve the timeliness and accuracy of alerts provided by ELTs and allow for new services to be provided (e.g. return link services). In conjunction with the new MEOSAR system, COSPAS-SARSAT is developing a new second-generation beacon specification system.

[131] Aircraft Tracking—Different Issues at Stake, Global Tracking 2014-WP/5, 7/5/2014, pp. 4–5.

France submitted to the meeting that short-term cost-effective solutions exist to address the issue of enhanced aircraft tracking and should be encouraged; vulnerabilities in relation to aircraft security have to be addressed; mid-/long-term solutions should be considered in terms of cost efficiency, having in mind the benefit in terms of safety and security; and mid-term solutions, as the activation at distance of new generation ELTs combined to airborne triggering in flight in case of unusual situations, is one example of solutions that should be envisaged.[132]

The International Council of Aircraft Owner and Pilot Associations (IAOPA) issued a note of caution against ICAO Standards and Recommended Practices (SARPs) that required the use of emergency locator transmitters (ELTs) as the approved equipage for SAR as being not suitable while at the same time claiming that mandating specific equipment would pose the trap of forcing aircraft owners to purchase new equipment each time the technology is updated or new technology is developed. Another danger envisioned by IAOPA is that such a move would preclude pilots and owners from using new technologies—even though the new technology may be better suited to their flight operations—simply because it "does not meet the regulatory requirement". As well, technology manufacturers are likely not to invest in research and development of new and more efficient technology when ICAO standards mandate the use of a specific and older technology.[133]

1.4.3 Legal Issues

1.4.3.1 Historical Background of ICAO's Work

In its Conclusions and Recommendations following the meeting on tracking aircraft, ICAO once again committed the usual blunder of not making any reference to the legal framework that might be developed simultaneously to implement some of the recommendations. For instance for the short term, ICAO is required to develop guidance material, based on available flight tracking best practices. Of what nature would these guidance material be and what compulsion or discretion of States might they carry? For the midterm, ICAO is expected to develop performance based "provisions". What legal effect or credibility would these provisions have unless their nature and application are laid out? At the least, ICAO could have declared that these guidelines and provisions would be developed, keeping in mind the relevant legal aspects.

The 31st Session of the ICAO Assembly, in December 1995, adopted Resolution A31-7 pursuant to which the Council established a Panel of Experts on the Establishment of a Legal Framework with Regard to Global Navigation Satellite Systems

---

[132] Aircraft Tracking, Global Tracking 2014-WP/8, 9/5/14, p. 3.

[133] General Aviation Experience with Aircraft Tracking, Global Tracking 2014-WP/9, 9/5/14, pp. 1–2.

(LTEP). The Panel was charged, *inter alia*, to develop a legal framework of conduct regarding the use of the GNSS. The framework developed by the LTEP took the form of a Charter containing various principles for the implementation and operation of GNSS, such as recognition of the paramount nature of safety in international civil aviation; non-discrimination and universal applicability and accessibility of GNSS; inviolability of States' sovereign rights; continuity, integrity, availability and reliability of services; and international cooperation. The Charter was recognized by the ICAO Council, at its 153rd Session in March 1998, as being worthy of submission to the 32nd Session of the Assembly, which in turn adopted the text of the Charter in a Resolution. The ICAO Assembly, at its 32nd Session in 1998, adopted Resolution A32-19 (Charter on the Rights and Obligations of States Relating to GNSS Services) containing fundamental principles of a Charter of Rights and Obligations of States in relation to GNSS Services.[134] In the Resolution, States, whilst recognizing that the primary use of GNSS services is to maintain safety in international civil aviation, reaffirm the principle that every State and aircraft of all States shall have access, on a non-discriminatory basis, and under uniform conditions to the use of GNSS services. The Resolution also grants every State authority and control over aircraft operations over their territory, and, *inter alia*, imposes obligations on provider States to ensure continuity, availability, integrity, accuracy and reliability of such services.

It is at this point that the problem arises, and legal discourse begins. The mere fact that the Charter is now an ICAO Assembly Resolution has prompted the comment:

> Adopted in the form of an Assembly Resolution, the Charter cannot be accorded any legal force and therefore must be regarded as legally not binding. Some commentators, having expressed serious doubts as to the usefulness of the instrument, seem to be somewhat displeased with the nomenclature employed which would be indicative of a legal instrument of fundamental importance.[135]

The significance of the legal status regarding the current principles on the conduct of States in using space based applications in civil aviation management lies in the compelling need to inquire as to whether rigid fragmentation of law and policy is really necessary, particularly in such an important area as aviation safety. Does one dismiss policy in this critical area purely on the inflexible notion that it is not enforceable? On the other hand, do States need to abandon rigid demarcations in instances such as these and agree to global adherence?

The 32nd Session of the ICAO Assembly also adopted Resolution A32-20 (Development and Elaboration of an Appropriate Long Term Legal Framework to Govern the Implementation of GNSS). This resolution, which recognizes that GNSS is an important element of the ICAO CNS/ATM System, is aimed at providing a framework governing safety critical services for aircraft navigation

---

[134] Res A32-19, Charter on the Rights and Obligations of States Relating to GNSS Services. Assembly Resolutions in Force (as of 5 October 2001) Doc. 9790, pp. V-3.

[135] Andrade (2001), p. 89.

with worldwide coverage. It also recognized, *inter alia*, the need for an appropriate long term legal framework to govern the implementation of GNSS and endorsed the Council decision taken earlier to authorize the ICAO Secretary General to establish a Study Group on Legal Aspects of CNS/ATM systems. The Assembly instructed the Council and the Secretary General to consider the elaboration of an appropriate long term legal framework to govern the operation of GNSS including consideration of an international convention.

Resolution A32-20 was a signal development in global recognition of the need to inquire into a liability regime regarding damage caused by the GPS process through signals transmitted to air navigation facilities. The Resolution resulted in an ICAO Secretariat Study Group being established to elaborate proposals for a liability framework. The Study Group reported to the 33rd Session of the Assembly (held in September–October 2001) in a somewhat divided way, some members recommending that the applicable regime under domestic law was adequate and appropriate to cope with the global navigation satellite system and others being of the view that a global international law instrument, such as a Convention, might be required to address issues of liability in the long term. A compromise between these two views, representing a model contractual framework, was also suggested to the Assembly as an alternative. The Assembly further remanded the matter to the Study Group to finalize the concept of a contractual framework as a first step, with a view to later considering the development of an international convention as a long term measure.

At the ICAO Eleventh Air Navigation Conference, held in Montreal from 22nd September to 3rd October 2003, the European Organization for the Safety of Air Navigation (EUROCONTROL), on behalf of its members and those of the European Civil Aviation Conference (ECAC), presented such a contractual framework to the Conference.[136] The framework contained main elements developed while taking into consideration regional requirements. The Conference was advised that, in the face of increased advancement and development of GNSS, there was a compelling need for an adequate legal and institutional framework to cover liability issues, *inter alia*, concerning GNSS. The main reason for this need, as was identified in support of the proposal, was that, as most States would not have direct involvement in the operation of the GNSS System or sub systems, air navigation services provided within their sovereign airspace will be compelled to rely on facilities which were beyond their control. It was also contended by EUROCONTROL and others presenting the contractual framework that States needed to be satisfied, *inter alia*, that there was an appropriate and adequate performance level of the GNSS signals and services offered within their airspace, ensuring their integrity, reliability, accuracy and continuity and that such services should come within a clearly defined liability framework.

[136] GNSS Legal Framework—Contractual Framework for the Implementation, Provision, Operation and Use of the Global Navigation Satellite System for Air Navigation Purposes, AN-Conf/11-WP/153, 18/9/03.

1.4.3.2 Positions of Other Key Players

The African States who presented an alternate proposal to the Conference of an international convention to cover liability issues of GNSS,[137] reiterated the European concern that there was a large constituent of user States who by themselves could not provide GNSS services (as required by Article 28 of the *Chicago Convention*) but were nonetheless responsible for the provision of those services under international treaty. As such, it was the view of the African States that an international convention was needed in order to establish and adopt principles regarding provider and user responsibility and rights in the event of damage. Such a Convention, according to the African States, would also serve to provide for an international institutionalized safety and security oversight process. The African States further suggested that the Convention should be based on ICAO's central responsibility to develop principles and techniques and foster planning and development of international air transportation, with safety as a paramount concern. The suggested Convention would involve mandatory submission to arbitration by parties concerned with an accident or damage and oblige providers to assure continuity, availability, accuracy and transparency while being liable for damage caused by GNSS services provided by them.[138]

It must be noted that Galileo, the European second-generation navigation satellite system—an outcome of collaborative work between the European Union (accomplished through the European Commission) and the European Space Agency, has done quite some work on developing an adequate legal regime. One suggestion has been to establish a Galileo Supervisory Authority (GSA) which would provide the political basis to enhance opportunities for the industry; introduce a fair and reasonable liability regime; ensure a proper certification scheme for providers; and establish safeguards against financial crises such as bankruptcy.[139]

The United States, on the other hand, recommended[140] to the Eleventh ICAO Air Navigation Conference that there was no need for either a new contractual framework or international convention. The United States was of the view that ICAO has flexible legal tools (such as SARPs) together with a sustained and long term institutional framework and experience that would enable the Organization to continue to serve without major problems, as it had, through such major aviation landmarks as the advent of the jet engine, radar and many other scientific advancements without legal or institutional problems. As such, the United States invited the Conference to agree that work on GNSS has progressed steadily over the past several years and no deficiencies had been found to impede technical implementation of CNS/ATM and that such work should not be impeded or delayed by work on legal and institutional issues.

---

[137] Legal Aspects of GNSS, AN-Conf/11-WP/143, 18/9/03.

[138] For an analysis of the liability of providers see generally, Abeyratne (1994), pp. 156–186.

[139] For an in depth discussion, see von der Dunk (2006), pp. 100–117 at pp. 101–103.

[140] Legal and Institutional Issues and the Status of CNS/ATM, AN-Conf/11-WP/160, 18/9/03.

Another school of thought, represented by the LTEP, was inclined to tie in the responsibility of States under Article 28 of the *Chicago Convention* to adherence to SARPs. The LTEP suggested that States providing signals in space or under whose jurisdiction such signals are provided should certify the signals in space by attesting that such provision is in conformity with SARPs. The LTEP went on to recommend that the States having jurisdiction under the *Chicago Convention* should ensure that avionics, ground facilities and training and licences requirements comply with ICAO SARPs.[141] These recommendations were based on the LTEP's considered view that, irrespective of whether or not a State handed over the provision of air navigation services to a privatized entity, ultimate responsibility continued to devolve upon the State for the provision of such services under Article 28 of the *Convention*.[142] The LTEP favoured an international convention on liability as an ultimate and long term measure, taking into consideration that a GNSS induced incident may, in certain circumstances, involve multiple and complex actions in several jurisdictions, thus requiring an internationally harmonized legal regime containing a simple, clear and speedy procedure.[143]

The LTEP presented its findings to the ICAO Council at the 170th Session of the Council in November 2003. In its Report,[144] the LTEP advised the Council that the Panel had finalized the draft text of a Contractual Framework Relating to the Provision of GNSS Services containing obligations of the GNSS signal provider as well as those of the air traffic service provider which makes use of the signal for the provision of its services. Essentially, the draft contractual framework foresees a series of contracts between the various stakeholders while stipulating that the signal provider is obligated to provide the signals with regularity, continuity, integrity, accuracy and uninterrupted availability.[145]

In 2004 the LTEP submitted its final Report to the Council of ICAO, concluding that there was no legal obstacle to the implementation of the CNS/ATM system and that it was not inconsistent with any of the provisions of the *Chicago Convention*. The LTEP has not been heard of since.

The trouble with ICAO is that it has artificially created individual turfs that consider silo interests. At the time of writing in early June 2014 it was still not too late to view the Conclusions and Recommendations relating to the global tracking of aircraft through a holistic approach that would consider those "provisions" and measures referred to in the Conclusions and Recommendations in the context of the legal implications and aspects involved. Admittedly, the preeminent considerations

---

[141] Recommendation 3 of LTEP. See Report of the Secretariat Study Group on Legal Aspects of CNS/ATM Systems, First Meeting, SSG-CNS/1-Report 9, April 1999, p. 4.

[142] *Id.*, p. 3.

[143] *Id.*, p. 5.

[144] Progress Report on the Work of the Secretarial Study Group on the Legal Aspects of CNS/ATM Systems, C-WP/12080, 7/11/03.

[145] *Id.*, p. 3.

would be technical but technical aspects could by no means be considered in isolation in the context of global policy.

It has already been established (within the scope of the Eleventh Air Navigation Conference of ICAO) that since the authorization of GNSS would require the involvement, participation and agreement by many States, both users and providers would have to agree to a binding legal agreement.[146] Furthermore, the Conference recognized that a contractual framework containing mandatory common elements could serve as the interim solution between the *status quo* and the long-term elaboration of a GNSS international convention. The framework would require, *inter alia*, a framework agreement among States at governmental level and that ICAO's long-term legal framework, namely the *Chicago Convention*, its Annexes and ICAO guidance material, was adequate and no deficiencies had been found to impede the technical implementation of CNS/ATM systems. While legal issues had been discussed in various bodies of ICAO, at no point did any ICAO body achieve consensus on a proposal for a new global conventional law.[147]

Since it has already been agreed that a legal framework could exist within the existing ICAO regime, it behoves ICAO to revisit the legal aspect of satellite communications and surveillance with a view to providing the ICAO High Level Safety Conference in February 2015 with a well rounded picture of both the technical, legal and policy aspects of global tracking of aircraft. The problem is that no one has agreed on what a "binding legal agreement" within the ICAO regime is. There are of course Assembly resolutions which are no more than results of political compromises to which no legal legitimacy can be ascribed[148]; procedures such as PANS-OPS[149]; SARPs which do not have any mandatory effect; Regional Air Navigation Plans; model legislation; model clauses and guidance material, all of which offer a whole range of useful purpose and direction that could result in the assurance of enforcement.[150] As to which of these, or a completely new animal that would define a "draft concept of operations" as referred to in the Conclusions and Recommendations of the ICAO meeting on global tracking of aircraft can be used, it is yet to be seen.

The philosophy of air law hinges on the fundamental premise that States have sovereignty over the airspace above its territory and adjacent territorial waters.[151]

---

[146] See AN-Conf/11-WP/143.

[147] AN-Conf/11-WP/160.

[148] Brownlie (1990), p. 691. Also Shaw (2003), p. 110.

[149] Procedures for Air Navigation Services—Aircraft Operations. PAN-OPS are rules for designing instrument approach and departure procedures. Such procedures are used to allow aircraft to land and take off when instrument meteorological conditions (IMC) impose instrument flight rules (IFR).

[150] Enforcement in this context does not necessarily result in punishment for non-compliance. It means justification, judgment and cooperation towards achieving a uniform global regime that ensures a common practice. See Brunnee (2005), pp. 3–13.

[151] Articles 1 and 2 of the Chicago Convention, supra.

Under this principle, seemingly a State is not answerable to their people for measures they take in the context of aviation within their territories. However, current approaches to international law brings to bear the primacy of humanity's law. Dr. Rudi Teitel, Professor of Comparative Law at New York Law School and Visiting Professor, London School of Economics, Global Governance, states:

> "sovereignty is no longer a self-evident foundation for international law. This shift is driving the move from the State-centric normative discourse of global politics – which had prevailed until recently – to a far ranging, transnational discourse in which references to changed subjectivity have consequences. That new discourse is constructed more among humanity law lines". This statement is consistent with the pronouncement of the International Criminal Tribunal for the former Yugoslavia which in its adjudication of *Prosecutor* v. *Dusko Tadic* said: "a state-sovereignty oriented approach has been gradually supplanted by a human being oriented approach[152]".

There are two broad reasons for this shift: the natural historical progression of world affairs which shifted trends chronologically; and the growing instances of torture, rape and killings in circumstances of internal strife and military warfare. Let us take the chronological evolution first. After World War II, the world has, over the past 70 years or so, gone through three global political and economic stages. The first was the modernist era, where naturally, State sovereignty was considered paramount over any other consideration in the immediate and proximate aftermath of the War where nations and States were attacked. This lasted until the 1960s.

From then on the trend shifted to the post-modernist era where the focus veered from the interests of the State to the welfare of the citizen. Post modernism, which was a characteristic of the 1960s and 1970s, progressed steadily toward the twenty-first century. Post modernist thinking was geared to accepting that human culture, as we knew it from a social and economic perspective, was reaching an end. This school of thought associated itself with the momentum of industrial society, drawing on an image of pluralism of cultures and a multitude of groups. The interaction between political modernism, which brought to bear the globalization of nations and deconstruction of separatism of human society, while at the same time ascribing to the individual certain rights at international law that transcended natural legislation parameters, has been symbiotic and essentially economic. In the post modernist era, the fundamental modernist philosophy of state sovereignty and peace gave way to an industrial culture that emphasized economic coexistence for the betterment of the global citizen.

The neo post modernist era was signalled as an immediate consequence of the attacks of 11 September 2001 where close to 3,000 human beings were decimated by terrorist activities within a few minutes. The principle of State Responsibility with regard to world peace and security which lies primarily in Article 24 of the United Nations Charter which calls upon all members to refrain in their international relations from the threat or use of force against the territorial integrity or

---

[152] Teitle (2011), p. 6.

political independence of any state, or in any other manner inconsistent with the purposes of the United Nations, was immediately resuscitated in the minds of the international community.

This principle is embodied in the work of the International Law Commission, through Article 2 of the Draft Articles on Responsibility of States for Internationally Wrongful Acts which provides that international responsibility of a State, which is referred to in Article 1, is attributable to that State if conduct of the State constitutes a breach of an international obligation of that State. The document also provides that the wrongfulness of an act of a State is precluded if the act constitutes a lawful measure of self defence taken in conformity with the Charter of the United Nations. The State responsible for an internationally wrongful act is under an obligation to compensate for damage caused, including reparation for financially assessable damage including loss of profits.

In addition to State responsibility for conduct attributable to that State, the International Law Commission has established that a crime against the peace and security of mankind entails individual responsibility, and is a crime of aggression. The Rome Statute of the International Criminal court, defines a war crime, *inter alia*, as intentionally directing attacks against civilian objects; attacking or bombarding, by whatever means, towns, villages, dwellings or buildings which are undefended and which are not military objects; employing weapons, projectiles, and material and methods of warfare that cause injury The Statute also defines as a war crime, any act which is intentionally directed at buildings, material, medical units and transport, and personnel using the distinctive emblems of the Geneva Conventions in conformity with international law.

Security of the human being, although the focus of international attention at the present time, has its roots in years gone by. The rule of law has always been intended to be a safeguard against arbitrary governance, whether by a totalitarian leader or by mob rule and is diametrically opposed to both dictatorship and anarchy.

Thomas Jefferson once wrote that the purpose of government is to enable the people of a nation to live in safety and happiness. Government exists for the interests of the governed, not for the governors. As Benjamin Franklin wrote, "In free governments the rulers are the servants and the people their superiors and sovereigns." The ultimate powers in a society, therefore, rest in the people themselves, and they should exercise those powers, either directly or through representatives, in every way they are competent and that is practicable.

The essence of a nation should be founded on human rights that are contrived from single instances of wrongs committed against the people. According to this principle, a right becomes something that is legitimately due to a person which he can justly claim as secured to him by law, and which ensures that some wrong committed in the past is effectively precluded by the right so secured. A right should not be confused with power, the former being based on moral justification and expectation and the latter being based on enforceability. Protection by the state of an individual, freedom to attend church or temple, and freedom to educate oneself are examples of a right where as sovereignty of State, authority to censor speech and enforce martial law are examples of power. A wise nation distinguishes between the two and maintains a balance.

Human rights should be viewed as something more than a concept which acts as a cultural artifact. They transcend fundamental rights, which are essentially political and civil rights, and expand to more basic rights such as the right to be equal to anyone with regard to the basic universal need for nourishment, shelter, clothing and education. In order to make sure that they are enjoyed by all of humanity, any community will have to make sure that human rights are a matter of course and are ensured by a guaranteed and contrived effort by all. Human rights and their worth cannot strictly be evaluated. Traditional modes of evaluation, with which the voter usually goes to the polls in a democratic environment to select the government, are "value for money", efficiency of service delivery and customer satisfaction. At best, these yardsticks have largely been political and economic abstractions which have prompted some academics and practitioners to consider the subject of governance-evaluation as being immeasurable or too much trouble. The issue is further aggravated by the fact that there is no scientifically approved or accepted model to assess the quality of public governance.

The bottom line is that human rights are enjoyed by the citizen through good governance. The first conclusion that one can reach is that good governance is no longer assessed by the provision of services by a government or other governing body but rather by the extent to which improvements were made possible to the quality of life of the individual. The second is that good governance has an international connotation, in that it should be assessed with the assistance and application of international standards. Also, good governance must be rewarded, for example through rewards along the lines of the Nobel Peace Price for "best practices" in good governance. Recognition should be given through "satisfaction surveys" where a direct causal nexus could be drawn between the manner in which the governed was enabled to reach a level of satisfaction with governance provided. Positive changes in expectation and results obtained should be weighed against perceived adequacies of government in the provision of services. Trust in government, through increased levels of health and well being (which must necessarily include a sense of security of life, habitation and movement) both from cultural and religious perspectives should be a primary indicator. The elimination of corruption is a key to good governance, and civil society, which has been overwhelmingly proactive in building awareness on human rights issues, has succeeded in persuading the international community of the value for transparency and honesty in public transactions. Arguably, the most important key to good governance is benevolence and understanding. A good government must assure its people that it has their well being at heart and pro actively move towards achieving that goal.

Alan Dershowitz, Felix Frankfurter Professor of Law at Harvard University, in his book "Rights from Wrongs"[153] states that rights do not come from nature, as nature is value neutral, nor do they come from logic or law alone because, if rights

---

[153] Dershowitz (2005).

emanated from law, there would be no basis to judge a given legal system. Dershowitz maintains that rights come from human experience, particularly experience with injustice. Our experience has taught us never again to tolerate a holocaust, never to curb freedom of expression and freedom of faith, and from that experience has stemmed the Universal Declaration of Human Rights and the United Nations Charter.[154] These documents, which embody fundamental rights are just pieces of paper if experience is not joined by logic. The marriage of logic and experience in the wisdom of human relations is ingrained in ancient Jewish philosophy, which, according to Isaac Abravanal, recognized that experience is more forceful than logic but logic and experience are not mutually exclusive. Without being applied to experience, logic tends to be hollow and directionless, but without the focus of logic, experience becomes multi directional and out of focus. Good decisions come from experience and experience comes from bad decisions. In other words, rights emerge from wrongs and not from ancient parchments or tomes of wisdom hidden away in a forgotten memory that is subsequently revived.

All this brings one to the irrefutable conclusion that the enforcement of human rights is a peremptory norm or *jus cogens* at international law. This means that the application of human rights takes precedence over any treaty, as enunciated by Article 53 of the Vienna Convention on the law of treaties, which states that whenever a *jus cogens* clashes with a treaty provision the former prevails. human rights.[155] Obligations pertaining to a *jus cogens* are applicable *erga omnes* or to the world at large irrespective of national or parochial interests. Aviation is no exception.

References

Abeyratne RIR (1992) The development of the machine readable passport and visa and the legal rights of the data subject. Ann Air Space Law (Annales de DroitArien et Spatial) XVII(Part II):1–31

Abeyratne RIR (1994) The evolution from FANS to CNS/ATM and products liability of technology Providers in the United States. Zeitschrift für Luft-und Weltraumrecht 43(2):156–186

Abeyratne R (2005) The E-passport and the public key directory – consequences for ICAO. Air Space Law XXX(4–5):255–268

Abeyratne R (2010a) The unlawful interference compensation convention of 2009 and principles of state responsibility. Ann Air Space Law XXXV:177–211 at 195–209

---

[154] *Id.*, p. 27.

[155] Article 53 states that a treaty is void if, at the time of its conclusion, it conflicts with a peremptory norm of general international law. For the purposes of the Convention, a peremptory norm of general international law is a norm accepted and recognized by the international community of states as a whole as a norm from which no derogation is permitted and which can be modified only by a subsequent norm of general international law having the same character.

Abeyratne R (2010b) The use of forged passports of foreign nationals in international criminal activity. J Transportation Secur 3(2):32–45
Abeyratne R (2013) The ePassport – new technology to counter security threats. J Transportation Secur 6(1):23–39
Abeyratne R (2014a) Integrity of travel documents: the wakeup call from flight MH 370. Zeitschrift fur Luft-und Weltraumrecht ZLW (German J Air Space Law ZLW) 63:238–249
Abeyratne R (2014b) Flight MH 17: the legal and regulatory fallout. Air Space Law 39 (6):329–342
Andrade AAL (2001) The global navigation satellite system. Ashgate, Aldershot, p 89
Brownlie I (1990) Principles of public international law, 4th edn. Clarendon, Oxford
Brunnee J (2005) Enforcement mechanisms in international law and international environmental law. Environ Law Netw Int Rev 3–13
Cancado Trinidade AA (2010) International law for humankind – towards a new Jus Gentium. Martinus Nijhoff, Leiden, p 281
Dershowitz A (2005) Rights from wrongs – a secular theory of the origin of rights. Basic Books, Cambridge
Drazenovich DA, Skovira RJ (2010) Mitigating risk in a high uncertainty avoidance culture – air traffic control. Issues Inf Syst XI(1):673–688
Hanlon M (2007) Attack on the cyber terrorists. Mail Online at http://www.dailymail.co.uk/sciencetech/article-457504/Attack-cyber-terrorists.html
Hurn J (1989) GPS – a guide to the next utility. Trimble Navigation, Sunnyvale, p 7
Jarvis DSL (2007) Risk, globalisation and the state: a critical appraisal of Ulrich Beck and the world risk society thesis. Global Soc 21(1):32
Leloudas G (2009) Risk and liability in air law. Informa, London, p 11
Margo RD (2000) Aviation insurance, 3rd edn. Butterworths, LexisNexis, p 338
Oster CB, Strong JS (2007) Managing the skies, public policy, organization and financing of air traffic management. Ashgate, Aldershot, pp 15–16
Petras CM (2004) An alternative proposal to modernize the liability regime for surface damage caused by aircraft to address damage resulting from hijackings or other unlawful interference. Gonzaga J Int Law. Tuesday 4 September 2004. http://www.gonzagajil.org/content/view/164/1/
Shaw MN (2003) International law, 5th edn. Cambridge University Press, Cambridge, p 110
Teitle R (2011) Humanty's law. Oxford University Press, Oxford, p 6
von der Dunk FG (2006) Towards Monitoring Galileo: the European GNSS supervisory authority in Statu Nascendi. Zeitschrift für Luft-und Weltraumrecht 100–117 at 101–103

Chapter 2
Aviation and Intervention

2.1 Unmanned Aircraft

In October 2013, CNN reported that 400 civilians had been killed in Pakistan by unmanned aircraft (UA)[1] since President Obama took office in 2009. Transnational legal theory suggests that such attacks are illegal and, in 2009, the United Nations special rapporteur on extra judicial killings suggested that the use of drones by the Obama administration in Afghanistan and Pakistan was untenable and contrary to international law.

The underlying concerns with regard to legal issues are state sovereignty, responsibility, and liability. One of the contentious issues would be whether unmanned aircraft flying outside their permitted areas violate state sovereignty, and if so—what can be done to prevent future UA violations of sovereignty of the airspace. Another would be, as earlier mentioned, whether the killings carried out by drones amount to extra judicial killings. The focus of this article is on legal challenges and whether military functions of UA can be brought under international regulation. Despite the inherent advantages of this relatively new mode of aviation—such as camouflage, low cost, and the avoidance of risk to the pilot on board in UA operations—serious concerns abound, which go to the heart of the law of humanity.

The problem of drones[2] has indeed become contentious. A compelling argument is that killing of humans, whether they fall into the category of terrorists or not, is

---

[1] Unmanned aircraft are sometimes referred to in this chapter as drones. The author gratefully acknowledges the contribution of Mr. Arman Khan LL.M to this chapter.

[2] The UA is not a new strategy and has nearly a century-old history in aviation. First included in Jane's All the World's Aircraft in 1920, UAs were tested during World War I, but not used in combat during that war. Germany's use of the simple yet deadly V-1 "flying bomb" during World War II laid the groundwork for post-war UA programs in the United States. However, it was not until the Vietnam War in the late 1960s and early 1970s that UAs such as the AQM-34 Firebee were used in a surveillance role (by the United States). The Firebee exemplifies the versatility of

R. Abeyratne, *Aviation and International Cooperation*,
DOI 10.1007/978-3-319-17022-0_2

premeditated murder. In many jurisdictions around the world premeditated murder is a crime, as it is in all 50 States of the United States. Additionally, political killings, which drone attacks often accomplish, are outlawed in the United States.[3] The crime is made more heinous when it is committed in foreign lands without the consent of the State concerned, and with callous disregard to the sovereignty of that State. The only exception to the use of drones in a foreign land (and this does not extend to arbitrary attacks that might involve non combatants) is in authorized armed conflict such as in instances where the international community, through the United Nations Security Council, authorizes such use in battle zones.

From an international perspective, the operative law with regard to victims of war is international humanitarian law. This limb of law is also known as the law of war, the laws and customs of war or the law of armed conflict. Basically, international humanitarian law encompasses four limbs, the first being that persons who are not, or are no longer, taking part in hostilities shall be respected, protected and treated humanely. They should be given appropriate care, without any discrimination. Secondly, captured combatants and other persons whose freedom has been restricted are required to be treated humanely. They should be protected against all acts of violence, in particular against torture and if they are brought to trial they have the right to enjoy the fundamental guarantees of a regular judicial procedure. Thirdly, the right of parties to an armed conflict to choose methods or means of warfare is not unlimited. No superfluous injury or unnecessary suffering should be inflicted. Finally, in order to spare the civilian population, armed forces are required at all times to distinguish between the civilian population and civilian objects on the one hand, and military objectives on the other. Neither the civilian population as such nor individual civilians or civilian objects should be the target of military attacks.

Certain circumstances over the past decade have made the world more cautious, leading it to guide its philosophy of mutual trust along a path which is now called "the precautionary approach". The world would no longer sit and wait, reacting only when a crisis causes massive human suffering and loss of lives. A new doctrine, propounded by a group of scholars at the behest of the United Nations Secretary General Kofi Annan in 2001 has come into being. Called "the responsibility to protect", this doctrine embraces the principle that all member States of the United Nations have a responsibility to protect the lives, liberty and basic human rights of their citizens, and that if they fail or are unable to carry it out, the international community has a responsibility to step in.

UAs. Initially flown in the 1970s, it was modified to deliver payloads and flew its first flight test as an armed UA on December 20, 2002.

[3] Executive Order 11905 [in Section 5(g)] issued by President Ford in 1976 explicitly provides that no employee of the United States shall engage in, or conspire to engage in political assassination. Later, President Reagan issued Executive Order 12333 which, by Section 2.11 provides that no person employed by or acting on behalf of the United States Government shall engage in or conspire to engage in assassination. The Order goes on to state that no agency of the Intelligence Community shall participate in or request any person to undertake activities forbidden by the Order.

2.1.1 *What Is an Unmanned Aircraft?*

Unmanned Aircraft [UA] is a pilotless aircraft capable of flying autonomously or semi-autonomously with some pilot assistance from a remote station and hence the term 'Remotely Piloted Aircraft' [RPA].[4] The United States Department of Defense defines UA as "a powered aerial vehicle that does not carry a human operator, uses aerodynamic forces to provide vehicle lift, can fly autonomously or be piloted remotely, can be expendable or recoverable, and carry a lethal or non-lethal payload."[5] In contrast, the ICAO circular has a simple definition of UA i.e. "An aircraft which is intended to operate with no pilot on board."[6] Unmanned Aircraft is one of the main components of the many sub-systems of Unmanned Aircraft system which includes an aircraft [unmanned air vehicle], payloads [dispensable or non-dispensable], communication and control systems, launching systems, transport systems etc.[7] Modern UA contain varying levels of autonomous decision making algorithms which enables UA to exchange data with its control base and to transfer thermal images along with its positioning, speed, altitude and other information it was designed for.[8] UA is also commonly known as 'Drone'. The leading aeronautical consultant Reg Austin reasons that UA is technically different than drone.[9] Another extension of UA is the Unmanned Combat Air Vehicles [UCAV] which is armed and specifically assigned to attack each target.[10] However, since the term drone has been popularized in the mainstream media, for the purposes of this paper the term UA will be interchangeably used with drones.

2.1.2 *Evolution of the Unmanned Aircraft*

From the humble beginnings of 'Aerial Torpedo', an unpiloted bomber plane made of wood, in the early twentieth century to the contemporary 'Predators' with the

---

[4] An Unmanned aircraft is a self piloted or remotely piloted aircraft that can carry cameras, sensors, communications equipment or other payloads. Therefore in its ultimate definition, a UA is an aircraft, which is defined by Annex 2 to the Convention on International Civil Aviation (Chicago Convention) as "any machine that can derive support in the atmosphere from the reactions of the air other than reactions of the air against the earth's surface." In addition to this quality, a UA is also a powered, aerial vehicle that does not carry a human operator, uses aerodynamic forces to provide vehicle lift, and can fly autonomously or be piloted remotely. A UA can be expendable or recoverable, and can carry a lethal or non lethal payload. UAs have been used to conduct reconnaissance and intelligence-gathering for nearly 60 years (since the 1950s).

[5] Bone and Bolkcom (2003), Congressional Research Service: The Library of Congress, CRS 1.

[6] ICAO (2011), p. x.

[7] Austin (2010), p. 3.

[8] *Ibid.*

[9] *Ibid.*

[10] See generally, Friedman (2010), p. 1.

capability of high resolution video cameras and payload weapons guided by the precision lasers, the metamorphosis of UA has come a long way.[11] UA come in various shapes, sizes, and capabilities' ranging from High altitude long endurance [HALE] which can fly up to 24 h and over 15,000 m altitude to Mini UAV [MUAV] which weigh less than 20 kg and is capable of flying up to 30 km range when hand-launched.[12] The history of UA changed forever when Predator made it first successful flight in 1994. Richard Newman notes that "Predator became an instant hit because it could transmit live video footage of enemy actions to commanders on the ground and aircrews above the battlefield. It illuminated targets for precision weapons fired from afar. It even, on occasion, fired its own weapons, a rarity for a UAV."[13] At present, Global Hawk is widely regarded to have taken over the Predator and has been flown over the Afghanistan, Bosnia, Iraq, Kosovo, Pakistan and Yemen. As one writer put it "Global Hawk does much broader surveillance. It flies at elevations of 60,000 feet or more, like the U-2, and has higher resolution cameras than the Predator. Global Hawk can fly for nearly day and a half without being refueled, meaning it can take off from a base 1,200 miles away, loiter for 24 hours, then fly back."[14] Today, the sky is no longer the exclusive domain of the birds or jumbo jets as thousands of UA are hovering all over the skies.

2.1.3 *Commercial Use of Unmanned Aircraft*

UA are primarily used for military purposes but their commercial use in on the rise. UA's are often used for 3D's i.e. 'dull, dirty, and dangerous jobs.'[15] Besides, UA are used in research and development of the aeronautical areas such as airborne testing under the real conditions.[16] There are many advantages of UA as compared to its counterpart manned aircrafts. Less operating costs, low fuel and maintenance costs, cheaper insurance and low salaries for the ground based personnel all translate into a lucrative monetary benefits to pursue UA for commercial purposes. Due to financial benefits associated with the UA, it is likely that UA rental business will soon sell their services.[17] Manned helicopters have been used in South Africa and Australia to monitor herds during hot weather and difficult terrains but it may save time and money if UA are used to keep track of these herds.[18] UA could be in hands of news channels and broadcasting companies for covering emergency events, real-

[11] Austin (2010), pp. 304–314.

[12] *Ibid.* at 4.

[13] See generally Newman (2002), p. 49.

[14] *Ibid.*

[15] *Ibid.* at 5.

[16] *Ibid.* at 6.

[17] Austin (2010), p. 273.

[18] *Ibid.* at 274.

time sports events, monitoring traffic accidents and war zones.[19] For the first time, the United Nations has ordered surveillance drones from an Italian firm for its peacekeeping missions in Congo and diplomats hope that if successful the drones could be deployed for missions in Ivory Coast and South Sudan.[20]

The advertisement line for drones is catchy, who wants to buy an expensive three million police helicopter when a much cheaper drone can do the same job more effectively. UAs have the potential of hunting down the suspects, criminals, and law-offenders. With the Global Positioning System [GPS] auto-pilot and a live video feed, one cannot hide from these omnipresent drones.

2.1.4 *State Utilization of Unmanned Aircrafts*

Approximately 70 countries now possess the UA however, only few possess the armed UA notably the US, UK, Israel, Italy, and China.[21] Although, the US is the leader when it comes to armed UA, it no longer has the monopoly over the expansion of drones. The militaries of various nations are quickly learning to develop UA capable of delivering lethal blow to the enemy taking advantage of the loopholes in the current international legal framework on the legitimate use of drones.[22] In 2001, a Predator evolved into a UCAV from a reconnaissance UA when it fired a Hellfire-C laser guided missile at a tank.[23] Meanwhile, the Israeli Defence Forces [IDF] have been using the UA for multiple intelligence and data gathering programs from 1980s.[24] Japan has been developing the UA for dual purposes i.e. both civilian and military uses since World War II.[25] The South Korean government has developed a smart vertical take-off and landing UA and is on the verge of developing an upgraded version of the US Eagle Eye tiltrotor UA.[26] The Italian Air Force has its hands on a precision strike UA and Rome is negotiating a "black" medium-altitude, long-endurance [MALE] project that would

---

[19] *Ibid.*

[20] The U.N. spokesman Martin Nesirky told reporters that "Unarmed UAVs (unmanned aerial vehicles) will allow our peacekeepers in the Democratic Republic of Congo to monitor the movements of armed groups and protect the civilian population more efficiently. The selected vendor is the Italian company Selex ES. The UAV is known as the Falco and is designed to be a medium altitude, medium endurance surveillance platform capable of carrying a range of payloads including several types of high resolution sensors." See Nichols (2013).

[21] Roberts (2013).

[22] *Ibid.*

[23] Lazarski (2002).

[24] Bone and Bolkcom (2003), p. 2, Washington/Congressional Research Service, RL31872 online: http://www.fas.org/irp/crs/RL31872.pdf.

[25] See generally Staff (2004), p. 138; Id. at 127.

[26] UAV Categorisation, in 2004 Yearbook: UAVs Global Perspective 156 (UVS International, Blyenburgh & Co.) at 156.

be flying in 2017–2018 told the Air Force Lt. Gen. Claudio Debertolis, Italy's secretary general of defence and national armaments director.[27] The Russian military has long collaborated with France's Dassault for manufacturing UAVs and is now acquiring the so-called mini-class of the drones.[28] Germany is in negotiations with Israel over the Heron UAV and with the U.S. for the Reaper UAV. "A Government spokesman told AFP: 'The intention is that we acquire 16 such devices in future and that the armed forces have them at their disposal from 2016, three years from now.'"[29] France made a $1.5 billion foreign military sales agreement with the US to buy 16 Reapers (UAVs) developed by General Atomics.[30] "Britain bought US Predator drones and rented Israeli drones for use in Britain's wars in Iraq and Afghanistan. . .the Watch keeper system for surveillance, and two separate BAE armed drones."[31] Besides conventional reconnaissance and surveillance the UCAVs can perform multiple offensive and defensive tasks such as "suppression of enemy air defenses [SEAD], close air support [CAS], defensive counter air [DCA], offensive counter air [OCA], and air interdiction [AI]."[32]

UA are increasingly used from reconnaissance to surveillance to security to border patrolling to law and order operations and every possible imaginable task. It is only a matter of time that drones will become a modern military necessity for each state.[33] One better pay heed to the words of General Petraeus, former head of the US Central Command, "[w]e can't get enough drones".[34] UA importance resonated with former CIA Director Leon Panetta affirming that the unmanned aerial vehicles are "[t]he only game in town"[35] when it comes to tackling Al-Qaeda.

---

[27] Butler (2013).

[28] Denis Fedutinov, a UAV expert described these new state of the art UAVs, "These are the short-range systems, with a range measuring from 5 kilometres to 25 kilometres, Denis Fedutinov says. These are small devices that are comparable to a model aero-plane. But they are very effective in ensuring the transmission of video- and photo-images in real time, and they can also reconnoitre over by the hill, in the close-in zone. These devices make it possible for the lowest-echelon units to quickly obtain intelligence and operational information and depend in no way on higher-echelon air-reconnaissance systems." **See** Nekhai (2013).

[29] Evans (2013).

[30] Harthy (2013).

[31] Cole and Wright (2010a).

[32] Peterson (2005), Published at 18.

[33] See Friedman (2010), p. 33, Friedman notes, "Militaries adopt revolutionary technology and tactics because they cannot go on the way they have been operating. In some cases, such as ours, circumstances make the existing way of war either literally impossible or unaffordable-or there are new requirements that existing forces cannot meet. In others the general adoption of some new technology makes it impossible to go on as before."

[34] See Capaccio (2010). "[t]he new aviation plan projects spending for all aircraft to increase to $29 billion by 2020 from about $22 billion in fiscal 2011, an average of 3 percent annual growth adjusted for inflation."

[35] Shachtman (2009).

2.1.5 *Legal Issues Related to Unmanned Aircraft*

2.1.5.1 State Sovereignty

Sovereignty here refers to air-space autonomy of each independent state as per Article 1 of the Chicago Convention stating that "[t]he contracting states recognize that every State has complete and exclusive sovereignty over the airspace above its territory."[36] Article 8 of the Chicago Convention titled '*Pilotless aircraft*' explicitly prohibits all States to fly UA without the prior consent of the other state.

UA used for surveillance and reconnaissance are in danger of breaching Article 36 of the Chicago Convention which gives authority to the contracting states to prohibit or regulate the use of photographic apparatus in aircraft over its territory.[37] Peterson calls this "an operational limitation" to UA reconnaissance functions and notes that the underlying principle is state sovereignty over airspace.[38] "Drones will no doubt make possible the dramatic expansion of the surveillance state. With the convergence of other technologies it may even make possible machine recognition of faces, behaviors, and the monitoring of individual conversations."[39] At present, the increased surveillance and unbridled reconnaissance by the major super-powers is breaching the principle of sovereignty. This embodies a legitimate concern of states. Moreover, UA carrying dangerous payloads or war munitions without authorization breach Article 35 of the Chicago Convention. With the rapid emergence of military drones, these UA carrying payloads, missiles and arsenals must be flown with the permission of the host State or else they will the state sovereignty.

Another paramount concern is when the State totally neglects the principle of sovereignty or blatantly breaches the terms and conditions of the parties' agreement? What should be the consequences for those states which habitually penetrate the foreign airspace? Milde suggests that such an aircraft may be "intercepted for purposes of identification; directed to leave the violated air space by a determined route; directed to land for the purpose of further investigation/prosecution; forced to land for further investigation/persecution."[40] Should the UA be shot down without any warning? Article 3 of the Chicago Convention prohibits the use of weapons against civil aircrafts—with a caveat that the same protection is not extended to the State aircraft in international law. In this context, one can draw a conclusion that UA do not enjoy any special immunity in international law. In reality, most states do not possess the technology to detect the stealthy UA,[41] even if, they detect

[36] Convention on International Civil Aviation, 7 December 1944, 15 UNTS 295, 61 US Stat 1180 (entered into force 4 April 1947).

[37] Article 36 of the Chicago Convention reads, "Each contracting State may prohibit or regulate the use of photographic apparatus in aircraft over its territory."

[38] Peterson (2005), p. 47.

[39] *See* Cole and Wright (2010b).

[40] Milde (2012), p. 65.

[41] The fact that majority of the UAs like Predator, Global Hawk, and Reapers fly above 70,000 ft and are beyond the technological or military capacity of most States to detect them, let alone shoot them.

foreign UA the fear of repercussions of shooting down a drone would make them think twice. But recently in May 2013, the Al-Shabaab militants shot a US drone in Somalia and posted pictures on the Twitter. The US officials confirmed the incident but did not provide any specifications to drone model believed to be one of *Schiebel* UAVs.[42] In an another case, the Russians in collaborations with Iranians observed the US drone's routes and its surveillance of Iranian skies, hacked into the drone system and forced it to land in Iran on 4 December 2011.[43] The Russians helped decode American drone RQ170s intercepted by the Iranians in exchange for anti-defense air system S300.[44] Despite the request by the US government to return it back, the Iranians have refused to comply and diplomatic tensions have escalated since then. Nonetheless, State reserves the right to regulate or prohibit the carriage in or above its territory in the interests of public order, safety, and above all its sovereignty.

2.1.5.2 Unmanned Aircrafts Operating Over High Seas

In accordance with Article 12 of the Chicago Convention, states cannot deviate from the rules of flight over the high seas which comprises for 70 % of airspace. Article 87 of the United Nations Convention on the Law of the Sea[45] of 1982 grants unrestricted freedom for the aircraft of all States to fly over the high seas. Abeyratne notes that "an important consideration in delineating territorial sovereignty lies in the expansion of Flight Information Regions [FIR] and the provision of air traffic management services by States particularly when such measures are influenced by the revenue generating capabilities."[46] Prof. Bin Cheng agrees that States must comply with the rules pertaining to the flight and maneuver of all aircrafts over the high seas as laid out in the Article 12 of the Chicago Convention.[47] Annex 2 of the Chicago Convention, further solidifies that State aircrafts over the high seas "[s]hall

[42] Laing (2013).

[43] Kahlili (2013). "A Russian military team from its air and space division had secretly entered Iran days after a strategic agreement between the two countries in 2007 and was stationed at Revolutionary Guard bases to help the Guards with its weapons program and access to modern U.S. technology. With this collaboration, all the information of the stealth drone is now in the hands of the Russians, and much has been given to Iran's Defense Ministry."

[44] *Ibid.*

[45] The Law of the Sea (1983). Moreover Article 29 gives the definition of a warship i.e. "a ship belonging to the armed forces of a state bearing the external marks distinguishing such ship of its nationality, under the command of an officer duly commissioned by the government of the State and whose name appears in the appropriate service list or its equivalent, and manned by a crew which is under regular armed forces discipline."

[46] Abeyratne (2010), p. 162.

[47] Cheng (1962), p. 148.

ensure that these operations do not compromise the safety, regularity and efficiency of international civil air traffic."[48]

2.1.5.3 Are Unmanned Aircraft a State or Civil Aircraft?

What is the status of UA under international air law? The Chicago Convention,[49] the *Magna Carta* for aviation and the primary bedrock source of the public international air law—in Article 3 distinguishes between civil and state aircraft. A distinct anomaly is observed in the question as to "how an international treaty, which on the one hand prescribes that it applies only to civil aircraft [Art. 3(a)], turns around and prescribes a rule for State aircraft [Art.3(c)]."[50] However, Article 3 does not provide a definition but at best gives a rebuttable presumption as those aircrafts utilized in furtherance of state activities such as search and rescue, natural disasters, governmental transport, and other state related services can be categorized as State aircraft.[51] This is far from a comprehensive definition and unreliable since the same Unmanned aircraft can transform into 'civil' or 'state' aircraft depending on the course of function.[52]

Whether UA is a 'civil' or 'state' aircraft raises different set of obligations and liabilities. If UA is a state aircraft, then UA would not come within the purview of Chicago Convention. If UA is a civil aircraft, then it would fall under the jurisdiction of ICAO and national laws. Some distinguishable characteristic of the Unmanned Aircraft for the purposes of interception and regulation can be the i) design of UA, ii) registration marks, iii) ownership, and iv) type of operation.

[48] Annex 2 to the Convention on International Civil Aviation, "Rules of the Air", 2.1.1 Note.

[49] Convention on International Civil Aviation also known as the Chicago Convention was signed at Chicago on December 7th, 1944. See ICAO doc 7300/9 Ninth Edition, 2006.

[50] Abeyratne (2010), p. 168.

[51] In the absence of any definition of 'military services' the interpretation must be done in accordance with the Article 31 of the Vienna Convention on the Law of Treaties that "ordinary meaning to be given to the terms of the treaty in their context and in the light of its object and purpose."

[52] **See**, Accident Investigation Board Report, August 1996, at 4 when USAF CT-43 A, a military equivalent of B-737 -200 carrying VIP Croatian officials crashed at Dubrovnik on 3 April 1996; the investigation report deemed the aircraft as civil aircraft per Article 3 of the Convention. **See also**, Resolution 1562 (2007) of the Parliamentary Assembly of the Council of Europe titled "Secret detention and illegal transfers of detainees involving Council of Europe member States" which was critical of countries involved with US authorities in transporting suspected terrorists in Gulfstream civilian aircrafts online at http://assembly.coe.int/main.asp?Link=/documents/adoptedtext/ta07/eres1562.htm. In the aftermath of 911, special rendition flights were performed to transport high-value terrorists for questioning in countries where they would not enjoy immunity against torture. The aircrafts identified were civilian B-737 and had US nationality marks but were used in a secret operative manner that would categorize them as State aircrafts. See also Milde (2008), pp. 477–486.

i) Design: In 2010, CIA deployed micro UA in Pakistan, approximately the size of a pizza platter specially designed to survey targets with more precision in order to minimize civilian causalities.[53] The drone operators also changed tactics by targeting terrorists in vehicles rather than residences to reduce collateral damage. However, the difficulty to distinguish UA lies in the fact when they can be easily re-modified for other purposes.
ii) Registration: The registration bearings can also distinguish an aircraft. However, the nationality and registration itself are insufficient proof that an aircraft is used solely for state or civil purpose. In a well-documented case of civil airplane B-737 MisrAir flight 2843 from Cairo to Tunis, suspected of carrying terrorists under Military Police escort was intercepted by US fighter planes and forced to land in Italy. The US explained its action stating that "it is our view that the aircraft was operating as a State aircraft at the time of interception. The relevant factors – including exclusive State purpose and function of the mission, the presence of armed military personnel on board and the secrecy under which the mission was attempted – compel this conclusion."[54]
iii) Ownership: The ownership may indicate that whether the aircraft is owned by the State or by a private entity. However, the problem arises where a state contracts out its military duties to private security contractors. For instance, the US State Department is operating surveillance drones to protect the US embassy in Iraq and is accepting bids from private security contractors to manage the drones program.[55] This has outraged the Iraqi officials who believe this program is an affront to Iraqi sovereignty.[56] Thus, ownership itself does not prove that the UA is used in state or a civil operation rather it further complicates the matter when the UA are contracted to private military security companies in the event liability arises.[57]

---

[53] Warrick and Finn (2010).

[54] ICAO document LC/29-WP/2-1 (1994) at pp. 11–12.

[55] As Schmitt and Schmidt reported, "American contractors say they have been told that the State Department is considering to field unarmed surveillance drones in the future in a handful of other potentially high-threat countries, including Indonesia and Pakistan, and in Afghanistan after the bulk of American troops leave in the next two years…The drones are the latest example of the State Department's efforts to take over functions in Iraq that the military used to perform. Some 5,000 private security contractors now protect the embassy's 11,000-person staff, for example, and typically drive around in heavily armored military vehicles." **See** Schmitt and Schmidt (2012).

[56] *Ibid.*

[57] 'The Montreux Document' is not legally binding however, it highlights the responsibilities of Contracting States that hire Private Military Security Companies (PMSCs) and emphasizes that States have an obligation to uphold international humanitarian law and have a duty to hold PMSCs accountable for criminal behavior. Both the US and Iraq have ratified 'The Montreux Document.' See "The Montreux Document on pertinent international legal obligations and good practices for States related to operations of private military and security companies during armed conflict" A/63/467- S/2008/636 (entered into force on 17 September 2008), online at www.icrc.org/eng/assets/files/other/icrc_002_0996.pdf.

In an another case, the Elbit Systems which is a defense electronics manufacturer, trades on NASDAQ and sells Unmanned Aircraft systems to many countries around the world. Although, 45 % shares are owned by Micheal Federmann, due to its close connections with the Israeli Ministry of Defense, Elbit Systems has been a target of the pro-Palestinian BDS (boycott, divestment, sanctions) movement due to the use of its Hermes 450 drone in West Bank and Gaza.[58]

iv) Type of Operation: The type of operation can prove the nature of either the UA flight for civil or secret state purpose. The drone operations in Yemen planned to go after Al-Qaeda in a way that they cannot escape, qualifies these drones as state aircrafts. For example, the CIA drone operation killed Anwar al-Awlaki, an American-born leader of a Yemen Al-Qaeda network without affording him an opportunity for a trial. Besides, these drone strikes have increased resentment both against the Yemeni and the US government.[59]

In the absence of any other regulatory guidance, the term 'used' in Article 3 (b) of the Chicago convention can determine the actual status of the UA. This functionality is the dispositive factor in ascertaining whether UA is a state or a civil aircraft. The CIA's use of drones in Pakistan, Yemen and Somalia is raising controversial issues. The CIA is not a military agency rather it is an independent civilian agency of the U.S.[60] This begs the question of whether the CIA drones operators who are civilians engaged in armed conflict can be categorized as unlawful combatants and subject to prosecution.[61] And whether the drone used for extra-judicial killing violate the UN Charter and state sovereignty.

2.1.6 Pros and Cons

For reasons adduced earlier in this article, drones would remain the popular weapon of choice despite calls to reduce its reliance on them. This incontrovertible fact is brought to bear by the fact that the Obama administration has used more than 400 drone strikes so far as against 50 strikes sanctioned by President George W. Bush during his 8 years in office.[62] On the positive side, drones have effectively rid the world of terrorist leaders and decimated terrorist organizations in foreign

---

[58] Cole and Wright (2010a).

[59] Smith (2012).

[60] Central Intelligence Agency, online: CIA https://www.cia.gov/about-cia/todays-cia/index.html.

[61] As Solis notes, "An unlawful combatant is one who takes an active and continuous part in armed conflict who therefore should be treated as a combatant in that he/she is a lawful target of attack, not enjoying the protections granted to civilians. Because unlawful combatants do not differentiate themselves from civilians and do not obey the laws of armed conflict they are not entitled to the privileges of combatants, for example, prisoner-of-war status". **See** Solis (2007), p. 130.

[62] Byman (2013), pp. 32–43 at 32.

countries that are infested with terrorist activity. One commentator unreservedly states that drones remain "a necessary instrument of counter-terrorism".[63] One of the effective achievements of drones is their ability to cut off communications of terrorist cells and destroy their training capabilities. However, a distinct danger is that along with the advantages to States fighting terrorism with drone attacks would be the tendencies to use such attacks at whim and indiscriminately on the tenuous justification that if it is good for the United States to use drones with impunity, they should also have the license to do so.

On the other side of the coin is a compelling argument against the use of drones, that the drone programme has taken a life of its own "to the point where tactics are driving strategy instead of the other way around"[64] resulting in "remote controlled repression".[65] Another grave disadvantage of drone attacks, particularly from a strategic point of view is that it kills individuals suspected of terrorism or proven terrorists without giving States the flexibility and advantage of questioning them. Therefore drone attacks are only good when a State decides that a terrorist would have to be eliminated rather than captured. However, the most compelling and indisputable argument against drone attacks is that drones have proved to be lethal to innocent civilians[66] whose killings go to the heart of the most important law of all brought to prominence in recent times with the growing instances of torture and killings of non-combatants in warfare and the solid grounding of international society—the law of humanity.

The use of drones brings to bear the age old maxim *Inter arma enim silent leges* attributed to Cicero, which translates as "In times of war, the laws are silent". In the twenty-first century, this maxim, which was purported to address the growing mob violence and thuggery of Cicero's time, has taken on a different and a more complex dimension, extending from the idealistic synergy between the executive and the judiciary in instances of civil strife, to the overall power, called "prerogative" or "discretion" of the sovereign, to act for the public good as the guardian of the rule of law.

One commentator has said:

> In recent years, international criminal lawyers have proclaimed the arrival of a new order where impunity for war criminals is extinguished or swept aside by crusading domestic prosecutors (e.g. Garzon in Spain), by national courts (e.g. the House of Lords in *Pinochet*), by bold legislators (e.g. the Belgian war crimes legislation in 1993 and 1999, notwithstanding subsequent amendments), and by proliferating . . .international tribunals.[67]

However, the good news is that the trend is shifting towards a preponderance of State responsibility. Dr. Rudi Teitel, Professor of Comparative Law at New York

---

[63] *Id.* 32.

[64] Cronin (2013), pp. 44–54 at 44.

[65] *Id.* 47.

[66] The Bureau of Investigative Journalism has reported that in 2011 alone 900 non combatants including almost 200 children were killed by drone strikes. See Byman (2013), *supra* note 62 at 37.

[67] Simpson (2007), p. 133.

Law School and Visiting Professor, London School of Economics, Global Governance, says:

> Sovereignty is no longer a self-evident foundation for international law. This shift is driving the move from the State-centric normative discourse of global politics – which had prevailed until recently – to a far ranging, transnational discourse in which references to changed subjectivity have consequences. That new discourse is constructed more among humanity law lines.[68]

This statement is consistent with the pronouncement of the International Criminal Tribunal for the former Yugoslavia which in its adjudication of *Prosecutor* v. *Dusko Tadic* said: "a state-sovereignty oriented approach has been gradually supplanted by a human being oriented approach".[69]

There are two broad reasons for this shift: the natural historical progression of world affairs which shifted trends chronologically; and the growing instances of torture, rape and killings in circumstances of internal strife and military warfare. Let us take the chronological evolution first. After World War II, the world has, over the past 70 years or so, gone through three global political and economic stages. The first was the modernist era, where naturally, State sovereignty was considered paramount over any other consideration in the immediate and proximate aftermath of the War where nations and States were attacked. This lasted until the 1960s.

From then on the trend shifted to the post-modernist era where the focus veered from the interests of the State to the welfare of the citizen. Post modernism, which was a characteristic of the 1960s and 1970s, progressed steadily toward the twenty-first century. Post modernist thinking was geared to accepting that human culture, as we knew it from a social and economic perspective, was reaching an end. This school of thought associated itself with the momentum of industrial society, drawing on an image of pluralism of cultures and a multitude of groups. The interaction between political modernism, which brought to bear the globalization of nations and deconstruction of separatism of human society, while at the same time ascribing to the individual certain rights at international law that transcended natural legislation parameters, has been symbiotic and essentially economic. In the post modernist era, the fundamental modernist philosophy of state sovereignty and peace gave way to an industrial culture that emphasized economic coexistence for the betterment of the global citizen.

The neo post modernist era was signalled as an immediate consequence of the attacks of 11 September 2001 where close to 3,000 human beings were decimated by terrorist activities within a few minutes. The principle of State Responsibility with regard to world peace and security which lies primarily in Article 24 of the United Nations Charter which calls upon all members to refrain in their

---

[68] *Humanity's Law,* Oxford University Press: 2011, at 35.

[69] Decision of 2 October 1995, Case No. IT-94-1-T 14 July 1997, United Nations International Tribunal for the Prosecution of Persons Responsible for Serious Violations of International Humanitarian Law Committed in the Territory of former Yugoslavia since 1991.

international relations from the threat or use of force against the territorial integrity or political independence of any state, or in any other manner inconsistent with the purposes of the United Nations, was immediately resuscitated in the minds of the international community.

This principle is embodied in the work of the International Law Commission, through Article 2 of the Draft Articles on Responsibility of States for Internationally Wrongful Acts which provides that international responsibility of a State, which is referred to in Article 1, is attributable to that State if conduct of the State constitutes a breach of an international obligation of that State. The document also provides that the wrongfulness of an act of a State is precluded if the act constitutes a lawful measure of self defence taken in conformity with the Charter of the United Nations. The State responsible for an internationally wrongful act is under an obligation to compensate for damage caused, including reparation for financially assessable damage including loss of profits.

In addition to State responsibility for conduct attributable to that State, the International Law Commission has established that a crime against the peace and security of mankind entails individual responsibility, and is a crime of aggression. The Rome Statute of the International Criminal court, defines a war crime, *inter alia*, as intentionally directing attacks against civilian objects; attacking or bombarding, by whatever means, towns, villages, dwellings or buildings which are undefended and which are not military objects; employing weapons, projectiles, and material and methods of warfare that cause injury The Statute also defines as a war crime, any act which is intentionally directed at buildings, material, medical units and transport, and personnel using the distinctive emblems of the Geneva Conventions in conformity with international law.

The law of war categorizes justification for killing into four groups: military necessity; distinction; proportionality and humanity. The last is what applies to drone killings. Targeting of civilians and medical personnel by no means comes even close to any of these categories. That drones—which some call "assassination machines"—in their missions of targeted killings, are used contrary to Article 51 of the United Nations Charter[70] make them clearly contrary to the principles of public international law. Drones are precision instruments which rarely miss their targets but unfortunately lack the means to distinguish between the innocent and the guilty.

Killings by drones come within the new heading under international law called the "conscience of mankind". It is a universal law which goes as far back as 1949 when the International Court of Justice (ICJ) recognized:

[70] Article 51 states: "Nothing in the present Charter shall impair the inherent right of individual or collective self-defence if an armed attack occurs against a Member of the United Nations, until the Security Council has taken measures necessary to maintain international peace and security. Measures taken by Members in the exercise of this right of self-defence shall be immediately reported to the Security Council and shall not in any way affect the authority and responsibility of the Security Council under the present Charter to take at any time such action as it deems necessary in order to maintain or restore international peace and security".

Certain well recognized principles, namely elementary considerations of humanity, even more existing in peace than in war.[71]

Again in 1951 the ICJ made a similar pronouncement which appeared once again in 1976 in the Draft Articles of the International Law Commission which recognized the existence of "international crimes". The principle of humanity law was immutably established with no room for doubt in 1986 when the ICJ in the *Nicaragua Case* evolved "elementary considerations of humanity" which essentially devolved responsibility of ensuring the safety and security of humanity on the State even outside treaty law, such as the provisions contained in the 1949 *Geneva Conventions on International Humanitarian Law.*[72]

The future of UA will largely depend on how technology, regulators, and states can work in tandem. As already discussed, the status of an unmanned aircraft can be ascertained by the actual function it is performing at the given moment. Any UA on a particular journey would be considered military if it is surveying for a combat mission, other time the same UA can be considered civil if it is surveying for a humanitarian relief. Hence, the status of each flight must be pre-approved by the approved flight plan accepted by the States to be flown above and should beforehand specify the nature of the flight. As history unfolds and technology advances the future of UAVs will largely rely on the governmental funding of its research and development. The advances in technology will not only decrease the costs as the demand increases but their easy availability may raise the problems of UA getting in wrong hands. Thus, the legislatures and engineers should work hands in hands to make sure that the use of unmanned aircrafts is for the good of humanity not for its destruction. In view of this article, the future regulation of Unmanned Aircrafts remains a step-by-step process.

Drones are retaliation to mass murder and the unacceptable abuse and torture of humanity. More than 12 years have passed since the outrageous and catastrophic events of 11 September 2001. Yet, none of us can get over the outrage we felt at the dastardly terrorist acts which killed nearly 3,000 people in a day. Most of the world that day felt an amalgam of concern, fear, and anger that motivated us to take precautions. Overnight, we converted ourselves from a society that was both laid back and reactive to a society that took pre-emptive and preventive action to make sure such things never happened again. For purposes of definition, Preemption is when an act, which is potentially harmful to a State and is imminent, is effectively precluded by military or other action. Prevention is when an act, which is potentially harmful to a State and is inevitable, is effectively precluded by military or other action.

---

[71] *Corfu Channel Case*, (*United Kingdom* v. *Albania*), ICJ Reports (1949) at 22.

[72] The Geneva Conventions and their Additional Protocols are at the core of international humanitarian law, the body of international law that regulates the conduct of armed conflict and seeks to limit its effects. They specifically protect people who are not taking part in the hostilities (civilians, health workers and aid workers) and those who are no longer participating in the hostilities, such as wounded, sick and shipwrecked soldiers and prisoners of war.

Alan Dershowitz, Professor of Law at Harvard University, asserts that "There is a desperate need in the world for a coherent and widely accepted jurisprudence of preemption and prevention, in the context of both self-defence and defense of others".[73] Preemption and prevention are necessary elements in today's political and military fabric, where legal legitimacy is ascribed to actions of States which act swiftly to avoid harm and protect its citizenry. These practices, however well meant from a strategic context, inevitably give rise to the law of outrage. Referring to the Holocaust during World War II President Roosevelt said on October 7, 1942: "It is our intention that just and sure punishment shall be meted out to the ringleaders responsible for the organized murder of thousands of innocent persons in the commission of atrocities which have violated every tenet of the Christian faith." The next year Roosevelt, Stalin and Churchill vowed, through the Moscow Declaration, to bring the perpetrators to justice. On 26 October 1943 the United Nations War Crimes commission composed of 15 allied nations met for the first time in London. On March 24, 1944 President Roosevelt again issued the warning: "None who participate in these acts of savagery will go unpunished. All who share in the guilt shall share the punishment". The last sentence was reiterated in a different way by Present George W. Bush in the aftermath of the events of 9/11:

> Every nation has a choice to make. In this conflict, there is no neutral ground. If any government sponsors the outlaws and killers of innocence, they have become outlaws and murderers themselves. And they will take that lonely path at their own peril.[74]

Although this is the ultimate justification for the eradication of the vermin of terrorism that nature ever suffered to crawl upon the surface of the world, historic responses such as drone attacks are extreme measures resonating killings of both the innocent and the guilty summarily, without arraignment or judgment. There are some societies that are terrorized by drone attacks where they occur. In instances of hooliganism and vandalism as in the London riots of 2011 the statement in Parliament of Prime Minister David Cameron during: "We will not allow a culture of fear to exist on our streets. And we will do whatever it takes to restore law and order and to rebuild our communities... anyone charged with violent disorder and other serious offences should expect to be remanded in custody not let back on the streets and anyone convicted should expect to go to jail" might resonate.[75] But whether this philosophy can be extended to civilians of Afghanistan or Pakistan or Yemen who, while celebrating their relatives wedding at a party find a drone attack descending on them killing the bride and most of the happy relatives, is entirely another matter.

---

[73] *Preemption – A Sword that Cuts Both Ways,* Norton: New York, 2006, 72.

[74] Joyner.

[75] UK riots: text of David Cameron's address to Commons, *The Telegraph*, 11 August 2011, at http://www.telegraph.co.uk/news/uknews/crime/8695272/UK-riots-text-of-David-Camerons-address-to-Commons.html.

2.2 Other Instances of Aggressive Intervention

2.2.1 Carriage by Air of Munitions of War

On 8 December 2014 in Chicago the ICAO Council adopted a Resolution[76] to mark the 70th anniversary of the signing of the Chicago Convention. The Resolution states inter alia that the Council is determined to ensure that international civil aviation will continue to contribute to the promotion of global peace and security, social integration among the peoples of the world, economic prosperity of nations, and sustainable development for future generations. It then resolved, again inter alia, to encourage all Member States of ICAO to continue to promote the ideals and principles of the *Convention on International Civil Aviation* and compliance with its provisions; and acknowledged the critical need for continued ICAO efforts aimed at identifying the challenges posed by increases in global air transport demand and capacity, as well as the opportunities offered by new and emerging technologies, and to address those challenges and take advantage of those opportunities in order to achieve the safe, secure and sustainable growth of the international civil aviation system. The Resolution also invited all stakeholders, including Member States and relevant organizations of the global civil aviation community, to continue sharing and promoting best practices and working together through ICAO in support of a worldwide air transport system, which serves and benefits all nations and peoples of the world.

Ironically, nowhere in this Resolution is mention that the Council was mindful of instances when aviation was used over the past 70 years to attack and destroy communities and societies. The following discussion addresses such instances.

The Universal Declaration of Human Rights of the United Nations underscores the compelling importance of human rights and calls upon every society and human to contribute towards ensuring that human rights are enjoyed by everyone. In this context one of the most blatant abuse of human rights in the field of aviation is the carriage of military equipment and arms as well as active involvement in the provision of military assistance to regimes that use military force on their communities and societies. The Chicago Convention, in Article 35 provides that no munitions of war or implements of war may be carried in or above the territory of a State in aircraft engaged in international navigation, except by permission of such State. Each State is required to determine by regulations what constitutes munitions of war or implements of war for the purposes of this Article, giving due consideration, for the purposes of uniformity, to such recommendations as the International Civil Aviation Organization may from time to time make. Article 35 also states that each contracting State reserves the right, for reasons of public order and safety, to regulate or prohibit the carriage in or above its territory of articles other than those

[76] Resolution Adopted at the Extraordinary Session of the Council on 8 December 2014 on the Occasion of the Seventieth Anniversary of the Signing of the Chicago Convention. See http://www.icao.int/Newsroom/Pages/ICAO-council-passes-chicago-convention-resolution.aspx.

enumerated above, provided that no distinction is made in this respect between its national aircraft engaged in international navigation and the aircraft of the other States so engaged; and provided further that no restriction are imposed which may interfere with the carriage and use on aircraft of apparatus necessary for the operation or navigation of the aircraft or the safety of the personnel or passengers.

The above facts make it imperative for air carriers to be consciously aware of the danger of their contributing to attacks on civilians. Munitions of war are considered dangerous goods when carried by air. The subject of carriage of dangerous goods by air is addressed in Annex 18 to the Chicago Convention. The material in this Annex was developed by the Air Navigation Commission in response to a need expressed by Contracting States for an internationally agreed set of provisions governing the safe transport of dangerous goods by air. At its 33rd session held in Montreal from 25 September to 5 October 2001, the ICAO Assembly adopted Resolution A33-1[77] which was a direct response to the terrorist acts of 9/11. The Resolution recognized that a new type of threat was posed to civil aviation which required new concerted efforts and policies of cooperation on the part of States. The Resolution also urges all ICAO member States to ensure, in accordance with Article 4 of the Chicago Convention, that civil aviation is not used for any purpose inconsistent with the aims of the Convention, and to hold accountable and punish severely those who misuse civil aircraft as weapons of destruction, including those responsible for planning and organizing such acts or for aiding, supporting or harbouring perpetrators. It also called upon States to cooperate with each other in this endeavour and to ensure that ICAO Standards and Recommended Practices (SARPs) relating to aviation security are adhered to. Finally the Resolution directed the Council of ICAO and the Secretary General to act urgently to address new and emerging threats to civil aviation, in particular to review the adequacy of existing aviation conventions on security.

Munitions of war are not only arms and ammunitions but are deemed to include noxious material and gases that, when released, could be dangerous to humans. The ICAO requirements for dangerous goods have been largely developed by a panel of experts which was established in 1976. This panel continues to meet and recommends the necessary revisions to the Technical Instructions. As far as possible the Technical Instructions are kept aligned with the recommendations of the United Nations Committee of Experts on the Transport of Dangerous Goods and with the regulations of the International Atomic Energy Agency. The use of these common bases by all forms of transport allows cargo to be transferred safely and smoothly between air, sea, rail and road modes.

The ICAO requirements for the safe handling of dangerous goods firstly identify a limited list of those substances which are unsafe to carry in any circumstances and then show how other potentially dangerous articles or substances can be transported

[77] Resolution A33-1, Declaration on misuse of civil aircraft as weapons of destruction and other terrorist acts involving civil aviation, Assembly Resolutions in Force (as of 8 October 2004) ICAO Doc. 9848, at VII-1.

safely. The nine hazard classes are those determined by the United Nations Committee of Experts and are used for all modes of transport.

Class 1 includes explosives of all kinds, such as sporting ammunition, fireworks and signal flares. Class 2 comprises compressed or liquefied gases which may also be toxic or flammable; examples are cylinders of oxygen and refrigerated liquid nitrogen. Class 3 substances are flammable liquids including gasoline, lacquers, paint thinners, etc. Class 4 covers flammable solids, spontaneously combustible materials and materials which, when in contact with water, exit flammable gases (examples are some powdered metals, cellulose type film and charcoal).

Class 5 covers oxidizing material, including bromates, chlorates or nitrates; this class also covers organic peroxides which are both oxygen carriers and very combustible. Poisonous or toxic substances, such as pesticides, mercury compounds, etc., comprise Class 6, together with infectious substances which must sometimes be shipped for diagnostic or preventative purposes. Radioactive materials are in Class 7; these are mainly radioactive isotopes needed for medical or research purposes but are sometimes contained in manufactured articles such as heart pacemakers or smoke detectors. Corrosive substances which may be dangerous to human tissue or which pose a hazard to the structure of an aircraft are dealt with in Class 8 (for example, caustic soda, battery fluid, paint remover). Finally, Class 9 is a miscellaneous category for other materials which are potentially hazardous in air transport, such as magnetized materials which could affect the aircraft's navigational systems.

In response to the requirement of A33-1, that ICAO act with some urgency to address new and emerging threats to civil aviation, an ICAO Special Sub Committee meeting of the Legal Committee on the subject of preparation of one or more instruments addressing new and emerging threats was held at ICAO Headquarters from 3 to 6 July 2007.[78] At this meeting, Australia submitted a proposal[79] to prohibit the intentional and unlawful transport by air of particularly dangerous goods and fugitives. In this paper, Australia quoted the Preamble[80] to the Chicago

[78] One of the terms of reference of the Sub Committee as agreed by the ICAO Council was: to prepare, in light of A33-1 and the guidance of the Council, one or more draft instruments addressing the new and emerging threats to civil aviation. See Special Sub Committee on the Preparation of One or More Instruments Addressing New and Emerging Threats, Introductory Note, *LC/SC-NET-WP/1*, 29/05/07 at p. 2.

[79] Proposal to Prohibit the International and Unlawful Transport by Air of Particularly Dangerous Goods and Fugitives, *LC/SC-NET-WP/3*, 5/07/07.

[80] The Preamble to the Chicago Convention recognizes that the future development of international civil aviation can greatly help to create and preserve friendship and understanding among the nations and peoples of the world, and yet its abuse can become a threat to the general security. It also states that it is desirable to avoid friction and to promote co-operation between nations and peoples upon which the peace of the world depends. In pursuance of these objectives, governments signed the Convention that contains certain principles and arrangements in order that international civil aviation may be developed in a safe and orderly manner and that international air transport services may be established on the basis of equality of opportunity and operated soundly and economically.

Convention and emphasized that ICAO was created to help ensure the safe and orderly growth of civil aviation and to encourage the operation of civil aircraft for peaceful purposes. It was also the view of Australia that there were gaps in the international legal framework with regard to the unlawful transport of biological, chemical and legal weapons and other dangerous material on board civil aircraft and that the international aviation community had a responsibility to address these lacunae and shortcomings, particularly when an opportunity such as the one presented through the ICAO meeting arose.

The Sub Committee meeting had the opportunity, through the Australian paper, to note other international legislation on the transportation of dangerous materials. For example, the 2005 Protocol to the Convention for the Suppression of Unlawful Acts Against the Safety of Maritime Navigation which underscores the extreme danger of use by unlawful activity of maritime transport of nuclear, chemical or biological weapons.[81] Additionally, there are other guidance material, such as those issued by the World Health Organization[82] which provide practical guidance to facilitate compliance with current international regulations for the transport of infectious substances[83] and patient specimens by all modes of transport, both nationally and internationally, and include the changes that apply from 1 January 2005.[84] The WHO regulations categorically state that The *Technical Instructions for the Safe Transport of Dangerous Goods by* are the legally binding international regulations.[85] IATA *Dangerous Goods Regulations* (DGR) that incorporate the ICAO provisions and may add further restrictions (where necessary such restrictions are included in these guidelines). The ICAO rules apply on all international flights. For national flights, i.e. flights within one country, national civil aviation authorities apply national legislation. This is normally based on the ICAO

---

[81] Wagner et al. (2006). Bio surveillance is the science of real-time disease outbreak detection. Its principles apply to both natural and man-made epidemics (bioterrorism). It is worthy of note that in addition to activity in this field in the United States, there is also work being done in Europe, where disease surveillance is beginning to be organized on a continent-wide scale needed to track a biological emergencies. The system not only monitors infected persons, but also attempts to discern the origin of the outbreak.

[82] Guidance on Regulation for the Transport of Infectious Substances, World Health Organization, September 2005, WHO/CDS/CSR/LYO/2005.22.

[83] For the purposes of transport, infectious substances are defined as substances which are known or are reasonably expected to contain pathogens. Pathogens are defined as microorganisms (including bacteria, viruses, rickettsiae, parasites, fungi) and other agents such as prions, which can cause disease in humans or animals. The definition is applied to all specimens except those explicitly excluded in the WHO Guidance Material.

[84] The international regulations for the transport of infectious substances by any mode of transport are based upon the Recommendations made by the Committee of Experts on the Transport of Dangerous Goods (UNCETDG), a committee of the United Nations Economic and Social Council. The Recommendations are presented in the form of Model Regulations. The United Nations Model Regulations are reflected in international law through international modal agreements.

[85] It is widely recognized that SARS is not a disease but a syndrome. See generally, Ruwantissa and Abeyratne (2002), pp. 53–80.

provisions, but may incorporate variations. State and operator variations are published in the ICAO Technical Instructions and in the IATA Dangerous Goods Regulations. The WHO guidelines also contain detailed packing instructions regarding infectious substances.[86]

Regionally, there is direct provision with regard to safety provisions that impact on protecting humanity at large. For instance, European Union Regulation EC 3922/1991[87] prescribes common standards for consistency and harmonization in the design, manufacture, operation and maintenance of aircraft. The United Kingdom has focused regulations, such as the Air Navigation Order 2005 (CAP 393), Article 69 of which prohibits aircraft from carrying munitions of war[88] unless they are carried under a letter of authority of the Civil Aviation Administration and the commander of the aircraft is advised in writing of all details of the consignment carried in his aircraft, such details to be given to the commander prior to take off.

In the United States The US Code of Federal Regulations (Title 18.922) permits only licensed manufacturers, importers or dealers to be involved in the dissemination of arms and ammunition by shipping or transportation ob surface or by air. There are other countries which have similar restrictions.[89]

---

[86] The system of packaging is recommended for use all infectious substances. It consists of three layers as follows: Primary receptacle—which is a primary watertight, leak-proof receptacle containing the specimen. The receptacle is packaged with enough absorbent material to absorb all fluid in case of breakage. Secondary packaging—which is a second durable, watertight, leak-proof packaging to enclose and protect the primary receptacle(s). Several cushioned primary receptacles may be placed in one secondary packaging, but sufficient additional absorbent material shall be used to absorb all fluid in case of breakage. Outer packaging—which are secondary packagings placed in outer shipping packagings with suitable cushioning material. Outer packagings protect their contents from outside influences, such as physical damage, while in transit. The smallest overall external dimension shall be 10×10 cm. Each completed package is normally required to be marked, labelled and accompanied with appropriate shipping documents (as applicable). *Supra*, note 82 (this chapter) at p. 6.

[87] Council Resolution EEC No. 3922/91 of 16 December 1991 on the Harmonization of the Technical Requirements and Administrative Procedures in the Field of Civil Aviation, Amended by Regulation EC No. 1899/2006 of 12 December 2006, Regulation EC No. 8/2008 of 11 December 2007, Regulation EC No. 859 of 20 August 2008.

[88] Any weapons or ammunition, any article carrying an explosive, noxious liquid or gas or any other thing that is designed to be made use of in war.

[89] In El Salvador, OPS 2065 prohibits the transport by air of weapons and munitions of war unless specifically permitted by the relevant authorities. In Barbados, Regulation 69 requires all States concerned and involved in the carriage and receipt of arms and ammunition carried by air to give their permission and agreement for such carriage before the consignments can be placed on board aircraft. In Jamaica, Air operators are not allowed to transport munitions of war by air unless prior approval has been granted by the Civil Aviation Authority: Munitions of war shall not be carried on an aircraft unless the munitions of war are carried with the written permission of the Authority and in accordance with the Eighteenth 667 and Twentieth 668 Schedules and any other conditions imposed by the Authority relating thereto; Weapons of war cannot be carried on an aircraft in any compartment to which passengers have access. In Trinidad and Tobago, Trinidad and Tobago no airlines, whether domestic or foreign, are allowed to transport weapons of war by air unless prior approval has been granted by the Civil Aviation Authority. In Nicaragua, Nicaragua, according to

2.2.2 Some Specific Instances

Countries such as Afghanistan, Sudan, Somalia, Nicaragua, Angola the Democratic Republic of Congo and a host of other States have suffered from the transportation of arms and ammunition with the use of commercial aircraft. In Nicaragua, in a conflict spanning from 1982 to 1990 where 42,000 perished, much of the damage on the ground was done by the transport of munitions by air into Nicaragua. Around the same time, support to the Contra groups in Honduras was given by anti Cuban regime groups in Miami when they provided by air loads of small arms and ammunition in 1981.[90] A commentary on the subject of carriage by air of weaponry states:

> Various airports and sea ports - located in 20 different countries - were used between 1981 and 1986 to carry out the logistic operations in support of the Contras and of the arms shipments to Iran. Some of the aviation companies serving the Iran-Contra operations in Europe, the Middle East and Central America were also engaged at the same time in arms and military supply operations in Angola and DR Congo... One of the first documented flights occurred on April 9, 1983, when Southern Air Transport 61 "Flight SJ1466 (carried out by an L-100), supposedly en route from Panama to Tegucigalpa, landed instead at Palmerola AFB, under secret instructions given to the pilot by SAT management. The plane transported 22 tons of small arms and ammunition picked up in Panama. Also in 1983, other companies, such as Summit Aviation, Investair and Westair Intl., provided the Contras with light planes, used for operations inside Nicaragua. In 1985 and 1986, several other arms shipments reached the Contras, mainly" own by Southern Air Transport (contracted by Richard Gadd's Eagle Aviation Service and Technology, EAST and AirMach), Arrow Air (subcontracted by SAT) and F. A. Conner Air. North's associates also organized large arms shipments from Portugal, through the shell companies ACE and Energy Resources as well as through arms dealers in Portugal and Canada. North's associates used forged Guatemalan Army end-user certi! cates and diverted the arms from their supposed destinations in Guatemala to Contra bases. The shipments included arms and ammunition worth about USD5.6 million altogether, with a volume that added up to some 800 tons.[91]

a 2007 Regulation, based upon the ICAO's Standards and Recommended Practices and the European Joint Aviation Requirements (JAR), transportation of munitions are generally forbidden. In India, India according to rule 8 of the Aircraft Rules, 1937, arms, ammunition, munitions of war, implements of war, explosives etc. shall be carried to, from, within or over India only with the written permission of the Central Government and subject to the terms and conditions of such permission. 675 The power to issue permission is delegated to officers of the Directorate General of Civil Aviation. Under the Civil Aviation Rules (1994) of Pakistan, the country prohibits the carriage of weapons or parts thereof, or ammunition or constituents thereof, which are designed for use in warfare, or against a person. 678 Only the Director- General of the Civil Aviation Authority can authorize the carriage of such items. Article 41 paragraph 1 of the Air Navigation Order of Singapore prohibits the carriage of munitions of war on aircraft. Munitions of war are defined as "such weapons and ammunition designed for use in warfare or against the person, including parts designed for such weapons and ammunition".

[90] See Danssaert and Finardi (2011), at note 58 where the authors say: "According to FBI investigators, Raul Arana, an agent for Contra leader Fernando Chamorro," flew "several large arms shipments from Miami to Honduras in 1981", see Walcott and Paztor (1987).

[91] *Id.* Danssaert and Finardi (2011), pp. 15 and 17.

Aviation, arms and ammunition and strikes against civilian targets are interlinked. The most ominous threat to attacks on civilians in the aviation context is military air strikes.

2.3 Military Air Strikes

2.3.1 The Gaza Airport Incident

On 4 December 2001, Israel Military forces attacked the Gaza International airport, destroyed the air navigation facilities and bombarded runways and taxiways until the airport became unserviceable. When the Palestinian Authority attempted a repair on 11 January 2002, the Israeli military forces bombarded once again the airport and its facilities by aircraft, artillery and tanks, thereby destroying the runway, the taxiways and all facilities.

Gaza International Airport (LVGZ), the only gateway for the Palestinians to and from Palestine, had been inaugurated for air navigation in November 1998 by President Arafat of Palestine and President Bill Clinton as a result of the Wye-River accord signed by Israel and the Palestine Liberation Organization. On 23 October 1998 and also in accordance with the Protocol regarding the establishment and operation of the International Airport in the Gaza Strip during the interim period. That Protocol, signed by Israel and Palestinian 20 November 1998 at Tel Aviv, reserved the right for the Palestinian Civil Aviation Authority to operate the airport 24 h a day around.

On 8 October 2000, the Israeli Government had ordered the closure of the airport's airspace, preventing Palestine Airlines, Egypt Air, Jordanian Royal Wings, Royal Air Maroc and any other airlines from carrying out operations between that airport and those of Cairo, Amman, Jeddah, Dubai, Abu Dhabi, Doha, Larnaca and Istanbul, and preventing any humanitarian flights, including the transportation of injured persons and medical assistance. This public punishment procedure, it was alleged in the ICAO Council when it deliberated on this issue, was without any excuse related to the airport operations, which had been functioning smoothly under joint security supervision by Palestinians and Israelis without dispute, since Israel had the overriding responsibility for security during implementation of joint activities such as aircraft arrival and departure procedures, passenger and luggage arrival and departure procedures, customs procedures, inbound and outbound procedures, aircraft security inspection before departure and after arrival etc.

The destroyed airport and air navigation facilities were used for the transportation of civilian passengers, search and rescue operations in case of emergencies, transportation of rescue material, including medical equipment, medicines and survival kits for safeguarding human lives. The observer from Palestine in the council for purposes of this discussion emphasized the need for a fully functional airport in the Palestine occupies territories, taking into account the prevailing tense and violent situation and the urgent need for humanitarian relief and medical assistance.

The airport was developed with voluntary contributions from a number of European countries, which recognized beyond doubt the urgent need for the airport. Nevertheless, the airport was destroyed without paying attention to any humanitarian consideration. This led the European Union to condemn the Israeli actions and reserve the right to demand compensation for the damages.

The ICAO Council was advised that the destruction of the civil airport in Gaza was an act deliberately perpetrated by a Contracting State. Such destruction took place under the watchful eyes of the international community and was widely covered by local and international media reports. Let alone the complaisance of the perpetrator, contempt of respect for human life, the disrespect of international laws, including the conventions on civil aviation security.

The issue brought to bear the relevance of the Chicago Convention which provides that:

> Each contracting state agrees not to use civil aviation for any purpose inconsistent with the aims of this Convention.[92]

Those aims, as enshrined in the Convention, include:

> Meet the needs of the peoples of the world for safe, regular, efficient and economical air transport.[93]

The States parties to the Montreal Convention (1971) considered that:

> "... Unlawful acts against the safety of civil aviation jeopardize the safety of persons and property, seriously affect the operation of air services, and undermine the confidence of the peoples of the world in the safety of civil aviation", that "the occurrence of such acts is a matter of grave concern" and that "for the purpose of deterring such acts, there is an urgent need to provide appropriate measures for punishment of offenders".

In its Resolution A20-1 Resolving Clause 3, The Assembly solemnly warned Israel that if it continues committing such acts the Assembly will take further measures against Israel to protect international civil aviation.

In its Resolution A33-2, the Assembly stated the following: "*Whereas* acts of unlawful interference against civil aviation have become the main threat to its safe and orderly development; *Recognizing* that all acts of unlawful interference against international civil aviation constitute a grave offence in violation of International law"; the Assembly: strongly condemned all acts of unlawful interference against civil aviation wherever and by whomever and for whatever reason they are perpetrated.

In the light of repeated military attacks by Israel against civil aviation and considering the framework of the Chicago Convention (1944), the civil aviation

---

[92] *See* Article 4 (Misuse of civil aviation).

[93] *See* Article 44 (Objectives).

security conventions, Assembly resolution A20-1 and Assembly resolution A33-2, the Council was requested by Algeria, Egypt, Lebanon and Saudi Arabia to accept its responsibilities and solemnly condemn the unlawful act perpetrated by Israel against the civil international Gaza Airport and issue a clear warning that such acts will not go unpunished.[94] It must be noted that Article 54(n) of the *Chicago Convention provides that*, "the Council shall ... consider any matter relating to this Convention which any Contracting State refers to it" and it was under this principle that the Council considered the Gaza issue.

It was claimed on behalf of the Palestinian National Authority that it had the right, under international law to demand compensation from Israel for all losses caused by their incursions on this civil airport for a total of US$14 million and a condemnation in the Council of Israel for the acts of unlawful interference with civil aviation.

The following incidents of Israeli aggression by air were cited to the ICAO Council:

1. The interception by Israeli fighter planes of a civil aircraft belonging to Lebanese airlines during its flight from Jerusalem to Beirut in 1949;
2. An attack by Israeli military planes and airborne command forces against the international airport of Beirut on 31 December 1968;
3. The shooting by Israeli military planes of a civilian aircraft belonging to Libya, which had lost track of its route above Egyptian territory in Sinai on 21 February 1973;
4. The interception by Israeli military aircraft of a Lebanese civil aircraft leased to Iraqi Airways on 10 August 1973;
5. A violation of the sovereignty of Tunisian airspace by Israeli military aircraft on 1 October 1985;
6. The interception by Israeli military aircraft of a Libyan civil aircraft on 4 February 1986 en route from Tripoli to Damascus.

The Council adoption a resolution which Stated as follows:

The Council,

Having considered the fact that Gaza International Airport facilities were destroyed by Israeli military forces on 4 December 2001 and 10 January 2002 so as to render the airport inoperable;

Having noted the facts presented and the views expressed by Israel and the Palestinian Authority;

Noting that such action is a violation of the principles enshrined in the *Convention on International Civil Aviation* (Chicago Convention, 1944);

Noting that such action is contrary to the principles of the Convention for the *Suppression of Unlawful Acts against the Safety of Civil Aviation* (Montreal Convention, 1971) and the *Protocol for the Suppression of Unlawful Acts of Violence at Airports Serving International Civil Aviation*, Supplementary thereto (Montreal Protocol, 1988), which consider

[94] C-WP/11791, 8 March 2002.

that unlawful acts of violence jeopardizing the safety of airports disturb the safe and orderly conduct of civil aviation for all States;

Considering that the occurrence of such acts is a matter of grave concern to the international community;

1. *Strongly condemns* all acts of unlawful interference against civil aviation, wherever, by whomsoever and for whatever reasons they are perpetrated;
2. *Strongly condemns* the destruction of Gaza International Airport and its air navigation facilities;
3. *Reaffirms* the important role of the International Civil Aviation Organization to facilitate the resolution of questions which may arise between Contracting States in relation to matters affecting the safe and orderly operation of international civil aviation throughout the world;
4. *Urges* Israel to fully comply with the aims and objectives of the Chicago Convention;
5. *Strongly urges* Israel to take the necessary measures to restore Gaza International Airport so as to allow its reopening as soon as possible;
6. *Requests* the President of the Council to attend to the implementation of this Resolution, in particular its preceding Clauses 3, 4 and 5, and to secure the full cooperation of the parties with respect to the application of the Chicago Convention and of the above-mentioned principles; and
7. *Requests* the Secretary General to inform all 187 Contracting States of this Resolution.

This issue was placed as an item for discussion at The High-level, Ministerial Conference on Aviation Security (AVSEC-Conf/02) held at ICAO Headquarters on 19 and 20 February 2002, which was attended by 717 participants from 153 Contracting States and 25 observers.[95] The Conference adopted a Declaration on aviation security which did not refer specifically to the Gaza issue but inter alia recognized the responsibility of States for the security and the safety of civil aviation, irrespective of whether the air transport and related services concerned are provided by Government, autonomous or private entities; The Declaration also noted the significant improvements in aviation security recently initiated in a large number of States and went on to recognize that a uniform approach in a global system is essential to ensure aviation security throughout the world and that deficiencies in any part of the system constitute a threat to the entire global system. It also affirmed that a global aviation security system imposes a collective responsibility on all States.

[95] The Council decided to place this matter before the High Level Safety Conference. See C-WP/11786, 27 February 2002.

2.3.2 NATO Strikes on Libya

During the "Arab Spring"[96] of 2011, the Libyan military, its equipment and personnel who were launching military attacks[97] on protesting civilians came under heavy attack from NATO as a consequence of United Nations Security Council Resolution 1973. This Resolution, which was adopted on 17 March 2011, demanded an immediate ceasefire in Libya, including an end to the current attacks against civilians, which it said might constitute "crimes against humanity", the Security Council this evening imposed a ban on all flights in the country's airspace—a no-fly zone—and tightened sanctions on the Qadhafi regime and its supporters.

Adopting Resolution 1973 by a vote of 10 in favour to none against, with 5 abstentions (Brazil, China, Germany, India, Russian Federation), the Council authorized Member States, acting nationally or through regional organizations or arrangements, to take all necessary measures to protect civilians under threat of attack in the country, including Benghazi, while excluding a foreign occupation force of any form on any part of Libyan territory—requesting them to immediately inform the Secretary-General of such measures.

Recognizing the important role of the League of Arab States in the maintenance of international peace and security in the region, and bearing in mind the United Nations Charter's Chapter VIII, the Council asked the League's member States to cooperate with other Member States in implementing the no-fly zone.

The Council stressed the need to intensify efforts to find a solution to the crisis that responded to the legitimate demands of the Libyan people, noting actions being taken on the diplomatic front in that regard. It further demanded that Libyan authorities comply with their obligations under international law and take all measures to protect civilians and meet their basic needs and to ensure the rapid and unimpeded passage of humanitarian assistance.

In that connection, the Council specified that the flight ban would not apply to flights that had as their sole purpose humanitarian aid, the evacuation of foreign nationals, enforcing the ban or other purposes "deemed necessary for the benefit of the Libyan people".

It further decided that all States should deny permission to any Libyan commercial aircraft to land in or take off from their territory unless a particular flight had been approved in advance by the committee that was established to monitor sanctions imposed by resolution 1970 (2011).

---

[96] The 2010–2011 "Arab Spring" is a revolutionary wave of demonstrations and protests and protests that has been taking place in the Middle East and North Africa. Since 18 December 2010. Prior to this period Sudan was the only Arab country to have successfully toppled dictatorial regimes.

[97] The uprising in Libya instantly became violent when the Libyan government reacted harshly towards peaceful protests. On February 18, 3 days after the protests began; the country erupted into an armed conflict when protesters executed policemen and men loyal to Colonel Muammar Gaddafi for killing protesters.

In tightening the asset freeze and arms embargo established by that resolution, the Council this evening further detailed conditions for inspections of transport suspected to be violating the embargo, requesting States enforcing the embargo to coordinate closely with each other and the Secretary-General on the measures they were taking towards implementation.

It requested the Secretary-Secretary to create an eight-member panel of experts to assist the Security Council committee in monitoring the sanctions.

Introducing the resolution, the Foreign Minister of France, Alain Juppé, said "the situation on the ground is more alarming than ever, marked by the violent re-conquest of cities that have been released". The Security Council could not stand by and "let the warmongers flout international legality". The world was experiencing "a wave of great revolutions that would change the course of history", but the will of the Libyan people had been "trampled under the feet of the Qadhafi regime". Earlier Council measures had been ignored and violence against Libyan civilians had redoubled.

He said that the urgent need to protect the civilian population had led to the elaboration of the current resolution, which authorized the Arab League and those Member States wishing to do so to take all measures to protect areas that were being threatened by the Qadhafi regime. "We have very little time left — perhaps only a matter of hours," he said, adding that each hour and day that passed "increased the weight" on the international community's shoulders.

Speaking after the vote, representatives who had supported the text agreed that the strong action was made necessary because the Qadhafi regime had not heeded the first actions of the Council and was on the verge of even greater violence against civilians as it closed in on areas previously dominated by opposition in the east of the country. They stressed that the objective was solely to protect civilians from further harm.

The full text of resolution 1973 (2011) reads as follows:

The Security Council,

Recalling its resolution 1970 (2011) of 26 February 2011,

Deploring the failure of the Libyan authorities to comply with resolution 1970 (2011),

Expressing grave concern at the deteriorating situation, the escalation of violence, and the heavy civilian casualties,

Reiterating the responsibility of the Libyan authorities to protect the Libyan population and *reaffirming* that parties to armed conflicts bear the primary responsibility to take all feasible steps to ensure the protection of civilians,

Condemning the gross and systematic violation of human rights, including arbitrary detentions, enforced disappearances, torture and summary executions,

Further condemning acts of violence and intimidation committed by the Libyan authorities against journalists, media professionals and associated personnel and *urging* these authorities to comply with their obligations under international humanitarian law as outlined in resolution 1738 (2006),

Considering that the widespread and systematic attacks currently taking place in the Libyan Arab Jamahiriya against the civilian population may amount to crimes against humanity,

Recalling paragraph 26 of resolution 1970 (2011) in which the Council expressed its readiness to consider taking additional appropriate measures, as necessary, to facilitate and support the return of humanitarian agencies and make available humanitarian and related assistance in the Libyan Arab Jamahiriya,

Expressing its determination to ensure the protection of civilians and civilian populated areas and the rapid and unimpeded passage of humanitarian assistance and the safety of humanitarian personnel,

Recalling the condemnation by the League of Arab States, the African Union and the Secretary-General of the Organization of the Islamic Conference of the serious violations of human rights and international humanitarian law that have been and are being committed in the Libyan Arab Jamahiriya,

Taking note of the final communiqué of the Organization of the Islamic Conference of 8 March 2011, and the communiqué of the Peace and Security Council of the African Union of 10 March 2011 which established an ad hoc High-Level Committee on Libya,

Taking note also of the decision of the Council of the League of Arab States of 12 March 2011 to call for the imposition of a no-fly zone on Libyan military aviation, and to establish safe areas in places exposed to shelling as a precautionary measure that allows the protection of the Libyan people and foreign nationals residing in the Libyan Arab Jamahiriya,

Taking note further of the Secretary-General's call on 16 March 2011 for an immediate ceasefire,

Recalling its decision to refer the situation in the Libyan Arab Jamahiriya since 15 February 2011 to the Prosecutor of the International Criminal Court, and *stressing* that those responsible for or complicit in attacks targeting the civilian population, including aerial and naval attacks, must be held to account,

Reiterating its concern at the plight of refugees and foreign workers forced to flee the violence in the Libyan Arab Jamahiriya, *welcoming* the response of neighbouring States, in particular Tunisia and Egypt, to address the needs of those refugees and foreign workers, and *calling on* the international community to support those efforts,

Deploring the continuing use of mercenaries by the Libyan authorities,

Considering that the establishment of a ban on all flights in the airspace of the Libyan Arab Jamahiriya constitutes an important element for the protection of civilians as well as the safety of the delivery of humanitarian assistance and a decisive step for the cessation of hostilities in Libya,

Expressing concern also for the safety of foreign nationals and their rights in the Libyan Arab Jamahiriya,

Welcoming the appointment by the Secretary General of his Special Envoy to Libya, Mr. Abdul Ilah Mohamed Al-Khatib and supporting his efforts to find a sustainable and peaceful solution to the crisis in the Libyan Arab Jamahiriya,

Reaffirming its strong commitment to the sovereignty, independence, territorial integrity and national unity of the Libyan Arab Jamahiriya,

Determining that the situation in the Libyan Arab Jamahiriya continues to constitute a threat to international peace and security,

Acting under Chapter VII of the Charter of the United Nations,

1. *Demands* the immediate establishment of a ceasefire and a complete end to violence and all attacks against, and abuses of, civilians;

2. *Stresses* the need to intensify efforts to find a solution to the crisis which responds to the legitimate demands of the Libyan people and *notes* the decisions of the Secretary-General to send his Special Envoy to Libya and of the Peace and Security Council of the African Union to send its ad hoc High-Level Committee to Libya with the aim of facilitating dialogue to lead to the political reforms necessary to find a peaceful and sustainable solution;

3. *Demands* that the Libyan authorities comply with their obligations under international law, including international humanitarian law, human rights and refugee law and take all measures to protect civilians and meet their basic needs, and to ensure the rapid and unimpeded passage of humanitarian assistance;

Protection of civilians

4. *Authorizes* Member States that have notified the Secretary-General, acting nationally or through regional organizations or arrangements, and acting in cooperation with the Secretary-General, to take all necessary measures, notwithstanding paragraph 9 of resolution 1970 (2011), to protect civilians and civilian populated areas under threat of attack in the Libyan Arab Jamahiriya, including Benghazi, while excluding a foreign occupation force of any form on any part of Libyan territory, and *requests* the Member States concerned to inform the Secretary-General immediately of the measures they take pursuant to the authorization conferred by this paragraph which shall be immediately reported to the Security Council;

5. *Recognizes* the important role of the League of Arab States in matters relating to the maintenance of international peace and security in the region, and bearing in mind Chapter VIII of the Charter of the United Nations, requests the Member States of the League of Arab States to cooperate with other Member States in the implementation of paragraph 4;

No-fly zone

6. *Decides* to establish a ban on all flights in the airspace of the Libyan Arab Jamahiriya in order to help protect civilians;

7. *Decides further* that the ban imposed by paragraph 6 shall not apply to flights whose sole purpose is humanitarian, such as delivering or facilitating the delivery of assistance, including medical supplies, food, humanitarian workers and related assistance, or evacuating foreign nationals from the Libyan Arab Jamahiriya, nor shall it apply to flights authorised by paragraphs 4 or 8, nor other flights which are deemed necessary by States acting under the authorization conferred in paragraph 8 to be for the benefit of the Libyan people, and that these flights shall be coordinated with any mechanism established under paragraph 8;

8. *Authorizes* Member States that have notified the Secretary-General and the Secretary-General of the League of Arab States, acting nationally or through regional organizations or arrangements, to take all necessary measures to enforce compliance with the ban on flights imposed by paragraph 6 above, as necessary, and *requests* the States concerned in cooperation with the League of Arab States to

coordinate closely with the Secretary General on the measures they are taking to implement this ban, including by establishing an appropriate mechanism for implementing the provisions of paragraphs 6 and 7 above,

9. *Calls upon* all Member States, acting nationally or through regional organizations or arrangements, to provide assistance, including any necessary overflight approvals, for the purposes of implementing paragraphs 4, 6, 7 and 8 above;

10. *Requests* the Member States concerned to coordinate closely with each other and the Secretary-General on the measures they are taking to implement paragraphs 4, 6, 7 and 8 above, including practical measures for the monitoring and approval of authorised humanitarian or evacuation flights;

11. *Decides* that the Member States concerned shall inform the Secretary-General and the Secretary-General of the League of Arab States immediately of measures taken in exercise of the authority conferred by paragraph 8 above, including to supply a concept of operations;

12. *Requests* the Secretary-General to inform the Council immediately of any actions taken by the Member States concerned in exercise of the authority conferred by paragraph 8 above and to report to the Council within 7 days and every month thereafter on the implementation of this resolution, including information on any violations of the flight ban imposed by paragraph 6 above;

Enforcement of the arms embargo

13. *Decides that* paragraph 11 of resolution 1970 (2011) shall be replaced by the following paragraph : "Calls upon all Member States, in particular States of the region, acting nationally or through regional organisations or arrangements, in order to ensure strict implementation of the arms embargo established by paragraphs 9 and 10 of resolution 1970 (2011), to inspect in their territory, including seaports and airports, and on the high seas, vessels and aircraft bound to or from the Libyan Arab Jamahiriya, if the State concerned has information that provides reasonable grounds to believe that the cargo contains items the supply, sale, transfer or export of which is prohibited by paragraphs 9 or 10 of resolution 1970 (2011) as modified by this resolution, including the provision of armed mercenary personnel, *calls upon* all flag States of such vessels and aircraft to cooperate with such inspections and authorises Member States to use all measures commensurate to the specific circumstances to carry out such inspections";

14. *Requests* Member States which are taking action under paragraph 13 above on the high seas to coordinate closely with each other and the Secretary-General and *further requests* the States concerned to inform the Secretary-General and the Committee established pursuant to paragraph 24 of resolution 1970 (2011) ("the Committee") immediately of measures taken in the exercise of the authority conferred by paragraph 13 above;

15. *Requires* any Member State whether acting nationally or through regional organisations or arrangements, when it undertakes an inspection pursuant to paragraph 13 above, to submit promptly an initial written report to the Committee containing, in particular, explanation of the grounds for the inspection, the results of such inspection, and whether or not cooperation was provided, and, if prohibited items for transfer are found, further requires such Member States to submit to the

Committee, at a later stage, a subsequent written report containing relevant details on the inspection, seizure, and disposal, and relevant details of the transfer, including a description of the items, their origin and intended destination, if this information is not in the initial report;

16. *Deplores* the continuing flows of mercenaries into the Libyan Arab Jamahiriya and *calls upon* all Member States to comply strictly with their obligations under paragraph 9 of resolution 1970 (2011) to prevent the provision of armed mercenary personnel to the Libyan Arab Jamahiriya;

Ban on flights

17. *Decides* that all States shall deny permission to any aircraft registered in the Libyan Arab Jamahiriya or owned or operated by Libyan nationals or companies to take off from, land in or overfly their territory unless the particular flight has been approved in advance by the Committee, or in the case of an emergency landing;

18. *Decides that* all States shall deny permission to any aircraft to take off from, land in or overfly their territory, if they have information that provides reasonable grounds to believe that the aircraft contains items the supply, sale, transfer, or export of which is prohibited by paragraphs 9 and 10 of resolution 1970 (2011) as modified by this resolution, including the provision of armed mercenary personnel, except in the case of an emergency landing;

Asset freeze

19. *Decides* that the asset freeze imposed by paragraph 17, 19, 20 and 21 of resolution 1970 (2011) shall apply to all funds, other financial assets and economic resources which are on their territories, which are owned or controlled, directly or indirectly, by the Libyan authorities, as designated by the Committee, or by individuals or entities acting on their behalf or at their direction, or by entities owned or controlled by them, as designated by the Committee, and *decides further* that all States shall ensure that any funds, financial assets or economic resources are prevented from being made available by their nationals or by any individuals or entities within their territories, to or for the benefit of the Libyan authorities, as designated by the Committee, or individuals or entities acting on their behalf or at their direction, or entities owned or controlled by them, as designated by the Committee, and directs the Committee to designate such Libyan authorities, individuals or entities within 30 days of the date of the adoption of this resolution and as appropriate thereafter;

20. *Affirms* its determination to ensure that assets frozen pursuant to paragraph 17 of resolution 1970 (2011) shall, at a later stage, as soon as possible be made available to and for the benefit of the people of the Libyan Arab Jamahiriya;

21. *Decides* that all States shall require their nationals, persons subject to their jurisdiction and firms incorporated in their territory or subject to their jurisdiction to exercise vigilance when doing business with entities incorporated in the Libyan Arab Jamahiriya or subject to its jurisdiction, and any individuals or entities acting on their behalf or at their direction, and entities owned or controlled by them, if the States have information that provides reasonable grounds to believe that such business could contribute to violence and use of force against civilians;

Designations

22. *Decides* that the individuals listed in Annex I shall be subject to the travel restrictions imposed in paragraphs 15 and 16 of resolution 1970 (2011), and *decides* further that the individuals and entities listed in Annex II shall be subject to the asset freeze imposed in paragraphs 17, 19, 20 and 21 of resolution 1970 (2011);

23. *Decides* that the measures specified in paragraphs 15, 16, 17, 19, 20 and 21 of resolution 1970 (2011) shall apply also to individuals and entities determined by the Council or the Committee to have violated the provisions of resolution 1970 (2011), particularly paragraphs 9 and 10 thereof, or to have assisted others in doing so;

Panel of Experts

24. *Requests* the Secretary-General to create for an initial period of one year, in consultation with the Committee, a group of up to eight experts ("Panel of Experts"), under the direction of the Committee to carry out the following tasks:

(a) Assist the Committee in carrying out its mandate as specified in paragraph 24 of resolution 1970 (2011) and this resolution;

(b) Gather, examine and analyse information from States, relevant United Nations bodies, regional organisations and other interested parties regarding the implementation of the measures decided in resolution 1970 (2011) and this resolution, in particular incidents of non-compliance;

(c) Make recommendations on actions the Council, or the Committee or State, may consider to improve implementation of the relevant measures;

(d) Provide to the Council an interim report on its work no later than 90 days after the Panel's appointment, and a final report to the Council no later than 30 days prior to the termination of its mandate with its findings and recommendations;

25. *Urges* all States, relevant United Nations bodies and other interested parties, to cooperate fully with the Committee and the Panel of Experts, in particular by supplying any information at their disposal on the implementation of the measures decided in resolution 1970 (2011) and this resolution, in particular incidents of non-compliance;

26. *Decides* that the mandate of the Committee as set out in paragraph 24 of resolution 1970 (2011) shall also apply to the measures decided in this resolution;

27. *Decides* that all States, including the Libyan Arab Jamahiriya, shall take the necessary measures to ensure that no claim shall lie at the instance of the Libyan authorities, or of any person or body in the Libyan Arab Jamahiriya, or of any person claiming through or for the benefit of any such person or body, in connection with any contract or other transaction where its performance was affected by reason of the measures taken by the Security Council in resolution 1970 (2011), this resolution and related resolutions;

28. *Reaffirms* its intention to keep the actions of the Libyan authorities under continuous review and underlines its readiness to review at any time the measures imposed by this resolution and resolution 1970 (2011), including by strengthening, suspending or lifting those measures, as appropriate, based on compliance by the Libyan authorities with this resolution and resolution 1970 (2011);

"29. *Decides* to remain actively seized of the matter."

The significant elements of this Resolution lie in the ban imposed on all flights over Libyan airspace (except for humanitarian flights) and in Resolving Clause 4 in terms of protection of civilians whereby States were authorized to take *all necessary measures*, notwithstanding paragraph 9 of resolution 1970 (2011),[98] to protect civilians and civilian populated areas under threat of attack in the Libyan Arab Jamahiriya, including Benghazi, while excluding a foreign occupation force of any form on any part of Libyan territory, and *requests* the Member States concerned to inform the Secretary-General immediately of the measures they take pursuant to the authorization conferred by this paragraph which shall be immediately reported to the Security Council. The phrase "all necessary measures" extended much flexibility for States to employ air power to strike the Libyan military, stopping short of foreign occupation of Libya.

Resolution 1973 followed Resolution 1970 on tough measures on the Libyan Regime in response to its crackdown on protesters.

Deploring what it called "the gross and systematic violation of human rights" in strife-torn Libya, the Security Council this evening demanded an end to the violence and decided to refer the situation to the International Criminal Court while imposing an arms embargo on the country and a travel ban and assets freeze on the family of Muammar Al-Qadhafi and certain Government officials.

Unanimously adopting resolution 1970 (2011) under Article 41 of the Charter's Chapter VII, the Council authorized all Member States to seize and dispose of military-related materiel banned by the text. It called on all Member States to facilitate and support the return of humanitarian agencies and make available humanitarian and related assistance in Libya and expressed its readiness to consider taking additional appropriate measures as necessary to achieve that.

Through the text, the Council also decided to establish a new committee to monitor sanctions, to liaison with Member States on compliance and to respond to violations and to designate the individuals subject to the targeted measures. Individuals and entities immediately subjected to the targeted sanctions were listed in an Annex to the resolution.

Regarding its referral of the situation in Libya since 15 February 2011 to the Prosecutor of the International Criminal Court, the Council recognized that States not party to the Rome Statute that established the Court had no obligations to it, but urged all States and concerned organizations to cooperate fully with the Court's Prosecutor.

The Council affirmed it would keep the actions of the Libyan authorities under continuous review and would be prepared to strengthen, modify, suspend or lift the prescribed measures in light of compliance or non-compliance with the resolution.

Following the adoption of the text, Secretary-General Ban Ki-moon welcomed the Council's "decisive" action. "While it cannot, by itself, end the violence and the repression, it is a vital step — a clear expression of the will of a united community of nations," he said.

---

[98] See below the complete text of Resolution 1970 for the text of Clause 9.

He expressed hope that the message that "gross violations of basic human rights will not be tolerated and that those responsible for grave crimes will be held accountable" would be "heard and heeded" by the Libyan regime and that it would bring hope and relief to those still at risk. He looked for similar action from the General Assembly and the international community as a whole, and warned that even bolder steps might be necessary.

In their explanations of vote, Council members welcomed the unanimity of the action and expressed solidarity with the people of Libya, hoping that their "swift and decisive" intervention would help bring them relief. Many expressed hope that the resolution was a strong step in affirming the responsibility of States to protect their people as well as the legitimate role of the Council to step in when they failed to meet that responsibility.

With the referral of the situation to the International Criminal Court, France's representative hoped the vote would open a new era in commitment to the protection of populations. Further to that goal, Brazil's representative expressed strong reservations to the provision in the resolution allowing for exemptions from jurisdiction of nationals from non-States parties, saying those were not helpful to advance the cause of justice and accountability.

Noting that five Council members were not parties to the Rome Statute that set up the International Criminal Court, including India, that country's representative said he would have preferred a "calibrated approach" to the issue. However, he was convinced that the referral would help to bring about the end of violence and he heeded the call of the Secretary-General on the issue, while stressing the importance of the provisions in the resolution regarding non-States parties to the Statute.

Some speakers, such as the representatives of Lebanon and the Russian Federation, stressed the importance of affirming the sovereignty and territorial integrity of Libya. The Chinese representative said he had supported the resolution taking into account the special circumstances in Libya.

Speaking last, Libya's representative said that the Council's action represented moral support for his people and was a signal that an end must be put to the fascist regime in Tripoli. He launched an appeal to all the officers of the Libyan armed forces to support their own people, and welcomed the referral to the International Criminal Court, as well as the decision not to impose sanctions on those who might abandon Mr. Al-Qadhafi in the end.

Also speaking this evening were the representatives of the United Kingdom, South Africa, Nigeria, United States, Colombia, Portugal, Germany, and Bosnia and Herzegovina and Gabon.

The full text of resolution 1970 (2011) reads as follows:

"The Security Council,

Expressing grave concern at the situation in the Libyan Arab Jamahiriya and condemning the violence and use of force against civilians,

Deploring the gross and systematic violation of human rights, including the repression of peaceful demonstrators, expressing deep concern at the deaths of civilians, and rejecting unequivocally the incitement to hostility and violence against the civilian population made from the highest level of the Libyan government,

Welcoming the condemnation by the Arab League, the African Union, and the Secretary General of the Organization of the Islamic Conference of the serious violations of human rights and international humanitarian law that are being committed in the Libyan Arab Jamahiriya,

Taking note of the letter to the President of the Security Council from the Permanent Representative of the Libyan Arab Jamahiriya dated 26 February 2011,

Welcoming the Human Rights Council resolution A/HRC/S-15/2 of 25 February 2011, including the decision to urgently dispatch an independent international commission of inquiry to investigate all alleged violations of international human rights law in the Libyan Arab Jamahiriya, to establish the facts and circumstances of such violations and of the crimes perpetrated, and where possible identify those responsible,

Considering that the widespread and systematic attacks currently taking place in the Libyan Arab Jamahiriya against the civilian population may amount to crimes against humanity,

Expressing concern at the plight of refugees forced to flee the violence in the Libyan Arab Jamahiriya,

Expressing concern also at the reports of shortages of medical supplies to treat the wounded,

Recalling the Libyan authorities' responsibility to protect its population,

Underlining the need to respect the freedoms of peaceful assembly and of expression, including freedom of the media,

Stressing the need to hold to account those responsible for attacks, including by forces under their control, on civilians,

Recalling article 16 of the Rome Statute under which no investigation or prosecution may be commenced or proceeded with by the International Criminal Court for a period of 12 months after a Security Council request to that effect,

Expressing concern for the safety of foreign nationals and their rights in the Libyan Arab Jamahiriya,

Reaffirming its strong commitment to the sovereignty, independence, territorial integrity and national unity of the Libyan Arab Jamahiriya.

Mindful of its primary responsibility for the maintenance of international peace and security under the Charter of the United Nations,

Acting under Chapter VII of the Charter of the United Nations, and taking measures under its Article 41,

1. *Demands* an immediate end to the violence and calls for steps to fulfil the legitimate demands of the population;

2. *Urges* the Libyan authorities to:

(a) Act with the utmost restraint, respect human rights and international humanitarian law, and allow immediate access for international human rights monitors;

(b) Ensure the safety of all foreign nationals and their assets and facilitate the departure of those wishing to leave the country;

(c) Ensure the safe passage of humanitarian and medical supplies, and humanitarian agencies and workers, into the country; and

(d) Immediately lift restrictions on all forms of media;

3. *Requests* all Member States, to the extent possible, to cooperate in the evacuation of those foreign nationals wishing to leave the country;

ICC referral

4. *Decides* to refer the situation in the Libyan Arab Jamahiriya since 15 February 2011 to the Prosecutor of the International Criminal Court;

5. *Decides* that the Libyan authorities shall cooperate fully with and provide any necessary assistance to the Court and the Prosecutor pursuant to this resolution and, while recognizing that States not party to the Rome Statute have no obligation under the Statute, urges all States and concerned regional and other international organizations to cooperate fully with the Court and the Prosecutor;

6. *Decides* that nationals, current or former officials or personnel from a State outside the Libyan Arab Jamahiriya which is not a party to the Rome Statute of the International Criminal Court shall be subject to the exclusive jurisdiction of that State for all alleged acts or omissions arising out of or related to operations in the Libyan Arab Jamahiriya established or authorized by the Council, unless such exclusive jurisdiction has been expressly waived by the State;

7. *Invites* the Prosecutor to address the Security Council within 2 months of the adoption of this resolution and every 6 months thereafter on actions taken pursuant to this resolution;

8. *Recognizes* that none of the expenses incurred in connection with the referral, including expenses related to investigations or prosecutions in connection with that referral, shall be borne by the United Nations and that such costs shall be borne by the parties to the Rome Statute and those States that wish to contribute voluntarily;

Arms embargo

9. *Decides* that all Member States shall immediately take the necessary measures to prevent the direct or indirect supply, sale or transfer to the Libyan Arab Jamahiriya, from or through their territories or by their nationals, or using their flag vessels or aircraft, of arms and related materiel of all types, including weapons and ammunition, military vehicles and equipment, paramilitary equipment, and spare parts for the aforementioned, and technical assistance, training, financial or other assistance, related to military activities or the provision, maintenance or use of any arms and related materiel, including the provision of armed mercenary personnel whether or not originating in their territories, and decides further that this measure shall not apply to:

(a) Supplies of non-lethal military equipment intended solely for humanitarian or protective use, and related technical assistance or training, as approved in advance by the Committee established pursuant to paragraph 24 below;

(b) Protective clothing, including flak jackets and military helmets, temporarily exported to the Libyan Arab Jamahiriya by United Nations personnel, representatives of the media and humanitarian and development works and associated personnel, for their personal use only; or

(c) Other sales or supply of arms and related materiel, or provision of assistance or personnel, as approved in advance by the Committee;

10. *Decides* that the Libyan Arab Jamahiriya shall cease the export of all arms and related materiel and that all Member States shall prohibit the procurement of such items from the Libyan Arab Jamahiriya by their nationals, or using their flagged vessels or aircraft, and whether or not originating in the territory of the Libyan Arab Jamahiriya;

11. *Calls upon* all States, in particular States neighbouring the Libyan Arab Jamahiriya, to inspect, in accordance with their national authorities and legislation and consistent with international law, in particular the law of the sea and relevant international civil aviation agreements, all cargo to and from the Libyan Arab Jamahiriya, in their territory, including seaports and airports, if the State concerned has information that provides reasonable grounds to believe the cargo contains items the supply, sale, transfer, or export of which is prohibited by paragraphs 9 or 10 of this resolution for the purpose of ensuring strict implementation of those provisions;

12. *Decides* to authorize all Member States to, and that all Member States shall, upon discovery of items prohibited by paragraph 9 or 10 of this resolution, seize and dispose (such as through destruction, rendering inoperable, storage or transferring to a State other than the originating or destination States for disposal) items the supply, sale, transfer or export of which is prohibited by paragraph 9 or 10 of this resolution and decides further that all Member States shall cooperate in such efforts;

13. *Requires* any Member State when it undertakes an inspection pursuant to paragraph 11 above, to submit promptly an initial written report to the Committee containing, in particular, explanation of the grounds for the inspections, the results of such inspections, and whether or not cooperation was provided, and, if prohibited items for transfer are found, further requires such Member States to submit to the Committee, at a later stage, a subsequent written report containing relevant details on the inspection, seizure, and disposal, and relevant details of the transfer, including a description of the items, their origin and intended destination, if this information is not in the initial report;

14. *Encourages* Member States to take steps to strongly discourage their nationals from travelling to the Libyan Arab Jamahiriya to participate in activities on behalf of the Libyan authorities that could reasonably contribute to the violation of human rights;

Travel ban

15. *Decides* that all Member States shall take the necessary measures to prevent the entry into or transit through their territories of individuals listed in Annex I of this resolution or designated by the Committee established pursuant to paragraph 24 below, provided that nothing in this paragraph shall oblige a State to refuse its own nationals entry into its territory;

16. *Decides* that the measures imposed by paragraph 15 above shall not apply:

(a) Where the Committee determines on a case-by-case basis that such travel is justified on the grounds of humanitarian need, including religious obligation;

(b) Where entry or transit is necessary for the fulfilment of a judicial process;

(c) Where the Committee determines on a case-by-case basis that an exemption would further the objectives of peace and national reconciliation in the Libyan Arab Jamahiriya and stability in the region; or

(d) Where a State determines on a case-by-case basis that such entry or transit is required to advance peace and stability in the Libyan Arab Jamahiriya and the States subsequently notifies the Committee within 48 h after making such a determination;

Asset freeze

17. *Decides* that all Member States shall freeze without delay all funds, other financial assets and economic resources which are on their territories, which are owned or controlled, directly or indirectly, by the individuals or entities listed in Annex II of this resolution or designated by the Committee established pursuant to paragraph 24 below, or by individuals or entities acting on their behalf or at their direction, or by entities owned or controlled by them, and decides further that all Member States shall ensure that any funds, financial assets or economic resources are prevented from being made available by their nationals or by any individuals or entities within their territories, to or for the benefit of the individuals or entities listed in Annex II of this resolution or individuals designated by the Committee;

18. *Expresses* its intention to ensure that assets frozen pursuant to paragraph 17 shall at a later stage be made available to and for the benefit of the people of the Libyan Arab Jamahiriya;

19. *Decides* that the measures imposed by paragraph 17 above do not apply to funds, other financial assets or economic resources that have been determined by relevant Member States:

(a) To be necessary for basic expenses, including payment for foodstuffs, rent or mortgage, medicines and medical treatment, taxes, insurance premiums, and public utility charges or exclusively for payment of reasonable professional fees and reimbursement of incurred expenses associated with the provision of legal services in accordance with national laws, or fees or service charges, in accordance with national laws, for routine holding or maintenance of frozen funds, other financial assets and economic resources, after notification by the relevant State to the Committee of the intention to authorize, where appropriate, access to such funds, other financial assets or economic resources and in the absence of a negative decision by the Committee within 5 working days of such notification;

(b) To be necessary for extraordinary expenses, provided that such determination has been notified by the relevant State or Member States to the Committee and has been approved by the Committee; or

(c) To be the subject of a judicial, administrative or arbitral lien or judgment, in which case the funds, other financial assets and economic resources may be used to satisfy that lien or judgment provided that the lien or judgment was entered into prior to the date of the present resolution, is not for the benefit of a person or entity designated pursuant to paragraph 17 above, and has been notified by the relevant State or Member States to the Committee;

20. *Decides* that Member States may permit the addition to the accounts frozen pursuant to the provisions of paragraph 17 above of interests or other earnings due on those accounts or payments due under contracts, agreements or obligations that arose prior to the date on which those accounts became subject to the provisions of this resolution, provided that any such interest, other earnings and payments continue to be subject to these provisions and are frozen;

21. *Decides* that the measures in paragraph 17 above shall not prevent a designated person or entity from making payment due under a contract entered into prior to the listing of such a person or entity, provided that the relevant States have determined that the payment is not directly or indirectly received by a person or entity designated pursuant to paragraph 17 above, and after notification by the relevant States to the Committee of the intention to make or receive such payments or to authorize, where appropriate, the unfreezing of funds, other financial assets or economic resources for this purpose, 10 working days prior to such authorization;

Designation criteria

22. *Decides* that the measures contained in paragraphs 15 and 17 shall apply to the individuals and entities designated by the Committee, pursuant to paragraph 24 (b) and (c), respectively;

(a) Involved in or complicit in ordering, controlling, or otherwise directing, the commission of serious human rights abuses against persons in the Libyan Arab Jamahiriya, including by being involved in or complicit in planning, commanding, ordering or conducting attacks, in violation of international law, including aerial bombardments, on civilian populations and facilities; or

(b) Acting for or on behalf of or at the direction of individuals or entities identified in subparagraph (a).

23. *Strongly encourages* Member States to submit to the Committee names of individuals who meet the criteria set out in paragraph 22 above;

New Sanctions Committee

24. *Decides* to establish, in accordance with rule 28 of its provisional rules of procedure, a Committee of the Security Council consisting of all the members of the Council (herein "the Committee"), to undertake to following tasks:

(a) To monitor implementation of the measures imposed in paragraphs 9, 10, 15, and 17;

(b) To designate those individuals subject to the measures imposed by paragraphs 15 and to consider requests for exemptions in accordance with paragraph 16 above;

(c) To designate those individuals subject to the measures imposed by paragraph 17 above and to consider requests for exemptions in accordance with paragraphs 19 and 20 above;

(d) To establish such guidelines as may be necessary to facilitate the implementation of the measures imposed above;

(e) To report within 30 days to the Security Council on its work for the first report and thereafter to report as deemed necessary by the Committee;

(f) To encourage a dialogue between the Committee and interested Member States, in particular those in the region, including by inviting representatives of such States to meet with the Committee to discuss implementation of the measures;

(g) To seek from all States whatever information it may consider useful regarding the actions taken by them to implement effectively the measures imposed above;

(h) To examine and take appropriate action on information regarding alleged violations or non-compliance with the measures contained in this resolution;

25. *Calls upon* all Member States to report to the Committee within 120 days of the adoption of this resolution on the steps they have taken with a view to implementing effectively paragraphs 9, 10, 15 and 17 above;

Humanitarian assistance

26. *Calls upon* all Member States, working together and acting in cooperation with the Secretary General, to facilitate and support the return of humanitarian agencies and make available humanitarian and related assistance in the Libyan Arab Jamahiriya, and requests the States concerned to keep the Security Council regularly informed on the progress of actions undertaken pursuant to this paragraph, and expresses its readiness to consider taking additional appropriate measures, as necessary, to achieve this;

Commitment to review

27. *Affirms* that it shall keep the Libyan authorities' actions under continuous review and that it shall be prepared to review the appropriateness of the measures contained in this resolution, including the strengthening, modification, suspension or lifting of the measures, as may be needed at any time in light of the Libyan authorities' compliance with relevant provisions of this resolution;

28. *Decides* to remain actively seized of the matter."

Jurists will argue on the legitimacy of this international move by the member States of the United Nations which are generally guided by the United Nations Charter, Article 2.4 provides that *All Members shall refrain in their international relations from the threat or use of force against the territorial integrity or political independence of any State, or in any other manner inconsistent with the purposes of the United Nations*. As a counter argument is the compelling need to protect persons who are attacked by a regime to quell a peaceful protest.

Arguably, an analogy to the incursion into Libyan airspace by NATO aircraft would be the United States (and allies) military offensives on Afghanistan and Iraq, both of which, in terms of Article 51 of the United Nations Charter[99] were

[99] Article 51 provides that nothing in the Charter shall impair the inherent right of individual or collective self-defence if an armed attack occurs against a Member of the United Nations, until the Security Council has taken measures necessary to maintain international peace and security. Measures taken by Members in the exercise of this right of self-defence shall be immediately reported to the Security Council and shall not in any way affect the authority and responsibility of the Security Council under the present Charter to take at any time such action as it deems necessary in order to maintain or restore international peace and security.

seemingly carried out outside the parameters of the United Nations Charter and prompted some jurists to allege breaches of sovereignty in both instances. State Sovereignty is all about control over a society. However, it is also about legitimacy, as the Montevideo principles enunciated. It has to be noted that, from a purely academic perspective, the concept goes back several centuries. The doctrine of sovereignty was introduced to the Western world by the French philosopher *Bodin*. At a time when political attitudes were in transition from the dominance of the universal church to a universal legal order, *Bodin* introduced sovereignty as a supreme power over citizens and subjects that was not itself bound by laws. *Bodin* elaborated that every independent community had to consider that while acknowledging the authority of the law, a State was above the law if it wished to govern successfully. Other jurists who supported the theory of exclusive sovereignty were Hugo Grotius, who maintained that sovereign States were independent of foreign control, and Thomas Hobbes, who said that sovereignty was absolute and its misuse was unthinkable. John Locke attempted to compromise the absolute quality of sovereignty by opining that sovereignty was not absolute and unquestionable in that it was an exchange of social trust between the government and the people. Accordingly, there was an inarticulate premise that a breach of the social trust between the two parties would erode the concept of sovereignty. This brings one to the inexorable conclusion that, in the ultimate analysis the sovereignty of a nation, and therefore its independence, lies in a stable population that subscribes to the aspirations of the people. This is the ultimate value of self determination, legitimacy and independence.

The above notwithstanding, Article 39 of the United Nations Charter empowers the Security Council to determine the existence of any threat to the peace, breach of the peace or aggression. The Security Council may make recommendations or decide on the measures that may be taken. Although practically such recommendations may not be adopted owing to the veto power of the permanent members of the Council no such veto was exercised in the case of Resolution 1973 against Libya.

United Nations General Assembly Resolution 3314 of 1974 provides some examples of aggression, among which is invasion or attack by armed forces of a State. One of the fundamental rules of humanitarian law in the context of armed conflicts was prepared by the International Committee of the Red Cross in 1978 requires parties to a conflict to distinguish at all times between the civilian population and combatants in order to spare civilian population and property. It goes on to say that neither the civilian population as such nor civilian persons shall be the object of attack. The 1981 Conventional Weapons Convention prohibits the use of mines, booby traps and other similar devices and incendiary weapons directed against the civilian population or used indiscriminately and the first Protocol of 1977 imposes very detailed target restraints with a view to protecting civilians.

2.4 Relief Flights

The operation of relief flights, either by States or such bodies as the United Nations, to alleviate human suffering in times of war, natural or manmade catastrophe, is yet another area in which the role of civil aviation is brought to bear in securing peace and security. There is a specific provision in Annex 9 to the Chicago Convention for provision by State of relief flights. Contracting States are required, by Standard 8.8 of Chapter 8 of the Annex, to facilitate the entry into, departure from and transit through their territories of aircraft engaged in relief flights performed by or on behalf of international organizations recognized by the United Nations or by or on behalf of States themselves and to take all possible measures to ensure their safe operation.[100] The relief flights referred to should be undertaken to respond to natural and man-made disasters which seriously endanger human health or the environment. An emergency is acknowledged in the Annex as "a sudden and usually unforeseen event that calls for immediate measures to minimize its adverse consequences". A disaster is described in the Annex as "a serious disruption of the functioning of society, causing wide spread human, material or environmental losses which exceed the ability of the affected society to cope using its own resources."[101]

The United Nations Charter lists the achievement of international cooperation in solving international problems of an economic, social, cultural or humanitarian character, as one of the purposes of the United Nations.[102] The problems that the United Nations is mandated by its Charter to solve should therefore be necessarily

[100] Standard 8.8 States: "Contracting States shall facilitate the entry into, departure from and transit through their territories of aircraft engaged in relief flights performed by or on behalf of international organizations recognized by the UN or by or on behalf of States themselves and shall take all possible measures to ensure their safe operation. Such relief flights are those undertaken in response to natural and man-made disasters which seriously endanger human health or the environment, as well as similar emergency situations where UN assistance is required. Such flights shall be commenced as quickly as possible after obtaining agreement with the recipient State.

Note 1.—According to its Internationally Agreed Glossary of Basic Terms, the United Nations Department of Humanitarian Affairs considers an emergency to be "a sudden and usually unforeseen event that calls for immediate measures to minimize its adverse consequences", and a disaster to be "a serious disruption of the functioning of society, causing widespread human, material or environmental losses which exceed the ability of the affected society to cope using only its own resources".

Note 2.— With respect to the application of measures to ensure the safe operation of relief flights, attention is drawn to Annex 11 — Air Traffic Services, *the* Manual Concerning Safety Measures Relating to Military Activities Potentially Hazardous to Civil Aircraft Operations *(Doc 9554) and the* Manual concerning Interception of Civil Aircraft *(Doc 9433)".*

Standard 8.9 States: "Contracting States shall ensure that personnel and articles arriving on relief flights referred to in 8.8 are cleared without delay".

[101] Annex 9, Facilitation, Thirteenth Edition – July 2011, Chapter 8, C, Standard 8.8, note 1 *Ibid.*

[102] *Charter of the United Nations and Statute of the International Court of Justice*, United Nations: New York, Article 1.3.

of an international nature. Article 2(7) of the Charter expands the scope of this philosophy further when it provides that the United Nations is not authorized to intervene in matters which are essentially within the domestic jurisdiction of any State, without prejudice to the right of the United Nations to intervene in matters which are within the domestic jurisdiction of any State, and apply enforcement measures where there is an occurrence of acts of aggression, a threat to the peace or breach thereof.[103] Therefore *stricto sensu*, the United Nations cannot intervene in instances where natural disasters such as famine, drought or earthquakes render the citizens of a State homeless, destitute and dying of starvation unless invited by the States concerned. The principle however cannot be too strictly interpreted, as natural disasters may usually lead to breaches of the peace. In such instances the United Nations Security Council may take such actions by air, sea or land as may be necessary to maintain or restore international peace and security.[104] For such instances, Article 43 of the Charter provides:

> All members of the United Nations, in order to contribute to the maintenance of international peace and security, undertake to make available to the Security Council, on its call and in accordance with a special agreement or agreements, armed forces assistance and facilities, including rights of passage necessary for the purpose of maintaining international peace and security.

Here again, action can only be taken for the maintenance of international peace, effectively precluding any direct intervention in a domestic issue.

Resolutions adopted by the United Nations Security Council relating to Somalia[105] and Bosnia and Herzegovina[106] clearly demonstrate the parameters of the scope of United Nations' intervention under its Charter.[107] In the case of Somalia, the United Nations Security Council recognized the unique character of the situation in the country, where conflict and violence demanded that all concerned take all necessary measures to facilitate the measures of the United Nations, its specialised agencies and humanitarian organizations to provide humanitarian assistance to the affected population in Somalia. In the case of Bosnia and Herzegovina the Security Council recognized in its Resolution that the situation in the two States constituted a threat to international peace and security. In both Resolutions, the Security Council had to function within its mandate of intervention only in instances of conflict and breaches of the peace.

The Geneva Conventions of 12 August 1949 for the Amelioration of the Wounded and Sick in Armed Forces in the Field contain provisions for facilitation

---

[103] *Id.* Chapter VII Articles 39, 41 and 42.

[104] *Id.* Article 42.

[105] *S/RES/794* (1992) 3 December 1992.

[106] *S/RES/770* (1992) 13 August 1992.

[107] See also the earlier Security Council Resolution 688 (1991), 5 April 1991 whereby the Security Council expressed grave concern at the repression of the Iraqi civilian population in parts of Iraq and insisted that Iraq allow immediate access by international humanitarian organizations to all parts of Iraq.

of aircraft for the removal of the wounded and sick and for the transport of medical personnel and equipment.[108] The thrust of the Conventions is that relief operations by air in the case of removing wounded and sick soldiers and civilians in instances of armed conflict are protected from attack and facilitated through. The Conventions also provide that medical aircraft of parties to the conflict may fly over the territories of neutral powers, land thereon in case of necessity, or use them as ports of call.[109] Medical aircraft are required by the Conventions to give neutral States previous notice of their passage through such States.

Humanitarian assistance *per se* is by no means contrary to the principles of international law. In *Nicaragua* v. *USA (Merits)*[110] the International Court of Justice (ICJ) held:

> there can be no doubt that the provisions of strictly humanitarian aid to persons or forces in another country, whatever the political obligations or objectives, cannot be regarded as unlawful intervention, or in any way contrary to international law.[111]

When the recipient State however, does not grant permission to a State or organization to operate relief flights into its territory, the *status quo* becomes different, and the legality of the relief flight becomes *prima facie* questionable. In such an instance, a relief flight would *ipso facto* be an intervention. The fact that such a flight does not obtain the permission to fly over or into the recipient State would clearly tantamount to a forcible entry. The question which emerges in this scenario is, when, if at all, does a unilateral humanitarian intervention such as a relief flight operated unilaterally, become legally justifiable?

The Court in the *Nicaragua* case, having concluded that the activities of the United States in assisting the *Contras* against the wishes of the Nicaraguan government constituted *prima facie* acts of intervention, held that the principle of non-intervention derives from customary international law, and a mere request for assistance by a person or group of persons or organization in a State would not justify at international law the unilateral intervention of State in the affairs of another.[112] Court relied heavily on the paramount authority of treaty provisions over the other sources of international law and applied the principle that the use of force or threat thereof by one State on another—to render assistance to the latter, without the latter's consent—was contrary to existing treaty law and therefore unacceptable. Judge Sette-Camara quoted the International Law Commission and observed:

[108] Geneva Conventions for the Amelioration of the Wounded and Sick in Armed Forces in the Field, *Geneva Conventions of August 12 1949*, International Committee of the Red Cross, Geneva: 1970, Article 36.

[109] *Id.* Article 37.

[110] *Military and Para Military Activities in and Against Nicaragua* (*Nicaragua* v. *US*) Merits *I.C.J. Rep.* 1986 at 14.

[111] *Id.* para. 242.

[112] *Id.* para. 246.

> the law of the Charter concerning the prohibition of the use of force in itself constitutes a conspicuous example of a rule of international law having the character of *jus cogens*.[113]

The Court in this case has therefore clearly arranged legal priorities in such a manner that the United Nations Charter is paramount over any other moral or ethical argument on humanitarian assistance.

Toussaint L'Ouverture International Airport at Port -au-Prince had only one runway which is 9,900 ft. in length. There is only one parking area at which only 44 aircraft can be accommodated at any given time. Because only one aircraft can land or depart from this runway at the same time, and because aircraft must be separated by a certain time to avoid collisions, the airport has a finite capacity. Air traffic control can also be limiting, since there are only so many aircraft an air traffic control unit can safely handle. Additionally, staff shortages, radar maintenance or equipment faults can lower the capacity of a unit. This can affect both airport air traffic control as well as en-route air traffic control centers.[114]

The fundamental principle with regard to the provision of air traffic services is contained in *Annex 11* to the *Chicago Convention* which provides that the need for the provision of air traffic services is determined by the types of air traffic involved; the density of air traffic,[115] the meteorological conditions and such other factors as may be relevant.[116] The main objective of the air traffic control service is to prevent collisions between aircraft and between aircraft and obstructions in the maneuvering area and to expedite and maintain an orderly flow of air traffic.[117] The objective of airspace management is to maximize, within a given airspace structure the utilization of available airspace by dynamic time-sharing and, at times, segregation of airspace among various categories of users based on short term needs.[118] The objective of air traffic flow management is to ensure an optimum flow of air traffic to or through areas during times, when demand exceeds or is expected to exceed available capacity of the air traffic floor system.

---

[113] *Id.* para. 199.

[114] http://www.airtrafficmanagement.net/view_news.asp?ID=1435 A seven member team from the US FAA is working with the Haitian government and the US Department of Defense (DOD) to supply technical assistance for air traffic management. The US Air force has successfully used the one-slot system which rapidly brings in an aircraft when one leaves.

[115] At the time of writing, the US Air Force was to open a second relief centre distribution point at Jacmel Airport in the south of Haiti to relieve the pressure on Toussaint Louverture International Airport at Port-au-Prince. Some 200 flights a day were proving a colossal strain at Port-au-Prince and Jacmel will immediately be able to receive C-130 Hercules deliveries initially supporting Canadian humanitarian assistance efforts. See http://www.airtrafficmanagement.net/view_news.asp?ID=1435.

[116] Annex 11 *Air Traffic Services*, Thirteenth Edition, July 2001, Standard 2.4.1.

[117] *Air Traffic Services Planning Manual, First (Provisional) Edition*, 1984, ICAO Doc 9486-AN/924, paragraph 1.2.2.1, at p. II-I-I-2.

[118] *Id.* Paragraph., 1.2.3.1.

One of the considerations in the context of relief flights in Haiti is that there could be both civil and military aircraft operating into Haiti.[119] From an aeronautical perspective, Annex 11 to the Chicago Convention, which deals with the subject of air traffic services, lays down requirements for coordination of activities that are potentially hazardous to civil aircraft. The International Standards and Recommended Practices in the Annex, Chapter 2, (2.17 and 2.18 in particular) contain provisions for co-ordination between military authorities and air traffic services and co-ordination of activities potentially hazardous to civil aircraft. These provisions specify that air traffic services authorities shall establish and maintain close co-operation with military authorities responsible for activities that may affect flights of civil aircraft. The provisions also prescribe that the arrangements for activities potentially hazardous to civil aircraft shall be co-ordinated with the appropriate air traffic services authorities and that the objective of this co-ordination shall be to achieve the best arrangements which will avoid hazards to civil aircraft and minimize interference with the normal operations of such aircraft. Standard 2.17.1 stipulates that arrangements for activities potentially hazardous to civil aircraft, whether over the territory of a State or over the high seas, shall be coordinated with the appropriate air traffic services authorities, such coordination to be effected early enough to permit timely promulgation of information regarding the activities in accordance with the provisions of Annex 15 to the Chicago Convention. Standard 2.17.2 of Annex 11 explains that the objective of the coordination referred to in the earlier provision shall be to achieve the best arrangements that are calculated to avoid hazards to civil aircraft and minimize interference with the normal operations of aircraft.

The above considerations of safety notwithstanding, it is incontrovertible that cooperation in the activities of military and civil aviation is not only about sharing airspace. It is also about the efficient allocation of airspace to both categories of activity in separating such flights, particularly in the context of military flights which operate in special use airspace and those proceeding to special use airspace across civilian air routes. This brings to bear the inevitable conclusion that there must essentially be coordination between military authorities and air navigation service authorities.

In the instance of a proliferation of aircraft in airspace, as the Haitian relief operations might entail, separations considerations also prove vital. Action on the part of air traffic services (ATS) to keep aircraft operating in the same general area at such distances from each other and to keep the risk of collision maintained at an acceptable level is critical to this process. In determining appropriate separation minima, the airspace planner must have a thorough knowledge of the existing airspace, the CNS/ATM[120] capabilities and the airspace characteristics, which

[119] There is also the issue of civil aircraft being deployed for military purposes in transport military personnel to Haiti. The responsibility of using civil aircraft and crew for military purposes rests with the State concerned. The fundamental legal premise which applies in such situations is that, in international relations, the erosion of one's legal interests by another brings to bear the latter's responsibility. See generally, Abeyratne (1997), pp. 1–23.

[120] For a discussion on CNS/ATM see O'Keeffe (2000), pp. 133–148.

may influence the safe separation minima. This involves a number of factors such as: the route structure; complexity of the airspace; traffic demand pattern; numbers and location of traffic tracks, amount of traffic operating on opposite direction tracks and the amount of traffic which is either climbing or descending.[121] However, the fundamental premise still remains in Article 12 of the Chicago Convention[122] which states that each contracting State undertakes to adopt measures to insure that every aircraft flying over or manoeuvring within its territory and that every aircraft carrying its nationality mark, wherever such aircraft may be, shall comply with the rules and regulations relating to the flight and manoeuvre of aircraft there in force. Each contracting State undertakes to keep its own regulations in these respects uniform, to the greatest possible extent, with those established from time to time under this Convention.

The final dimension, without which relief flights would lose their purpose, is the dissemination of relief which forms the final consideration of relief by air. At the initial stages, it was reported that in Haiti, although medical and food supplies arrived at the airport there was no infrastructure to distribute them to those in need. The coordination of relief in armed conflicts or natural or manmade disasters raises very real problems. In many instances, the lack of coordination in relief operations often results in an imbalance in consignments, foodstuffs perishing in large warehouses and the lack of adequate transport to provide relief in areas which need assistance. Paragraph 5 of Article 70 of Protocol 1 to the Geneva Convention[123] lays down the principle of *effective* international coordination of relief. This provision lays down obligations of all parties concerned i.e. donors, transit countries and beneficiaries.

In 1969 the XXIst International Conference of the Red Cross adopted a Resolution whereby States were requested to exercise their sovereign and legal rights so as to facilitate the transit, admission and distribution of relief supplies provided by impartial international humanitarian organizations for the benefit of civilian populations in disaster areas when disaster situations imperil the life and welfare of such populations.[124] The United Nations subsequently announced that this Resolution would also apply to situations arising from armed conflict.[125]

---

[121] *Manual on Airspace Planning Methodology for the Determination of Separation Minima*, ICAO Doc. 9689-AN/953, First Edition, 1998, Paragraphs 3.1 and 3.2 at p. 14.

[122] *Supra*, note 116 (this chapter).

[123] Supra, note 108 (this chapter).

[124] Resolution XXVI, sub para (5), Declaration of Principles for International Humanitarian Relief to Civilian Populations in Disaster Situations, XXIst International Conference of the Red Cross, Istanbul, 1969, International Committee of the Red Cross (1987), p. 661.

[125] Resolution 2675 (XXV) of the United Nations, Principle 8 wherein the United Nations declared:

> The provision of international relief to civilian populations is in conformity with the humanitarian principles of the Charter of the United Nations, the Universal Declaration of Human Rights and other international instruments in the field of human rights. The Declaration of Principles for International Humanitarian Relief to the Civilian Population Disaster Situations, as laid down in Resolution XXVI adopted by the twenty first

The above discussion is not intended to obfuscate the fact that the provisions of the Geneva Conventions and the United Nations Charter do not admit of the operation of relief flights at the will and pleasure of the benefactor, without the permission of the recipient State. It is also not intended to circumvent the fact that States are primarily responsible for organizing relief. Relief societies such as the Red Cross and the Red Crescent Organizations are merely called upon to play a supplementary role by assisting the authorities of the States concerned in their task.

Since it is clear that the intervention of the United Nations Security Council in a matter lying within the domestic jurisdiction of a State can only be justified in instances where there is a threat to international peace and security, a breach of the peace within a State or an act of aggression, a question which arises when a relief flight is operated as a part of a humanitarian project is whether the operation of such a flight could be considered a legitimate unilateral action by States. The Question would essentially be ground in a legal analysis of the principles of humanitarian law and State sovereignty. On the one hand, everyone has the right to life, liberty and security of person[126] and the right to a standard of living adequate for the health and well being of himself and of his family, including food, clothing, housing and medical care.[127] On the other, there is overall recognition of the fact that every State has complete and exclusive sovereignty over the airspace above its territory. Except for the Paris Convention of 1956,[128] which provides for civil aircraft registered in a member State of the European Civil Aviation Conference (ECAC) to fly freely into member States for the purposes of discharging or taking on traffic where such aircraft are engaged *inter alia* in non-scheduled flights for the purpose of meeting humanitarian or emergency needs,[129] there is no multilateral or bilateral agreement that admits of unilateral intervention of a State in another for humanitarian purposes, where the intervening State does not obtain permission of the recipient State. In fact, Resolution 46/182[130] explicitly provides in the Annex to the Resolution that the sovereignty, territorial integrity and national unity of States must be fully respected in accordance with the Charter of the United Nations and that in this context, humanitarian assistance should be provided with the consent of the affected country and in principle on the basis of an appeal by the affected country. These conflicting principles, although not bestowing legal authority on the United Nations to intervene in a State with relief flights, at least give some degree of justification to the United Nations' efforts to mediate with States concerned in the

International Conference of the Red Cross, shall apply in situations of armed conflict, and all parties to a conflict should make every effort to facilitate this application.

[126] *Universal Declaration of Human Rights*, United Nations Department of Public Information, Article 3.

[127] *Id.* Article 25.

[128] *Multilateral Agreement On Commercial Rights of Non-Scheduled Air Services In Europe*, signed at Paris on 30 April 1956, *See* Weishaupt (1979), p. 409.

[129] *Id.* Articles 1 a) and b) and 2.1.a).

[130] *Infra*, note 137 (this chapter).

promotion of relief operations and to seek the support of other States, with the concurrence of affected States.

The author subscribes to the view that if humans are dying, one has got to help at all costs.[131]

2.4.1 The Haitian Earthquake

The earthquake which devastated the capital of Haiti and much of its environs in January 2010, causing massive fatalities and damage to property, could be distinguished from the Tsunami of December 2004 in that the latter left the government infrastructure and machinery intact. The earthquake, on the other hand, crippled the Haitian government and infrastructure, rendering government authorities weak in the running of the country. From an aeronautical perspective, this brought to bear issues of sovereignty within the parameters of relief flights and humanitarian law. Another devastation which was unique to the earthquake was that although the only runway at the airport was undamaged, the rest of the aviation infrastructure lay in a shambles. The flow of the numerous relief flights that came into Haiti after the fact had therefore to be managed with caution and diligence.

On Tuesday 12 January 2010, at 16.53 local time, an earthquake with a magnitude of 7.0 Mw on the Richter scale[132] hit Haiti at a depth of 13 km. Its epicenter was near Leogane, which was approximately 25 km or 16 miles west of the country's capital Port-au-Prince.[133] The International Red Cross estimated the people affected by the earthquake at around three million.[134] It was estimated that around 100,000–200,000 people would have perished as a result of the catastrophe.[135]

The quake damaged many important buildings including the Presidential Palace, The National Assembly and the Cathedral of Port-au-Prince. The headquarters of the United Nations Stabilization Mission in the city collapsed killing many UN

[131] See letter to the Editor by Professor Leff, Yale Law School, *New York Times* October 4 1968 at 46 Column 3, cited in Lillich (1973), p. 151.

[132] The Richter scale is the best known scale for measuring the magnitude of earthquakes. The magnitude value is proportional to the logarithm of the amplitude of the strongest wave during an earthquake. A recording of 7, for example, indicates a disturbance with ground motion 10 times as large as a recording of 6. The energy released by an earthquake increases by a factor of 30 for every unit increase in the Richter scale. See http://www.matter.org.uk/schools/content/seismology/richterscale.html.

[133] Wikipedia records that the United States Geological Survey counted at least 33 aftershocks, 14 of which were between magnitudes 5.0 and 5.9. See http://en.wikipedia.org/wiki/2010_Haiti_earthquake.

[134] CBS News. 13 January 2010. http://www.cbsnews.com/stories/2010/01/13/world/main6090601.shtml?tag=cbsnewsSectionContent.4. Retrieved 13 January 2010.

[135] Haiti Quake Death toll may hit 200,000-Minister, *Reuters Alertnet* (Reuters). http://www.alertnet.org/thenews/newsdesk/N15143632.htm. Retrieved 15 January 2010.

workers including the Chief of Mission. The disaster also caused severe damage to communication systems, air, land, and sea transport facilities, hospitals, and electrical networks, drastically hampering rescue and aid efforts.

Many countries responded to appeals for humanitarian aid, pledging funds and dispatching rescue and medical teams, engineers and support personnel. Communication systems, air, land, and sea transport facilities, hospitals, and electrical networks had been damaged by the earthquake, which hampered rescue and aid efforts.[136] From an aeronautical perspective, there was considerable confusion over who was in charge of the relief efforts and of flights in and out of Haiti, and the problem was compounded by air traffic congestion and problems with prioritization of flights which in turn hampered relief work.

2.4.2 The Aeronautical Intervention

Several nations that pitched in with relief in the early aftermath of the disaster may not have pondered the legalities concerned with entering the airspace over Haiti or landing therein. This would doubtless have provoked some members of the legal profession to question issues of sovereignty and legal rights of Haiti in this instance. Technically, although the Chicago Convention of 1944 provides that Member States of ICAO recognize that every State has complete and exclusive sovereignty over the air space above its territory,[137] the general consensus of contemporary

---

[136] Initially, relief flights from around the world were delayed as a two-tier system was put in place to make sure that the most urgently needed supplies and personnel were getting in first, and the rest followed using a pre-arranged arrival and take-off schedule. Once a ground control team from the Air Force's first Special Operations Wing began landing aircraft on January 13, a team from the U.S. Federal Aviation Administration and Haitian authorities began working to restore navigation and communication capabilities. A team of U.S. military and civilian aviation experts working with Haitian officials began sorting flights, allowing some to land immediately based on what was needed most.

[137] As per Article 1 of the Convention. The Convention applies only to State aircraft and not to military aircraft. However, it states, in Article 3 that no aircraft used in military, customs or police services shall fly over the territory of another State or land thereon without authorization by special agreement or otherwise, and in accordance with the terms thereof. It must also be borne in mind that United Nations General Assembly Resolution *A/RES/46/182* of 19 December 1991 provides, in resolving clause 3 that the sovereignty, territorial integrity and national unity of States must be fully respected in accordance with the Charter of the United Nations. In this context, humanitarian assistance should be provided with the consent of the affected country and in principle on the basis of an appeal by the affected country. In April 1992, the Secretary-General established the Department of Humanitarian Affairs (DHA), incorporating UNDRO, various UN units that had been dealing with specific emergency programs, and the secretariat for the International Decade for Natural Disaster Reduction. The Secretary-General appointed an Emergency Relief Coordinator to head the new department. The DHA has its headquarters in New York and an office in Geneva.

international law is that this is not an absolute and pristine principle anymore. One commentator states very aptly:

> The role of the State in the modern world is a complex one. According to legal theory, each State is sovereign and equal. In reality, with the phenomenal growth in communications and consciousness, and with the constant reminder of global rivalries, not even the most powerful of States can be entirely sovereign. Interdependence and the close knit character of contemporary international commercial and political society ensures that virtually any action of a State could well have profound repercussions upon the system as a whole and the decisions under consideration by other States.[138]

Therefore, although sovereignty is the basic constitutional doctrine of the law of nations, modern exigencies of international relations, particularly in the case of disasters, whether man made or natural, have not regarded this principle as inflexible. The essence of intervention is compulsion. The legal question with regard to the inviolability of the sovereignty of a State is not whether the intervention concerned was an armed or unarmed one, but whether it was effected unilaterally under compulsion or threat by the intervening State.[139] Starke is inclined to stretch the principle of sovereignty to accommodate external involvement by a State in the affairs of another in special circumstances:

> ..."Sovereignty" has a much more restricted meaning today than in the eighteenth and nineteenth centuries when, with the emergence of powerful highly nationalised States, few limits on State autonomy were acknowledged. At the present time there is hardly a State which, in the interests of the international community, has not accepted restrictions on its liberty of action. Thus most States are members of the United Nations and the International Labour Organization (ILO), in relation to which they have undertaken obligations limiting their unfettered discretion in matters of international policy. Therefore, it is probably more accurate today to say that the sovereignty of a State means the *residuum* of power which it possesses within the confines laid down by international law. It is of interest to note that this conception resembles the doctrine of early writers on international, law, who treated the State as subordinate to the law of nations, then identified as part of the wider "law of nature".[140]

Oppenheim holds a similar view that the "traditional" law of humanity is incorporated into contemporary international law. He views this attitude as:

> recognition of the supremacy of the law of humanity over the law of the sovereign State when enacted or applied in violation of human rights in a manner that may justly be held to shock the conscience of mankind.[141]

Some authorities in international law also believe that intervention should, if absolutely necessary, be effected when there is cogent evidence of a breakdown in the minimum guarantees of humanity.[142] Accordingly, it may be argued that any act

---

[138] Shaw (2003), p. 120.

[139] de Lima (1971), p. 16.

[140] Starke (1977), p. 106.

[141] Oppenheim (1955), pp. 312–320.

[142] Hyde (1945), p. 253. *Stowell's Intervention at International Law* 1921 at 126 and 350. Also, Wehberg (1938), p. 115.

of intervention aimed at saving the lives of human beings which are in danger, would be legally and morally justifiable. Fernando Teson[143] argues that since the ultimate justification for the existence of States is the protection and enforcement of the natural rights of the citizens, a government that engages in substantial violations of human rights betrays the very purpose for which it exists and so forfeits not only its domestic legitimacy, but also its international legitimacy as well. He goes on to say:

> I suggest that from an ethical standpoint, the rights of States under international law are properly derived from individual rights. I therefore reject the notion that States have any autonomous moral standing - that they hold international rights that are independent from the rights of individuals who populate the State.[144]

Schwarzenberger analyses the concept somewhat clinically and concludes that in the absence of an international *jus cogens* which corresponds to municipal *jus cogens* of advanced communities, where the latter prevents the worst excesses of inequality of power, the supremacy of the rule of force would prevail.[145]

There is also a contrasting view that humanitarian intervention is generally resorted to by States only in instances of serious abuses of human rights by one State upon its people or others. Dr Michael Akehurst argues that if a State intervenes forcibly on the territory of another in order to protect the local population from serious human violations, such an armed intervention could inevitably constitute a temporary violation *de facto* of the territorial integrity of the latter State, and to an extent of its political independence, if carried out against its wishes.[146] Dr. Akehurst goes on to assert:

> Any humanitarian intervention, however limited, constitutes a temporary violation of the target State's political independence and territorial integrity if it is carried out against the State's wishes.[147]

The doctrine of humanitarian intervention is thought of by some commentators as an invention of strategy to circumvent the strong *jus cogens* nature of the principle of sovereignty and inviolability of States which Dr. Akehurst refers to. Professor Brownlie is of the view that States have generally invoked the doctrine to give support to their commercial and strategic considerations.[148] The United Kingdom legislature recently considered the view of the British Minister of State who was of the view:

> When members of the United Nations act in a forcible manner either they should do so within and under the authority of the United Nations or that which they do should be authorised by the principles of international law.[149]

---

[143] Teson (1956), p. 5.

[144] *Id.* at 16.

[145] Schwarzenberger (1971), p. 63.

[146] Akehurst (1977), p. 16.

[147] Akehurst (1984), p. 105.

[148] Brownlie (1963), pp. 338–340.

[149] *Hansard* H.C, Vol 219, col.784 (23 Feb 1993).

Clearly, this statement establishes the view that international law in the context of intervention is *jus cogens*. The British Foreign Office has supported this position in the following language:

> the best case that can be made in support of humanitarian intervention is that it cannot be said to be unambiguously illegal but the overwhelming majority of contemporary legal opinion comes down against the existence of a right of humanitarian intervention.[150]

Despite this strong alignment towards *anti* humanitarian intervention, it is believed that there is a school of thought within the British legislature that is prepared to accept unilateral intervention as justifiable under customary international law in cases of "extreme humanitarian need."[151]

The view that despite these divergent views, the non-intervention principle remains sacrosanct as a contemporary postulate of international law and deviations from the principle, although recognized as ethical and moral in certain instances by scholars, would be justified only in extreme cases[152] sounds logical and practically acceptable.

2.5 Aviation and Rights to Environmental Protection

2.5.1 Noise

Aircraft noise is generated whenever the passage of air over the structure of the aircraft or flowing through its power plants cause fluctuating pressure disturbances which transform into auditory impulses in the human being.[153] Such impulses, called unwanted sound, not desired by the recipient,[154] is measured using a logarithmic unit called the decibel (dB) in terms of pressure exerted in the ear.[155] In the case of jet aircraft, two distinct kinds of engine noise affect the human ear adversely, namely the roar of the jet exhaust and the whine of the compressor fan.[156] The roar of the jet exhaust occurs mainly in the take-off phase, where the engines are at maximum power to enable the aircraft to be airborne.[157] The whine of the compressor occurs mainly during the landing approach where interaction of the

[150] UKMIL (1986) 57 *B.Y.I.L.* 619.

[151] See Lowe and Warbrick (1993), Part 4, 938 at 944.

[152] See Vincent (1974), p. 313.

[153] See Rayleigh (1945), p. 21. For more recent discussions on aircraft noise see Frey and Kinsler (1982), p. 82 and Nelson (1987), p. 321.

[154] Kryfer (1985), p. 1.

[155] Harper (1988), p. 121.

[156] *Id.* at p. 120.

[157] Grey (1975), p. 72.

fan blades with turbulence and wakes results in a whine of high frequency in the compressor of the engine.[158]

There is definite medical evidence to prove that exposure to aerial noise causes deleterious psychological effects. For instance, a study made in relation to admissions to a psychiatric hospital in England draws a distinct relationship between the disturbed mental state of the admitted inmates of the hospital and high intermittent noise levels from Heathrow Airport, to which they were exposed.[159] One writer states that airport and industrial noise are the causes of mental stress and maladjustment, increases in chronic fatigue and neurotic complaints in some cases, although different personalities may underlie individual differences in noise effects.[160]

Aircraft noise is at best a nuisance and at worst a gross infringement of a fundamental right. The issue of aircraft noise becomes even more critical in the face of noise curfews imposed by certain jurisdictions that force some States to have their carriers depart at night in order to beat night-time curfews imposed by the State at which the aircraft arrives. Several African States have raised this issue in the Council of ICAO in the context of African carriers facing night curfews in Europe—an issue brought up also by States such as India in the Council.

The term "noise" has so far not been examined in depth by the courts specially in relation to tolerance levels. It is clear that courts have not had occasion to consider noise as a significant factor in litigation which calls for detailed consideration in terms of medical and scientific pronouncements to ascertain what degree of noise is detrimental to the human being. Courts have not been able to do so due to two significant factors.

The first is that the issue of noise as pollution has surfaced rarely. The second is that even if it did surface, courts have always had to question noise as it is referred to in regulations or municipal by-laws, the validity of which has been the primary factor of consideration. It can be seen that most by-laws refer generally to noise and the prohibition of it. It is not surprising therefore, that courts have never attempted to define noise empirically or medically and instead preferred to observe the nature of noise on a very general basis.

2.5.1.1 Regulatory Issues

The Chicago Convention is the fundamental source which grants regulatory powers to the international community on matters relating to international civil aviation. The Convention in its Preamble proclaims that the governments that are parties to it agree on certain principles and arrangements in order that international civil

---

[158] *Ibid.*

[159] Abey-Wickrama et al. (1969), pp. 1275-1277.

[160] Cohen (1968), 25 pp.

aviation may be developed in a safe and orderly manner. Public safety is a feature of major concern to the Convention which requires that:

> Each contracting State may, for reasons of military necessity or public safety, restrict or prohibit uniformly the aircraft of other States from flying over certain areas of its territory. . .Such prohibited areas shall be of reasonable extent and location so as not to interfere unnecessarily with air navigation. Descriptions of such prohibited areas in the territory of a contracting State, as well as any subsequent alterations therein, shall be communicated as soon as possible to the other contracting States and to the International Civil Aviation Organization.[161]

The Convention also established the International Civil Aviation Organization which is required as one of its objectives to foster the planning and development of international air transport so as *inter alia* to ensure the safe and orderly growth of international civil aviation throughout the world,[162] meet the needs of the people of the world for safe, regular, efficient and economic air transport,[163] promote safety of flight in international air navigation,[164] and promote generally the development of all aspects of international civil aviation.[165] To this end, ICAO is mandated by the Convention to adopt and amend from time to time as may be necessary, international Standards and Recommended Practices and Procedures (SARPS) dealing *inter alia* with characteristics of airports and landing areas and such matters concerned with the safety, regularity and efficiency of air navigation as may from time to time be appropriate.[166] Each contracting State has undertaken to collaborate in securing the highest practicable degree of uniformity in the above regulations, standards and procedures of the Organization.

In order to carry out its policies in environmental issues related to civil aviation, ICAO established the Committee on Environmental Protection (CAEP) in 1983.[167] CAEP is a technical committee reporting to the ICAO Council and consisting of 14 members nominated by their States. The members are experts in the field of aviation and the environment. The Committee is charged with addressing the main environmental problems affecting civil aviation, which are pollution caused by aircraft engine emissions and aircraft noise. The three main approaches adopted by CAEP in its work on aircraft noise concern reduction of noise at source; use of noise abatement operating measures; and land use planning. One of the CAEPs positive contributions to ICAO's environmental programme has been the development of a new chapter for Volume 1 of Annex 16, which contains a noise certification scheme for light helicopters as well as a number of modifications to the technical specifications of the Annex. CAEP believes that although ICAO initiatives on Chapter 2

[161] Article 9.

[162] Article 44(a).

[163] Article 44(d).

[164] Article 44(n).

[165] Article 44(i).

[166] Article 37(b).

[167] For more details on CAEP see Mortimer (1992), p. 6. See also, Crayston and Hupe (1999), p. 5.

aircraft withdrawal will ameliorate the problem of noise near airports, it will be short lived, as the rapidly increasing proportions of international air travel in the next few years will render the benefits brought about by the initiative, nugatory.

At its Fourth Meeting held in April 1998, CAEP focussed *inter alia* on the subject of nuisance and noise from light aircraft and recommended that pilots and operators be aware of the nuisance their operations may cause and take steps to minimize noise levels of their aircraft.

In the light of developments concerning the phasing out of Chapter 2 aircraft wholly by Chapter 3 aircraft, CAEP has been charged with an extension to its mandate, to undertake additional work on reviewing Chapter 3 noise standards with a view to identifying a standard even more stringent than the Chapter 3 representation.[168] The Noise Scenarios Group, established within CAEP for the purpose of addressing increased options for possible aircraft noise restriction levels.[169]

ICAO's active involvement in aviation related environmental issues is not without good reason, as 9 of the 27 principles contained in the Declaration of the United Nations Conference on Environment and Development (UNCED)—also known as the "Earth Summit"—held in Rio de Janeiro in June 1992. They are:

1. "States have ... the responsibility to ensure that activities within their jurisdiction or control do not cause damage to the environment of other States or of areas beyond the limits of national jurisdiction." (Principle 2).
2. "The special situation and needs of developing countries, particularly the least developed.... shall be given special priority. International actions in the field of environment and development should also address the interests and needs of all countries." (Principle 6).
3. "... In view of the different contributions to global environmental degradation, States have common but differentiated responsibilities. The developed countries acknowledge the responsibility that they bear in the international pursuit of sustainable development in view of the pressures their societies place on the global environment and of the technologies and financial resources they command." (Principle 7).
4. "To achieve sustainable development and a higher quality of life for all people, States should reduce and eliminate unsustainable patterns of production and consumption ..." (Principle 8).
5. "States shall enact effective environmental legislation. Environmental standards, management objectives and priorities should reflect the environmental and developmental context to which they apply. Standards applied by some countries may be inappropriate and of unwarranted economic and social cost to other countries, in particular developing countries." (Principle 11).
6. "Environmental measures addressing trans-boundary or global environmental problems should, as far as possible, be based on an international consensus." (Principle 12).

---

[168] Only two aircraft that will meet standards beyond those prescribed in Chapter 3 of Annex 16 to the Chicago Convention will be in production after the year 2000, namely the Boeing 747-400 and the Airbus A321. It is expected that by the year 2007, more than 70 % of the world airline fleet would meet the most stringent of the proposals under consideration by CAEP. See Chapter 3½ *The Avmark Aviation Economist*, January/February 1995, at p. 10.

[169] See Crayston and Hupe (1999), *supra,* note 167, at p. 6.

7. "In order to protect the environment, the precautionary approach shall be widely applied by States according to their capabilities. Where there are threats of serious or irreversible damage, lack of full scientific certainty shall not be used as a reason for postponing cost-effective measures to prevent environmental degradation." (Principle 15).
8. "National authorities should endeavour to promote the internalization of environmental costs and the use of economic instruments, taking into account the approach that the polluter should, in principle, bear the cost of pollution, . . ." (Principle 16).
9. "Environmental impact assessment, as a national instrument, shall be undertaken for proposed activities that are likely to have a significant adverse impact on the environment and are subject to a decision of a competent national authority." (Principle 17).

Standards and Recommended Practices for aircraft noise were first adopted by ICAO on 2 April 1971 and designated as Annex 16 to the Convention. The development of the Annex originated in September 1968 at the 16th Session of the ICAO Assembly which adopted Resolution A 16-3 which recognized that the problem of aircraft noise was so serious in the vicinity of many of the world's airports that public reaction was mounting to a degree that gave cause for great concern and required urgent solution. The Assembly also noted that noise concerned the public and civil aviation and was becoming a matter for concern with the increase in air traffic and that the introduction of future aircraft types could increase and aggravate the problem further unless action was taken to alleviate the problem. Accordingly, the Assembly resolved to instruct the ICAO Council to call an international conference within the machinery of ICAO as soon as possible; establish international specifications and associated guidance material relating to aircraft noise; and, to include, in the appropriate Annexes and other relevant ICAO documents and possibly in a separate Annex on noise, such material as the description and methods of measurement of aircraft noise and suitable limitations on the noise caused by aircraft that was of concern to communities in the vicinity of airports.

In response to the Assembly Resolution, a Special Meeting on Aircraft Noise in the Vicinity of Aerodromes was convened in Montreal in November–December 1969 to examine the following aspects related to the problems of aircraft noise:

a) procedures for describing and measuring aircraft noise;
b) human tolerance to aircraft noise;
c) aircraft noise certification;
d) criteria for establishment of aircraft noise abatement operating procedures;
e) land use control; and,
f) ground run-up noise abatement procedures.

Based on the recommendations of the Special Meeting on Aircraft Noise in the Vicinity of Aerodromes, draft International Standards and Recommended Practices for Aircraft Noise were developed and, after amendment following the usual consultation with the Contracting States of the Organization, were adopted by the Council to form the text of Annex 16.

With the development of Standards and Recommended Practices dealing with the control of aircraft engine emissions, it was felt that all provisions relating to environmental aspects of aviation should be included in a single document.

Accordingly, Annex 16 was re- titled as "Environmental Protection". Volume I of the Annex contains the existing provisions relating to aircraft noise and Volume 11 contains the provisions related to aircraft engine emissions.[170] The much dreaded sonic boom[171] See *ICAO Doc 8894*, SPB/11, at 1-5. caused by supersonic aircraft was broadly discussed during the Second Meeting of ICAO's Sonic Boom Committee in Montreal in June 1973.[172] The Committee found that The sonic boom had no ill effects on the human eye and ear. As for psychological effects of the boom, it was noted by the Committee that it caused a startle reaction in the human being, in addition to a slight increase in heart rate which returned to normalcy within a very short period. It was also the Committee's view that the startle effects of the *Concorde* and *TU-144* aircraft did not give rise to any significant circulatory effects. With regard to the sociological effects of the sonic boom, the Committee considered a study of the *Concorde* aircraft over Australia in June 1972 and concluded that this aspect be best studied as both a public and private law issue by the Legal Committee of ICAO.

The ICAO Assembly has, in addition, adopted several Resolutions concerning aviation and the environment. At its 22nd Assembly held in September/October 1977 the ICAO Assembly adopted Resolution A 22-12 which recognized *inter alia* the following:

1) advancing technology has caused aviation to become a significant influence in the environment;
2) many of the adverse environmental effects of civil aviation activity can be reduced by, the application of integrated measures embracing technological improvements, appropriate noise abatement operating procedures, proper organization of air traffic and the appropriate use of airport planning and land use control mechanisms;
3) other international organizations are becoming involved in activities relating to noise abatement policies;
4) in fulfilling its role, ICAO strives to achieve a balance between the benefit of accruing to the world community through civil aviation and the harm caused to the human environment in certain areas through the progressive advancement of civil aviation;

---

[170] See *Annex 16 to the Convention on International Civil Aviation*, Second Ed. 1988 Vol 1 Foreword.

[171] Sonic boom was identified in 1970 as:

> ... a phenomenon peculiar to supersonic flight. It is caused by the mach waves that an aircraft inevitably generates aerodynamically when it flies at a speed greater than that of sound. Roughly speaking these waves take the form of compression waves from the bow and the stern separated by expansion waves...these waves extend from the aircraft as an audible pattern of roughly conical shape, much as the water waves from a boat extend from it as a visible pattern of roughly v-shape, and just as the water waves from a boat cause a disturbance that often extends to and travels along a neighbouring shore as the boat passes by, so do these shock waves from an aircraft cause a disturbance that normally extends to and travels along the ground as the aircraft flies over., The passage of these shock waves is perceived as a sonic boom.

[172] See *ICAO Doc 9064*, SBC/2, at 4-1.

The Assembly therefore declared:

1) that ICAO is conscious of the adverse environmental impacts that may be related to aircraft activity and of its responsibility and that of its Contracting States to achieve maximum compatibility between the safe and orderly development of civil aviation and the quality of the human environment;
2) that the Council Should maintain its vigilance in the pursuit of aviation interests related to the human environment and also maintain the initiative in developing policy guidance on all aviation matters related to the human environment, and not leave such initiatives to other organizations;

The Assembly also invited States to continue their active support for ICAO's Action Programme Regarding the Environment on all appropriate occasions as their participation in civil aviation's contribution to the United Nations Environment Programme (UNEP) and authorised the ICAO Council, if and when it deems this desirable, to enter into cooperative arrangements with the United Nations Environment Programme for the execution of environmental projects financed by the United Nations Environment Fund. The Assembly urged States to refrain from unilateral measures that would be harmful to the development of international civil aviation.

At the same Session, the Assembly adopted Resolution a 22-13 on airports and the environment, observing *inter alia*:

1) the compatibility between the airport and its environment was one of the elements to be taken into account in long-term systems planning;
2) the problem of aircraft noise in the vicinity of many of the world's airports continued to arouse public concern and required appropriate action;
3) the introduction of future aircraft types could increase and aggravate this noise unless action was taken to alleviate the situation.

The Assembly therefore requested the council to continue its work on establishing Standards and Recommended Practices relating to the alleviation of the problem and urged contracting States to adopt, where appropriate, the ICAO measures and procedures applicable.

In the following Session (September/October 1980), the Assembly adopted Resolution A 23-10 on aircraft noise and engine emissions from subsonic aircraft and requested contracting States not to allow the operation of foreign registered subsonic jet planes that did not conform to ICAO's specifications on noise certification standards as specified in Annex 16 until 1 January 1988.[173] At the 28th Assembly Sessions held in October 1990, the ICAO Assembly observed that while certification standards for subsonic jet aircraft noise levels are specified in Volume 1, Chapter 2 and Chapter 3 of Annex 16 and that environmental problems due to aircraft noise continued to exist in the neighbourhood of many international airports, some States were consequently considering restrictions on the operations of aircraft which exceed the noise levels in Volume I, Chapter 3 of Annex 16. The

[173] See *Assembly Resolutions in Force* (as of 6 October 1989), *ICAO Doc 9558* at II-18.

Assembly also recognized that the noise standards in Annex 16 were not intended to introduce operating restrictions on aircraft and that operating restrictions on existing aircraft would increase the costs of airlines and would impose a heavy economic burden, particularly on those airlines which do not have the financial resources to re-equip their fleets. Therefore, considering that resolution of problems due to aircraft noise must be based on the mutual recognition of the difficulties encountered by States and a balance among their different concerns, the Assembly, by Resolution A 28-3, urged States not to introduce any new operating restrictions on aircraft which exceed the noise levels in Volume I, Chapter 3 of Annex 16 before considering:

a) whether the normal attrition of existing fleets of such aircraft will provide the necessary protection of noise climates around their airports;
b) whether the necessary protection can be achieved by regulations preventing their operators from adding such aircraft to their fleets through either purchase, or lease/charter/interchange, or alternatively by incentives to accelerate fleet modernization;
c) whether the necessary protection can be achieved through restrictions limited to airports and runways the use of which has been identified and declared by them as generating noise problems and limited to time periods when greater noise disturbance is caused; and,
d) the implications of any restrictions for other States concerned, consulting these States and giving them reasonable notice of intention

The Assembly further urged States:

a) to frame any restrictions so that Chapter 2 compliant aircraft of an individual operator which are presently operating to their territories may be withdrawn from these operations gradually over a period of not less than 7 years;
b) not to begin the above phase-in period for any restrictions before I April 1995;
c) not to restrict before the end of the phase-in period the operations of any aircraft less than 25 years after the date of issue of its first individual certificate of airworthiness;
d) not to restrict before the end of the phase-in period the operations of any presently existing wide-body aircraft or of any fitted with high by-pass ratio engines;
e) to apply any restrictions consistently with the non-discrimination principle in Article 15 of the Chicago Convention so as to give foreign operators at least as favourable treatment as their own operators at the same airports; and,
f) to inform ICAO, as well as the other States concerned, of all restrictions imposed.

The Assembly also strongly encouraged States to continue to cooperate bilaterally, regionally and inter-regionally with a view to:

a) alleviating the noise burden on communities around airports without imposing severe economic hardship on aircraft operators; and
b) taking into account the problems of operators of developing countries with regard to Chapter 2 aircraft presently on their register, where they cannot be replaced before the end of the phase-in period, provided that there is proof of a purchase order or leasing contract placed for a replacement Chapter 3 compliant aircraft and the first date of delivery of the aircraft has been accepted;

The Assembly, while urging States, if and when any new noise certification standards are introduced which are more stringent than those in Volume I,

Chapter 3 of Annex 16, not to impose any operating restrictions on Chapter 3 compliant aircraft, urged the Council to promote and States to develop an integrated approach to the problem of aircraft noise, including land-use planning procedures around international airports, so that any residential, industrial or other land-use that might be adversely affected by aircraft noise is minimal. The Assembly further urged States to assist aircraft operators in their efforts to accelerate fleet modernization and thereby prevent obstacles and permit all States to have access to lease or purchase aircraft compliant with Chapter 3, including the provision of multilateral technical assistance where appropriate. This Resolution superseded Resolution A23-10, which was discussed above.

Resolution A 28-3 represents a cautious balance between the concerns of the aircraft manufacturers, the airline industry and developing States who do not wish to lose in the near future, the services of Chapter 2 aircraft which are already in use and service. Although aircraft manufactured prior to October 1977 that are included in Chapter 2 of Annex 16 and called "Chapter 2 aircraft" are required to be phased out, the compromise in Resolution A 28-3 allows States that have noise problems at airports to start phasing out operations by Chapter 2 aircraft from the year 1995 and to have all of them withdrawn by the year 2002, with some exceptions. The Resolution envisages that by the year 2002 only aircraft manufactured after October 1977 and described in Chapter 3 of Annex 16 (called "Chapter 3 aircraft") would be in operation. Following this resolution, a number of developed States have already started to phase out Chapter 2 aircraft, while giving due recognition to the compromise reached in Resolution A28-3.

At its 32nd Assembly, held in September 1998, Assembly Resolution A32-8[174] containing a consolidated statement of continuing ICAO policies and practices related to environmental protection was adopted, making current the regulatory policies relating to aviation and the environment. Appendix B to the Resolution cites Annex 16 Volume 1 as comprising, *inter alia,* noise certification standards for future subsonic aircraft and mentions that aircraft manufacturers and operators need to note that future generations of aircraft have to be so designed as to operate efficiently and with the least possible environmental disturbance. Appendix C calls upon Contracting States and international Organizations to recognize the leading role of ICAO in dealing with aircraft noise and requests the former to work closely together to ensure the greatest harmonization of work in the area of environmental protection as related to air transport. In Appendix G, which relates to the problem of sonic boom, the Assembly reaffirms the importance attached to ameliorating problems caused to the public by sonic boom as a result of supersonic flight, invites States involved in the manufacture of supersonic aircraft to furnish ICAO with proposals that would meet specifications established by ICAO on the subject.

The most topical issue addressed by Resolution A32-8, is in its Appendix D, which, whilst reiterating the time limits specified for the phasing out of Chapter 2

[174] *Assembly Resolutions in Force*, (As of 2 October 1998), *ICAO Doc 9730*, ICAO: Montreal, at I-36.

Aircraft and related dates, strongly encourages States to continue to co-operate bilaterally, regionally and inter-regionally with a view to alleviating the noise burden on communities and also to take into account the problems that may be faced by some operators in phasing out their Chapter 2 aircraft before the end of the period specified. The Resolution also urges States, if and any noise certification standards are introduced new which are more stringent than those in Volume 1, Chapter 3 of Annex 16 not to impose any operating restrictions on Chapter 3 compliant aircraft. More importantly, States are urges to assist operators in their efforts of fleet modernization with a view to preventing obstacles and permit all States to have access to lease or purchase aircraft compliant with Chapter 3.

The qualification in Resolution A32-8 seemingly admits of Chapter 2 aircraft which are converted to be compliant with Chapter 3 noise levels being considered for operation at least until 1 April 2002. The Resolution urges States to consider the difficulties faced by operators of Chapter 2 aircraft who are unable to make them Chapter 3 compliant by the given date, implying that it would be in the economic interests of such operators to be given additional time in order to make the necessary replacements. Chapter 2 aircraft could be made Chapter 3 compliant whereby the aircraft can be re-certified to Chapter 3 standards through re-engining or hush kitting. Chapter 2 aircraft which are likely to be re-engined or hush-kitted are Boeing 727s and 737s, DC-9s, BAC1-11s and some Boeing 747-100s that need hush-kitting.

There is an attempt on the part of the European Union to limit and eventually eliminate Chapter 3 compliant aircraft from operating within countries of the European Union. This ban would also be calculated to affect the importation of such aircraft into the region. Legislation passed by the Union in April 1999 was intended to bar Chapter 3 compliant aircraft from European registries from 4 May 2000 (originally 1 April 1999, which is 3 years before the date specified in Resolution A32-8, namely, 1 April 2002) and to prohibit their operation into the countries of the European Union after 1 April 2002. Such action has been reportedly criticized by the Air Transport Association which claims that the inflexibility of such a deadline "will severely undercut, if not destroy entirely - ICAO's efforts to address environmental issues on a uniform international basis".[175]

The action of the European Union seeks justification on the basis that the exponential air traffic growth in Europe will increase noise around European airports, requiring stringent noise standards. A related fear is reportedly that hush-kitted aircraft, which are rare in Europe, will find a new home in the Continent. Unlike in Europe, airlines in the United States have been somewhat prolific in the use of hush-kits in aircraft[176] and many United States carriers operate Stage

[175] Flint (1999), p. 29.

[176] It was reported that American Airlines is installing the Raisbeck System on 52 B 727 aircraft and hush-kitting 20 more. United is installing hush-kits on 75 B 727 aircraft and 24 B 737-200s. Delta is hush-kitting 104 727s and 54 737-200s. Southwest, TWA, Alaska Airlines and US Airways are other carriers who plan to hush-kit their Chapter 2 aircraft. See Flint (1999), *Id.* at p. 34.

2 hush-kitted aircraft into Europe and even have based equipment in the Continent. At the time of writing, although the U.S. State Department had reacted forcefully to the European Union's hush-kitted aircraft ban, the application of which was extended by the Union until May 2000, it was reported that both the United States and the European Union were working together on a possible new ICAO Standard (presumably to be called Chapter 4).[177]

The issue is a "double edged sword" involving two distinct disciplines. As discussed earlier, the European contention is based on the strictly legal issue of noise pollution and overtones of the tort of nuisance committed by operators whose aircraft are not compliant with Chapter 3 standards to the satisfaction of the European Union. Others who oppose what they claim to be a premature enforcement of ICAO standards, as contained in Resolution A32-8, argue that the European hush-kit rule would cost the manufacturing industry significant losses. It is reported that the United States industry would lose $2 billion if the ban is enforced in Europe as scheduled.[178]

The aircraft noise issue has entered a phase where trading and environmental issues are at a delicate balance. On the one hand, competition issues pertaining to the sale of hush-kits and other equipment calculated to reduce aircraft engine noise to levels prescribed by ICAO are quite significant from the perspective of international trade, and on the other hand, environmental issues that have been addressed by the ICAO Council through CAEP and also by the Assembly should also be given careful consideration. As discussed earlier, ICAO Assembly Resolution A32-8 urges States to give consideration to the economic difficulties that some States may have to face in phasing out Chapter 2 aircraft by the year 2002. However, the Resolution does not leave room for States to claim that pure economic factors would effectively preclude them from phasing out Chapter 2 aircraft by the date stipulated in the Resolution.

The 33rd Session of the ICAO Assembly (Montreal, 25 September–5 October 2001), Resolution A33-7 urged Contracting States to adopt a balanced approach to noise management, taking full account of ICAO guidance, applicable legal obligations, existing agreements, current laws and established policies, all to be given due consideration when addressing noise problems at international airports. Contracting States are urged to adopt appropriate mechanisms in implementing this balanced approach, notably: in establishing a transparent process based on objective, measurable criteria for the assessment of the noise problem; in evaluating likely costs and benefits of various measures with a view to achieving maximum environmental benefit; and in providing for dissemination of the evaluation results that may be used in consultation with stakeholders and dispute resolutions.

As part of the balanced approach, Resolution A33-7 encourages States, *inter alia*, to conduct studies, support research and technology programmes aimed at reducing noise at source and to work closely with each other to ensure that their

---

[177] Esler (1999), pp. 52 and 53.

[178] Europe Considers Delaying Ban (1999), p. 1.

noise management programmes are harmonized taking into consideration the particular economic exigencies of developing countries and also taking particular care so as not to derogate the non-discrimination principle enshrined in Article 15 of the Chicago Convention.[179]

Resolution A33-7, in Appendix D, urges Contracting States not to introduce any phase-outs of subsonic jet aircraft which exceed noise levels contained in Volume 1 of Annex 16 to the Chicago Convention before giving consideration to a clear determination as to whether normal attrition of existing fleets of such aircraft will provide the necessary protection of noise climates around airports and whether necessary protection can be achieved by regulators preventing their operations from adding such aircraft to their fleets through either purchase, or lease/charter/interchange or alternatively by incentives to accelerate fleet modernization. Before phasing out subsonic aircraft that may exceed the above mentioned noise levels, the Resolution also calls upon Contracting States to give careful thought to whether alternatively effective noise management can be achieved by applying regulations preventing operations of such aircraft through restrictions limited to airports and runways, the use of which has been identified and declared by these States as generating noise problems and limited to time periods when greater noise disturbance is caused. Finally, Contracting States are called upon to consider implications of any restrictions for other States concerned, through consultation and reasonable notification of the application of restrictions.

One of the most significant achievements of Resolution A33-7 lies in the skilful balance achieved in offering a compromise to Contracting States which, despite the above mentioned criteria, decide to phase out aircraft which comply with noise certificates Standards in Volume 1, Chapter 2 of Annex 16 but which exceed the noise levels in Volume 1, Chapter 3 of Annex 16. This compromise recommends that such States frame any restrictions so that Chapter 2 compliant aircraft of an individual operator which are operating at present to their territories may be withdrawn from such operations gradually, over a period of not less than 7 years. The Resolution also requests Contracting States not to restrict before the end of the above period the operations of any aircraft less than 25 years after the date of issue of its first individual certificates of airworthiness, and, not to restrict before the end of the period the operations of any presently existing wide-body aircraft or of any fitted with engines that have a by-pass ratio higher than 2 to 1. Contracting States are required to inform ICAO, as well as the other States concerned, of all restrictions imposed.

Finally, on the subject of noise, the Resolution urges States not to introduce measures to phase out aircraft which comply, through original certification or re-certification, with the noise certification standards in Volume 1, Chapter 3 or

---

[179] Article 15, *inter alia*, generally requires that any charges that may be imposed or permitted to be imposed by a Contracting State for the use of airports and air navigation facilities by aircraft shall not be higher than those that are payable by national aircraft of the State imposing such charges.

4 of Annex 16 and, in particular, not to impose any operating restrictions on Chapter 3 compliant aircraft, except as part of the balanced approach to noise management developed by ICAO and in accordance with Appendices C and E to the Resolution which address issues pertaining to phase-out of subsonic aircraft and local noise related operations at airports.

At the 38th Session of the ICAO Assembly held in September/October 2013, the Assembly adopted Resolution A38-17 (*Continuous statement of continuing ICAO policies and practices related to environmental protection - general provisions, noise and local air quality*) Appendix C of which is on policies and programmes based on a "balanced approach" to aircraft noise management. The Resolution Calls upon all ICAO Member States and international organizations to recognize the leading role of ICAO in dealing with the problems of aircraft noise, and urges States to: adopt a balanced approach to noise management, taking full account of ICAO guidance (Doc 9829), relevant legal obligations, existing agreements, current laws and established policies, when addressing noise problems at their international airports; institute or oversee a transparent process when considering measures to alleviate noise, including:

a) assessment of the noise problem at the airport concerned based on objective, measurable criteria and other relevant factors;
b) evaluation of the likely costs and benefits of the various measures available and, based on that evaluation, selection of measures with the goal to achieve maximum environmental benefit most costeffectively; and
c) provision for dissemination of the evaluation results, for consultation with stakeholders and for dispute resolution.

Resolution A38-17 also encourages States to: promote and support studies, research and technology programmes aimed at reducing noise at source or by other means taking into account interdependencies with other environmental concerns; apply land-use planning and management policies to limit the encroachment of incompatible development into noise-sensitive areas and mitigation measures for areas affected by noise, consistent with Appendix F to the Resolution; apply noise abatement operational procedures, to the extent possible without affecting safety and considering interdependencies with other environmental concerns; and not apply operating restrictions as a first resort but only after consideration of the benefits to be gained from other elements of the balanced approach and in a manner which is consistent with Appendix E to the Resolution and taking into account the possible impact of such restrictions at other airports.

States are requested by the Resolution to work closely together to ensure the harmonization of programmes, plans and policies to the extent possible; ensure that the application of any measures to alleviate noise are consistent with the non-discrimination principle in Article 15 of the Chicago Convention; and take into consideration the particular economic conditions of developing countries. States are requested to keep the Council informed of their policies and programmes to alleviate the problem of aircraft noise in international civil aviation.

The Resolution requests the Council to: assess continuously the evolution of the impact of aircraft noise; ensure that the guidance on the balanced approach in Doc 9829 is current and responsive to the requirements of States; and promote the use of the balanced approach, for example through workshops and calls upon States to provide appropriate support for this work on ICAO guidance and any additional work on methodologies, and for the assessment of the impact or effectiveness of measures under the balanced approach as necessary.

It is noteworthy that at the 38th ICAO Assembly, South Africa, while recalling Assembly Resolution A27-11 which is still in force, (still in force), called upon to consider the possible relaxation of operating restrictions for aircraft meeting the requirements of Chapter 3 of Annex 16—Environmental Protection, including the easing of night curfews and/or quotas for off schedule arrivals by such aircraft, submitted that the issues arising from night curfews that have existed for many years continue to remain despite the fact that advancement in aircraft engine technology has contributed considerably to noise abatement in the past two decades. South Africa noted that although the measure of night curfews can help reduce the aircraft noise problem at the airport, it can also have a deleterious effect on the operation of air services, particularly international services to and from the airport, and the economic well-being of the local community and the country at large.[180]

2.5.1.2 *Cursus Curiae*

The only Common Law cause of action which merits detailed discussion in this chapter is nuisance. The word "nuisance" remains the most uncertain word in legal terminology even though it has been put to diverse use by the judiciary. Its tentacles lie spread-eagled on a multitude of wrongs ranging from the noises made by animals to obnoxious smells. Perhaps the reason for this is due to the spontaneous human response of identifying any interference, obstruction or damage caused to an individual as a nuisance. In attempting to interpret the word "nuisance" the judiciary has been in a quandary. The meaning of the word "nuisance" has been stretched from "a potential source of annoyance"[181] to "liability for any act of obstruction of the enjoyment of a legal right".[182] Though the very concept of nuisance originated to retaliate against offences of a public nature, it has since been categorized into public and private nuisance. English common law rigidly classifies a public nuisance as a crime and a private nuisance as a civil wrong.[183] It is submitted that in the environmental law area it is the concept of private nuisance

---

[180] South Africa's Approach and Recommendations on Aircraft Noise Curfews *A38-WP/240 EX/81*.

[181] *Per* Lord Atkin in *Sedleigh-Denfield* v. *O'Callaghan* [1940] A.C. 880, 886.

[182] *Per* Dixon J. in *Torette House Pty, Ltd.* v. *Berkman* (1940) 62 C.L.R. 637, 652.

[183] *Salmond on Torts* (17 ed., 1977, ed., R.F.V. Heuston) 49.

which is most relevant. The concept of private nuisance can be divided into a wrongful disturbance of land and a wrongful interjection of deleterious matter into land belonging to another. Therefore, private nuisance is particularly appropriate to claims founded upon excessive noise because it is directed at the interference with a person's interest in the peaceful use and enjoyment of his property. The basis of this branch of the law is the necessity of maintaining a balance between the right of an occupier to use and enjoy his property and the right of an occupier to be protected from interference or injury.[184] Therefore, courts when adjudicating on environmental disputes and determining whether the tort of private nuisance is present, have to balance the equities between the parties. This chapter will attempt to trace the manner in which the Anglo Australian judiciary has dealt with nuisance in noise pollution disputes.

Howlett v. *McCarthy*[185] is an early case which exemplifies the difficulty faced by the courts in this respect. In this instance, a New South Wales court examined a resolution of a council which ordered the occupiers of premises to cause "no noise" in the use of such premises, between specified hours. In the context of the phrase "no noise" it was extremely difficult for Judge Betts, who delivered judgment, to examine whether the sound caused by machinery was noise which would amount to a breach of such by-law. The court cautiously observed that a liberal meaning should be attributed to "noise" as a strict interpretation of the term would hinder anyone in carrying out a business or industry.[186] Therefore, "noise" was identified as that which would carry beyond the confines of the premises it emanated from to such a degree or of such a nature in quality or volume as to interfere with the life of the residents of normal temperament in the neighbourhood. The interference referred to was further elaborated to be worry, inconvenience, annoyance, discomfort ir loss of sleep during the prohibited hours.

Judge Betts faced the dilemma of being confronted with two conflicting interests—that of the plaintiff who alleges the injury or annoyance and of the business man who carries on an industry as a means of livelihood. Eventually, His Honour introduced an effective compromise by stating that any sound does not necessarily become noise unless it causes the discomfort specified above, and if it does, the person who suffers the annoyance is entitled to seek redress in order that the noise be abated by the person who causes it. Although the judgment seems fair and just for the time it was delivered, it does not suffice that such a general observation be applied in the present day when industry and population grow simultaneously and the consideration of sustainable development has become a pre-eminent factor in human development.

The judgment was fully endorsed by Richardson J. in a case decided 20 years later[187] where His Honour did not deviate from the general stance taken by Judge

[184] Per Cleary J. in *Morgan* v. *Kyatt* [1962] N.Z.L.R. 791, 794 (C.A.).

[185] (1936) 13L.G.R. 73.

[186] *Id.* 75.

[187] *Williams* v. *Storey* (1957) 2L.G.R.A. 226.

Betts. The same criteria were used in different terminology to state that noise should be modulated so as not to inconvenience others and cause discomfort[188] and that noise caused by such things as hammers, building implements and loud speakers should not hinder the comfortable enjoyment of life by neighbours.[189] The judgment was followed verbatim by Manning J. in *Bankstown Municipal Council* v. *Berzins*[190] in which His Honour stated that noise is distinguishable from sound, in that noise, unlike sound, materially interferes with the ordinary comfort and convenience of residents of the neighbourhood.[191]

The early attitude towards noise as environmental pollution does not differ in any way to that which was taken by the courts. The courts too were faced with the problem of being confronted by inadequately framed by-laws and the paucity of precedent which adequately defined noise. In *Leslie* v. *City of Essendon,*[192] a case decided in Victoria, Australia in 1902, Sholl J. dealt casually with noise by calling it "loud sound"[193] whereas Copell A.J., quoting from them Oxford English Dictionary preferred to expand the definition to a "loud or harsh sound of any kinda din".[194] It is not unreasonable to assume that in the light of the few cases decided, noise was always considered subjectively which made it difficult for courts to ascertain whether in any given instance a person aggrieved had a cause of action. What was lacking was a definite idea of what the basic human tolerance level was. At that time courts did not even mention the possibility of excessive noise causing injury to the human ear. However, as was mentioned earlier, the above cases were decided between 1936 and 1957 during which time medical science, though not totally unaware of the detrimental effects of noise on the human ear, had not published sufficient material on the subject.

Although ample evidence was available in the 1960s on noise pollution and human physiology, a case decided in 1967 in Victoria, Australia[195] decided to follow precedent and appeared to agree with the general view that noise becomes intolerable when it causes discomfort.[196] Although the courts only examined the validity of a by-law prohibiting noise on a highway, it did mention noise in general without attempting to assess modern scientific findings on the subject.

The British case of *Regina* v. *Fenny Stratford Justices, ex parte Watney Mann (Midlands) Ltd.*[197] decided in 1976 adopted a more positive attitude even though it did not consider the effects of noise in terms of scientific findings. Watkins

---

[188] *Id.* 233. See also 236.

[189] *Ibid.*

[190] [1962] N.S.W.R. 641.

[191] *Id.* 644.

[192] [1952] V.L.R. 222.

[193] *Id.* 231.

[194] *Id.* 241.

[195] *Schofield* v. *City of Moorabbin* [1967] V.R. 22.

[196] *Id.* 24.

[197] [1976] 1 W.L.R. 1101.

J. expressing his opinion of a jukebox as a modern appliance was both sincere and unequivocal. His Lordship said:

> …at any rate it is my opinion that excessive noise is one of the curses of the modern age. One of the reasons why this is so is that somebody invented the amplifier, which serves at least one purpose, namely, very nearly to blast out the eardrums of anyone within its proximity.[198]

This statement is certainly more expressive than those exemplifying the previous attitude, although noise related discomfort and injury was not dealt with specifically. At least the plaintiff is given the assurance that even though the courts have still not considered modern medical findings on the subject, they consider any kind of noise with the same apprehension as when they consider any other assault on the human body. In addition, one can be comforted by the thought that it will not be long before the courts take full cognizance of the impact of modern scientific findings.

Pollution has been defined as:

> the introduction by man, directly or indirectly, of substances or energy into the environment resulting in deleterious effects of such nature as to endanger human health, harm living resources and ecosystems, and impair or interface with amenities and other legitimate uses of the environment.[199]

The definition covers human health which has been a major concern of environmentalists for some time. It also accommodates the role played by the United Nations under its Charter. Article 1 of the United Nations Charter identifies as one of the purposes of the United Nations, the achievement of international cooperation in solving international problems *inter alia* of a human character[200] and charges the General Assembly of the United Nations to promote international cooperation *inter alia* in the health fields.[201] Article 55 of the Charter requires the United Nations to promote higher standards of living[202] and arrive at solutions concerning *inter alia* health problems.[203] For this purpose of the Economic and Social Council of the United Nations (ECOSOC) may make or initiate studies and reports *inter alia* with respect to international health matters.[204] The role of the United Nations in preserving the environment is thereby clearly entrenched in its Charter. International regulation, as discussed above, are but a corollary to this *status quo*.

---

[198] *Id.* 1103.

[199] *OECD and the Environment*, Paris:1979 at 108. This definition has also been adopted by the *Law of the Sea Conference*. See Article 1(4), A/CONF.62/WP 10/Rev.3.

[200] *Charter of the United Nations and Statute of the International Court of Justice*, United Nations: New York, Article 1.3.

[201] *Id.* Article 13.1.b.

[202] Article 55(a).

[203] Article 55(b).

[204] Article 62(1).

The tort of nuisance is caused by an unprivileged interference by a person of another's enjoyment of his or her private property, causing discomfort to the latter, and invariably causing the property to diminish in value.[205] There are two instances, however, where recovery against the tort of nuisance is not possible, namely, where a State can invoke sovereignty; and where the defence of pre-emption can be successfully claimed.[206] In all other instances, where nuisance is alleged to have been committed by aircraft noise, particularly where a State run airport is held answerable, a successful legal approach for the plaintiff would lie in the theory of inverse condemnation. The principle of inverse condemnation has been identified as:

> the popular description of a cause against a governmental defendant to recover the value of property which has been taken in fact by the governmental defendant even though no formal exercise of the power of eminent domain has been attempted by taking the agency.[207]

and was brought to light by the seminal decision in the case of *United States* v. *Causby,*[208] decided in 1946, which involved repeated flights over the plaintiff's property by military aircraft, held that there had been a compensable interference with property and consequent taking down of its value which was at variance with the Fifth Amendment of the United States' Constitution.[209]

The 1962 case of *Thornburg* v. *Port of Portland*[210] made a significant deviation from the usual principle that compensation for a nuisance caused by noise produced by aircraft could only be awarded if the aircraft was directly above the property concerned and ruled that a trespass resulting in nuisance could occur even if the noise was coming from an aircraft flying beside the property. The *Thornburg* principle was followed some years later in *Martin* v. *Port of Seattle*[211] which rejected the rigidity of the principle that, in the words of the court, insisted upon "the wingtip of the aircraft passing through some fraction of an inch of the airspace directly above the plaintiff's land".[212]

The early English case of *Walter* v. *Selfe*[213] was the *tour de force* in awarding the remedy of private nuisance in the environmental law area. The famous dictum of

---

[205] Stein (1991), p. 555.

[206] *Ibid.*

[207] See. *Thornburg* v. *Port of Portland*, (1962) 233 Or. 178, 376 P.2d. 100 at 101, note 1.

[208] 328 U.S. (1946) at p. 256.

[209] The Fifth Amendment provides that private property shall not be taken for public use without payment of compensation. See also the cases of *Griggs* v. *Allegheny County* 326 U.S. at 84 (1962) and *Batten* v. *United States* 306 F.2d. 580 (10th Cir. 1962) which awarded damages for diminishing of value of property under the United States' Constitution.

[210] *Supra,* note 207 (this chapter).

[211] 391 P.2d. 540 (1964).

[212] *Id.* at p. 545. See also *Aaron* v. *City of Los Angeles*, 40 Cal.App. 3d. 471, 115 Cal. Rptr. 162 (1974) which held that compensation should be awarded to persons for property values affected by aircraft noise even though the property was not located within a direct flight path of the Los Angeles airport.

[213] 4 De G & Sm 315; [1851] 64 E.R. 849.

Knight Bruce V.C. in this case concerned the test of whether emanations from land constitute a nuisance to persons in possession of other land. The criterion was whether they create:

> . . . an inconvenience materially interfering with the ordinary comfort physically of human existence, not merely according to elegant or dainty modes and habits of living, but according to plain and sober and simple notions among the English people.[214]

Another early case which is frequently considered as formulating the legal principles on nuisance as it applies to noise pollution is *Gaunt* v. *Fynney*. In this case Lord Selborne L.C. states:

> A nuisance by noise (supposing malice to be out of the question) is emphatically a question of degree. If my neighbour builds a house against a party-wall, next to my own, and I hear through the wall more than is agreeable to me of the sounds from his nursery or his music room, it does not follow (even if I am nervously sensitive or in infirm health) that I can bring an action or obtain an injunction. Such things, to offend against the law, must be done in a manner which, beyond fair controversy, ought to be regarded as exceptive and unreasonable.[215]

Later cases in determining whether noise amounts to a nuisance have applied these principles. In determining what is a reasonable noise and what are plain, sober and simple notions the courts have had to consider the general nature of the neighbourhood, the nature of the location at which the alleged nuisance took place, the character, duration of time of occurrence of the noise and the effects of the noise.[216]

In an early case[217] decided by the Supreme Court of Victoria, Australia, the ringing of a church bell in the early morning hours of Sunday and public holidays in such a manner as to disturb persons residing in the neighbourhood was considered a nuisance. The court took into account, among other things, the degree of the disturbance caused. This case could be contrasted with a South Australian case decided a few years later.[218] The plaintiff who lived 17 ft. away from the defendant's premises claimed an injunction and £500 in damages for depreciation of the value of his house by reason of the noises made by the defendant who conducted religious services on Sundays, commencing at 7 a.m. The noises the plaintiff complained of were singing, band playing, hand clapping and walking about the hall. It was held that singing in a place of religious worship is a sound considered to be ordinarily proceeding from such a place. Although the noises complained of in the two cases are similar, the judges in the South Australian case distinguished that case from the Victorian case on the basis that the necessary inference that can be drawn from the Victorian case is that the ringing of a church bell is not considered

[214] *Id.* 322. See also [1851] 64 E.R. 849, 852.

[215] *Id.* 12.

[216] *Per* Harris J. in *Oldham* v. *Lawson* [1976] V.R. 654, 655.

[217] *Haddon* v. *Lynch* [1911] V.L.R. 230.

[218] *McKenzie* v. *Powley* [1916] S.A.L.R. 1.

part of the service, whereas singing is a necessary part of religious worship. Whether such a distinction is logical is a moot point. Therefore, on similar facts the judges in the two cases came to different conclusions.

In a New South Wales case[219] decided in 1928, the plaintiff who was the defendant's neighbour, complained that the carrying on of building operations by the defendants by the use of mechanical drills, caused a continuous deafening noise during normal business hours. The plaintiff contended that this noise prevented business being carried on in a building owned by the plaintiff. Despite the protests of the company no effort had been made by the defendants to conduct their operations during hours when the plaintiff's business would not be affected. The court granted an injunction restraining the defendants from causing the nuisance. It was held that an alternative method of conducting their operations was available to the defendants even though such method was more costly and time consuming. Long Innes J. distinguished this case from an early English case[220] in which on similar facts the court held that an alternative method of operation would have endangered the lives of the employees.

However, in the later case of *Painter* v. *Reed*[221] the Supreme Court of South Australia held that the alterations made by the defendant to make his business as unobjectionable as possible was not sufficient to prevent the plaintiff from succeeding in an action for nuisance. In contrast, a 1953 case[222] determined by the Supreme Court of South Australia found the alterations made by the defendant's operations, which substantially reduced the noise were sufficient to no longer make the noise an actionable nuisance. In this case the plaintiff complained of the noise (and also the dust and the fumes) from a nearby sports ground which was also used as a speedway. Napier C.J. relying on *Gaunt* v. *Fynney*[223] and *Rushmer* v. *Polsue & Alfieri Ltd.*[224] took into account the nature of the trades usually carried on in the locality and the benefit to the public by the entertainment provided by the defendant's business. His Honour found a considerable interference by the defendant with the peace and comfort of the plaintiffs, but concluded that the nuisance no longer existed due to the alterations made by the defendant.

In a later case[225] the court took into account the fact that the plaintiff's sole occupation was poultry farming and that the use of explosives by the defendant to clear his land of stumps, which had adverse effects on the plaintiff's poultry, was a serious nuisance which warranted the granting of an injunction to prevent the

[219] *Daily Telegraph Co. Ltd.* v. *Stuart* (1928) 28 S.R. (N.S.W.) 291.

[220] *Id.* 295 where His Lordship cited *Harrison* v. *Southwark and Vauxhall Water Company* [1891] 2 Ch. 409.

[221] [1930] S.A.S.R. 295.

[222] *Field and Others* v. *South Australian Soccer Association (Incorporated) And Shields* [1953] S.A.S.R. 224, hereafter referred to as *Field*'s case.

[223] (1872) L.R. 8 Ch. App. At p. 8. See also *Field*'s case *Id.*

[224] [1906] 1 Ch. 234. See 231 of *Field*'s case.

[225] *McMahon* v. *Catanzaro* [1961] Qd.R. at 22 Q.W.N. 28.

nuisance. It was held that the defendant could use as an alternative, a reasonably practicable method of removing the trees, even though such a method would be more expensive. In a recent case determined by the Supreme Court of Western Australia[226] even though the defendants had taken considerable steps to reduce the noise (and the dust) from their quarrying operations, the plaintiffs succeeded in their action in nuisance as the noise interfered with the use and enjoyment of their land.

In analysing the above cases it is clear that in *Daily Telegraph Co. Ltd.* v. *Stuart* and in *McMahon* v. *Catanzaro* the courts granted injunctions to restrain the nuisance and ordered the defendants to use alternative methods even though such methods were expensive. In *Field*'s case an injunction was not granted as the nuisance had already been abated by an alternative method. In all three cases the courts took into consideration the alternative methods of operation that the defendants could use and had used. However, in *Painter* v. *Reed* although it was held by Richards J. that "the defendant had done very much to make the manner of carrying on his business unobjectionable"[227] this was not considered enough to have abated the nuisance. In *Aldridge* v. *Clough*, Burt C.J. found that "[E]ach step taken no doubt had some beneficial effect" but His Honour too came to the same conclusion as Richards J. Therefore, what is an unreasonable act to the plaintiff in nuisance disputes in the noise pollution area has to be determined according to the facts of each case, following the basic idea given in *Walter* v. *Selfe*. An objective standard cannot be used.

In continuing to analyse the judicial dicta on private nuisance as a remedy for noise pollution, the decision in *Ruthning* v. *Ferguson*[228] took into account the test of whether the noise of the defendants' birds interfered with the ordinary physical comfort of human existence according to the plain, sober and simple notions obtaining among the Australian people. An injunction had been granted restraining the defendant from keeping roosters in such a manner as to cause a nuisance to the plaintiffs. This decision was reversed on appeal because it was found in evidence that the complaints of the plaintiffs were grossly exaggerated.

A case decided in 1937 by the High Court of Australia[229] held that mere competition which is offered by the defendants does not amount to an actionable nuisance. The facts of the case briefly were that descriptions of races taking place on the plaintiff's land were broadcast by the defendants on adjoining land, without the plaintiff's permission. As a result of such broadcasting, attendance at the race course diminished. The plaintiff asked for an injunction to restrain such broadcasting as it amounted to a nuisance. It was held that although the plaintiff suffered a

[226] *Aldridge* v. *Clough*, Unreported S.C. Decisions of Western Australia No. 1246 of 1979.

[227] [1930] S.A.S.R. 295 at p. 297.

[228] [1930] St. R. Qd. 325.

[229] *Victoria Park Racing and Recreation Grounds Company Ltd.* v. *Taylor and Others* (1937) 58 C. L. R. 479.

substantial monetary loss due to the defendants' actions, yet it was not considered as an interference with the use and enjoyment of the plaintiff's land.

Spencer v. *Silva*[230] was a case where the plaintiff complained of the noise caused by the defendant's use of saws in his business. Mayo J. in determining whether the noise infringed the plaintiff's right to comfort considered whether the plaintiff was peculiarly susceptible to noise. His Honour held the view that the test of determining such susceptibility was objective, i.e. it is to be measured according to the plain, sober and simple notions among ordinary people. However, Mayo J. held that although the measure must be objective, the application of the test is subjective. The test is subjective to the extent that in some circumstances, which do not include cases where the discomfort is purely personal, only very slight evidence of damage need be proved.[231] Concluding that the plaintiff was entitled to an injunction His Honour held that:

> When that noise occurs daily for long periods, I think, it becomes a menace to reasonable comfort. It may be true that some people can readily adapt themselves to it without effort, others can with the exercise of determination accustom themselves to it,..but that does not necessarily mean, because that is so, if it is, that the law requires all persons to enure themselves to the unpleasantness or quit.[232]

In the case of *Dunstan* v. *King*[233] the prolonged and excessive noise emitted from a sawmill and a diesel engine and by smoke and oil fumes therefrom was the alleged nuisance complained of by the plaintiffs who were four residents of that locality. The court followed the rationale in the earlier case of *Spencer* v. *Silva* and decided that the plaintiffs had truthfully described the effects of the mill upon themselves.

Similarly, a case decided 2 years later[234] considered whether the plaintiff was peculiarly sensitive to the noise complained of. Roper C.J. asked the question:

> Are they fastidious or neurotic or oversensitive to something which according to simple and sober notions is not an annoyance at all, or not an annoyance to any appreciable degree?[235]

and came to the conclusion that the noise of the defendant's pigeons flying over the plaintiff's land would be an intolerable nuisance to reasonable normal people.

In *Clarey* v. *Principal and Council of the Women's College*,[236] the High Court of Australia had to consider whether a landlord who lets a portion of his residence for the accommodation of university students, could complain that the conduct of the students in making noises which are incidental to the occupation of premises as a dwelling place such as walking about, having a bath, laughing and talking,

---

[230] [1942] S.A.S.R. 213 [S. Ct.].

[231] *Id.* 219.

[232] *Id.* 220.

[233] [1948] V.L.R. 269.

[234] *Fraser* v. *Booth* (1950) 50 S.R. (N.S.W.) 113.

[235] *Id.* 116.

[236] (1953) 90 C.L.R. 170.

constitute a nuisance to adjoining occupiers within the meaning of sub-s. 41 (5) (d) of The Landlord and Tenant Acts of 1948–1950. The court using Pollock on Torts as an authority came to the conclusion that the use of a dwelling house in an ordinary and accustomed manner, is not a nuisance though it may produce more or less noise and inconvenience to a neighbour.[237]

In *Munro* v. *Southern Dairies Ltd.*[238] the plaintiff complained of the noise and smell from the defendant's dairy. The court in determining whether the noise and the smell amounted to a nuisance took into consideration *inter alia*, the particular sensitivity if any, of the plaintiff. Sholl J. held that although a man is not entitled to relief merely because he may happen to be unduly sensitive to noise, yet a man who lives next door to premises from which the noise emanates may get relief when another man some distance away will fail to obtain it.[239]

The American case of *Boomer* v. *Atlantic Cement Company*[240] is an example of a case in which the court was swayed by the economic position of the defendant company.[241] Here the plaintiff alleged property damage caused by the dirt and vibrations resulting from the defendant's industrial complex. The court recognized a substantial interference of the plaintiff's rights. Although the court cites the law in New York, which is to the effect that a nuisance will be found to exist even though there is marked disparity in economic consequence between the effect of the injunction and the effect of the nuisance, yet it strangely decided to ignore it on the basis that the rule should not be followed too "literally".[242] The court granted damages to the plaintiff instead of an injunction obviously because they were favourably impressed by the vast wealth of the defendant company and knew that a permanent injunction would have caused a significant loss to the company. The court may have been influenced in its decision by the loss of employment to the 300 workers of the defendant company if the court granted an injunction. However, the ability of the common law in protecting the ordinary citizen is illustrated in the English case of *Halsey* v. *Esso Petroleum Co. Ltd.*[243] The plaintiff in this case lived in an area zoned for residential purposes. The defendant, who was a mammoth oil company, used a land on the opposite side as an oil storage depot. The plaintiff alleged nuisance resulting from the noise and smell of the defendant's operations. In this case the court refused to be awed by the economic position of the defendant and seemed to base the judgment on the principle that the standards binding on both parties are those of the ordinary reasonable man. Although the defendant company had taken a number of steps to abate the nuisance, the court found that an actionable nuisance existed and granted an injunction restraining the defendant from carrying

---

[237] *Id.* 176.

[238] [1955] V.L.R. 332.

[239] *Id.* 335.

[240] 257 N.E. 2d. 870.

[241] The defendant's investment in the plant was in excess of $45,000,000.

[242] *Boomer*, supra, note 240 (this chapter) at pp. 872–873.

[243] [1961] 1 W.L.R. 683.

on the activities which caused the nuisance. Although in both cases the court held that a nuisance existed, in the English case the court by granting an injunction prevented the recurrence of the nuisance. These two cases illustrate the subjective approach adopted by the judiciary in nuisance disputes. A general principle cannot be culled from these decisions.

The case of *Aldridge* v. *Clough*[244] echoes the time honoured words in *Walter* v. *Selfe*. Burt C.J. considered carefully whether the plaintiffs were specially sensitive to the noise and dust from the quarry. His Honour came to the conclusion that the evidence of the plaintiffs was not only honest but also fairly objective. The earlier New South Wales case of *Farley and Lewers Ltd.* v. *Attorney General*[245] was cited as authority for the proposition that the plaintiffs' reaction should represent the reaction of the ordinary person. Burt C.J. in the course of his judgment referred to the Noise Abatement Act of 1972 which attempts to prescribe noise levels which if exceeded, constitutes an offence, and held that on the evidence of the plaintiffs the latter had made good the claim in nuisance.[246] Although the judgment is clearly based on the rationale in *Walter* v. *Selfe*, Burt C.J. determined the case on the evidence, according to his individual view of what constitutes an interference with "plain and sober notions among our people".[247]

2.5.1.3 Relevance of Locality

A frequent factor that courts take into consideration in determining whether a particular activity is a nuisance is the locality of the alleged nuisance. For instance, in *Painter* v. *Reed*,[248] in determining whether the noises caused in the early morning by the movement and stamping of horses which disturbed the sleep of the plaintiff constituted an actionable nuisance and should be restrained by injunction, Richards J. considered carefully the locality in which the alleged nuisance occurred. His Honour was of the opinion that although the defendant's business was situated in a very suitable place in the city, yet the noise complained of was not as reasonable as the law would require. In *Dunstan* v. *King*,[249] in coming to the conclusion that the noise amounted to an actionable nuisance, the court considered the type of locality in question. Although it was adduced in evidence that there were other sawmills in the district, the court held that there was no other industrial undertaking within reasonable distance of the particular locality, which had the character of a quiet countryside.[250] In *Munro* v. *Southern Dairies Ltd.*[251] too, in

---

[244] (1979) W.A.L.R. 771.

[245] [1963] N.S.W.R. 1624. The facts of this case are similar to the facts of *Aldridge* v. *Clough*.

[246] *Aldridge* v. *Clough*, above 11.

[247] *Id.* 5.

[248] *Supra*, note 221 (this chapter).

[249] [1948] V.L.R. 269.

[250] *Id.* 272.

[251] [1955] V.L.R. 332.

considering whether the noise complained of was characteristic of the particular neighbourhood, Sholl J. held that it was not so.[252]

In another case[253] the defendant was a haulage contractor and the noise and the dust caused by his business constituted a substantial nuisance to the plaintiff. The court cited a great deal of judicial authority in this area of the law. After considering the pecuniary loss resulting to the defendant if the court granted a permanent injunction, the fact that no one else living in the neighbourhood supported the plaintiff and the fact that the plaintiff had a motive for complaining about the noise (there was ill-feeling between the two families) the court came to the conclusion that in a residential suburb a man is not at liberty habitually to start a noisy motor vehicle at very early hours of the morning when his neighbours are usually asleep. Upon the defendant undertaking to abate the nuisance, the action was dismissed. This case underscores the preoccupation of judges in considering the locality of the alleged nuisance important, even in circumstances where there is evidence in favour of the defendant.

The relevance of locality has been considered in situations where the defendant avers that the plaintiff came to the nuisance. However, "coming to the nuisance" is not a defence to an action in nuisance. Certain cases have discussed this non-defence. One of the earliest cases[254] which has done so held that the plaintiff's having come to the nuisance did not dis-entitle him to equitable relief. In *Barber* v. *Penley*[255] the offending noise came from the crowd at a theatre in the locality where the plaintiff lived. It was held that a person who deliberately goes next door to a theatre cannot expect to have precisely the same amenities that the lessee of a private house in a street occupied by private houses only in the West End would expect. However, the court held that such a lessee still has the right to expect that no nuisance shall be committed to the prejudice of such person.[256] In the recent case of *Miller* v. *Jackson*[257] Geoffrey Lane L.J. posed the question:

> . . . can the defendants take advantage of the fact that the plaintiffs have put themselves in such a position by coming to occupy a house on the edge of a small cricket field, with the result that what was not a nuisance in the past now becomes a nuisance?[258]

His Lordship held that although he was inclined to find for the defendants he could not do so as he was bound by judicial precedent. The case of *Sturges* v. *Bridgman*[259] was cited as authority for the proposition that it is no answer to a

---

[252] *Id.* 337.

[253] *Kidman* v. *Page* [1959] Qd. R. 53.

[254] *Tipping* v. *St. Helen's Smelting Co.* [1865] 1Ch. App. 66.

[255] [1893] 2 Ch. 447.

[256] *Id.* 449.

[257] [1977] Q.B. 966. (C.A.).

[258] *Id.* 986.

[259] (1879) 11 Ch. D. 852. In this case the plaintiff who was a physician was hindered in his work by the noise and vibrations from the neighbouring confectionery shop.

claim in nuisance for a defendant to show that the plaintiff brought the trouble upon himself by building and coming to live in a house so close to the defendant's premises that he would inevitably be affected by the defendant's activities.[260] Geoffrey Lane L.J. thought that this rule works injustice. He reluctantly held for the plaintiff. His Lordship seemed to have been much impressed by the fact that the locality was a country area in which cricket was an important activity. Therefore, courts throughout the ages have been influenced by the locality in question in considering whether the plaintiff came to the nuisance.

2.5.1.4 The Public Benefit Criterion

Certain cases have considered whether the benefit to the public by the defendant's activity would override the nuisance such activity would cause to the plaintiff. *Munro* v. *Southern Dairies Ltd.*[261] was such a case. In this case Sholl J. cited the early case of *Shelfer* v. *City of London Electric Lighting Company*[262] in support of his contention that the public benefit resulting from the defendant's activity was not sufficient to prevent the plaintiff from succeeding in the nuisance action. In the more recent English case of *Kennaway* v. *Thompson and another*[263] the Court of Appeal considered the jeopardy caused to the public interest in granting an injunction prohibiting the defendant, a boating club, from carrying on noisy activities which prevented the plaintiff from using and enjoying her land. Lawton L.J. held that although the plaintiff was not entitled to an injunction restraining all of the club's activities because there had to be a reasonable amount of give and take, those activities which caused a serious nuisance to her should be restrained by injunction.[264] His Lordship was of the opinion that the public benefit generated by the defendant's activities outweighed the inconvenience to the plaintiff who had to absent herself from her house for many days in order to avoid the noise.[265] In dismissing the case of *Miller* v. *Jackson*[266] as not binding on the present case, His Lordship held that the statement of Lord Denning M.R. in that case, that the public interest should prevail over the private interest, runs counter to the principles enunciated in the early case of *Shelfer* v. *City of London Electric Lighting*

---

[260] [1977] Q.B. 966 at 986,987.

[261] [1955] V.L.R. 332.

[262] [1895] 1 Ch. 287. Lindley L.J. stated at 316:

> Neither has the circumstance that the wrongdoer is in some sense a public benefactor (e.g., a gas or water company or a sewer authority) ever been considered a sufficient reason for refusing to protect by injunction an individual whose rights are being persistently infringed. Cited at 337 in *Munro* v. *Southern Dairies Ltd.*

[263] [1980] 3 All. E. R. 329.

[264] *Id.* 333.

[265] *Ibid.*

[266] [1977] Q.B. 966. (C.A.).

Company.[267] *Kennaway* v. *Thompson* reiterated the attitude of the courts in considering a person's right to use and enjoy his land as sacrosanct as opposed to the benefit to the public that may accrue from the defendant's activities.

2.5.1.5 Remedies for Nuisance

Damages and injunctions are the two important remedies[268] granted to a successful plaintiff in a nuisance action. Judges have had difficulty in determining whether damages or injunctions should be granted and if damages are to be awarded, the test in determining reasonable damages. The loss need not be of a pecuniary nature but must be material or substantial. The loss complained of in most of the cases on nuisance by noise is the interference with the use and enjoyment of the complainants premises.

The decision of *Oldham* v. *Lawson*[269] examined the question of whether damages should be granted for non-pecuniary loss. Harris J. held that until the 1975 case of *Bone* v. *Seale*[270] there was no test applicable that would warrant the award of damages in the instance where the nuisance complained of does not affect the plaintiff's property. Harris J. relied on the dictum of Scarman L.J. in *Bone* v. *Seale*. Scarman L.J. held:

> Nuisance is a wrong to property, but it is well recognised that even when there is no physical damage to property it may cause annoyance inconvenience and discomfort to the occupier of the property in his enjoyment of it.[271]

Scarman L.J. relied on McGregor on Damages to buttress his proposition that damages are allowable for non-pecuniary loss caused by nuisance and that such damages should be regarded as part of the normal measure of damages; in determining what the normal measure of damages are, His Lordship held that the general principle in cases where there is no financial yardstick in awarding damages were formulated by Lord Morris of Borth-y-Gest in *H. West & Son Ltd.* v. *Shephard*,[272] when His Lordship said that all that judges and courts can do is to award sums which must be regarded as giving reasonable compensation.

On the question of what is reasonable Scarman L.J. held that there was very little guidance in the law, which led His Lordship to formulate the following principle:

> One must bear in mind also a further general principle, that, when one is removed from the world of pecuniary loss and is attempting to measure damages for non-pecuniary loss, an element in reasonableness is the fairness of the compensation to be awarded. There must be

[267] [1895] 1 Ch. 287.

[268] The other remedies are abatement without recourse to legal proceedings, summary proceedings for penalties or abatement of statutory nuisances and criminal proceedings.

[269] [1976] V.R. 654.

[270] [1975] 1 All.E.R. 787.

[271] *Id.* 794., cited at 658 in *Oldham* v. *Lawson*.

[272] [1964] A.C. 326., cited at 658 in *Oldham* v. *Lawson*.

moderation; some attention must be paid to the rights of the offending defendant as well as the rights of the injured plaintiff.[273]

Although Harris J. accepted this proposition as an accurate statement of the law, it is submitted that Scarman L.J. seemed to have experienced the same difficulty as his predecessors in formulating a test for non-pecuniary damages. His Lordship too does not formulate a comprehensive test.

In *Spencer* v. *Silva*[274] Mayo J. discussed the instances when damages should be granted in lieu of injunctions.[275] Damages instead of an injunction might be given if the injury to the plaintiff's legal right was small, was capable of being estimated in money, could be adequately compensated by a small money payment or if the injunction would be oppressive to the defendant. Mayo J. discusses another instance when damages should be granted instead of an injunction. This is the instance when considerable time has elapsed before the plaintiff instituted legal proceedings.[276] It is submitted that although such a clear statement of the law has been formulated, judges have persisted in applying subjective criteria in determining whether damages or injunctions should be granted. For instance, in *Boomer* v. *Atlantic Cement Company*,[277] the judge refused to grant an injunction preventing the defendant from continuing his commercial activities which was a considerable nuisance to the plaintiff, because the judge was swayed in his judgment by the wealth of the defendant company.

In *Vincent* v. *Peacock*[278] the court discussed the instances when an injunction should not be granted. In this regard the court considered the statement of Lord Evershed M.R. in *Pride of Derby and Derbyshire Angling Association LD.* v. *British Celanese LD.*[279] important. According to Lord Evershed M.R., if A proves that his property rights are being wrongfully interfered with by B, and that B intends to continue his wrong, then A is *prima facie* entitled to an injunction, and he will be deprived of that remedy only if special circumstances exist, including the circumstance that damages are an adequate remedy for the wrong that he has suffered.[280] The court in *Vincent* v. *Peacock* held that instances of special circumstances would be extreme hardship, where the effect of granting an injunction would be to order someone to perform an impossibility or to abstain from doing something which is impossible for him not to do. The Supreme Court overruled the decision of the lower court which failed to grant an injunction to restrain the defendant's drunken behaviour which prevented the plaintiffs from the reasonable

---

[273] *Bone* v. *Seale*, *Supra*, note 270 (this chapter), at p. 795, and cited at 658 in *Oldham* v. *Lawson*, *supra*, note 109 (Chap. 1).

[274] [1942] S.A.S.R. 213 [S. Ct.].

[275] These instances were first discussed by Smith (1895), 1 Ch. 287.

[276] *Spencer* v. *Silva* [1942] S.A.S.R. 213 [S. Ct.]. at 222.

[277] 257 N.E. 2d. 870.

[278] [1973] 1 N.S.W.L.R. 466.

[279] [1953] Ch. 149.

[280] *Id.* 188, cited at 467–468 in *Vincent* v. *Peacock*, *supra* at n. 259 (this chapter).

enjoyment of their home. The lower court seemed to have considered the facts that the offensive behaviour occurred only when the defendant was drunk and the admission of the defendant that he was an alcoholic sufficed as special circumstances for an injunction not to be granted.[281] The reasoning of the lower court was based on the premise that it was futile to issue an injunction against an offender who cannot control the commission of the offence, being in a perpetual state of intoxication. It is apparent that these were not special circumstances within the rule mentioned above. Special circumstances could be interpreted by different judges in different ways as the above case illustrates.

The courts may not only grant a remedy to a plaintiff subsequent to the commission of a nuisance from which he suffers; they may also issue an injunction to prohibit any activity which could cause a nuisance. The latter is called a *quia timet* injunction at common law. This effectively precludes the commission of an act which would result in a nuisance. A *quia timet* injunction should be granted only if three factors are present. They are:

1. Would an act considered in a *quia timet* action if not prohibited, cause a nuisance to the plaintiff?
2. If it would, how imminent is the nuisance?
3. Would the nuisance caused to the plaintiff result in substantial damage to him?

In determining whether these three factors are present, the onus of proof on the plaintiff was considered to be an important consideration in the case of *Grasso* v. *Love*[282] determined by the Full Court of the Supreme Court of Victoria, Australia. Disapproving of the expression "moral certainty" used in an earlier case[283] to describe what the plaintiff must prove, as suggesting a criminal onus of proof, the court held that there is no third onus of proof required in *quia timet* injunctions, lying between the civil and criminal onus. It was held by the Full Court that the plaintiff must prove that there is a real probability that the defendants' activities would cause a nuisance.[284] Citing academic authority in support of their view,[285] the judges held that in determining the real probability of the alleged nuisance, the relevant circumstances of the case should be taken into account, to determine whether the issuance of the injunction would be unfair to either party to the action.[286] For example, in *Grasso* v. *Love* on an appeal against the granting of an injunction against the construction of a drive-in cinema, the appellants argued mainly on the grounds that evidential facts were insufficiently considered in the court of first instance which resulted in a permanent injunction against them. It was

[281] *Vincent* v. *Peacock*, at 468.

[282] [1980] V.R. 163.

[283] Per Farwell J. in *Royal Insurance Co. Ltd.* v. *Midland Insurance Co. Ltd.* (1908) 2 R.P.C. 95 at 100. This case is also cited in *Grasso* v. *Love* at 167.

[284] [1980] V.R. 163 at 167.

[285] I.C.F. Spry, *Equitable Remedies*, (1971) 342–343.

[286] Ibid. See also *Grasso* v. *Love*, [1980] V.R. 163 at p. 167.

held on appeal that although the issuance of the injunction was justified to the extent that substantial damage would have been caused to the respondents who resided in a portion of the land had the injunction not been issued, it need not have applied to the entire land as there was insufficient evidence to show that the proposed drive-in cinema would have caused a nuisance to the respondents had it been erected on a portion of the land a considerable distance away from the respondents' home. Therefore, the Supreme Court considered the rights of both parties in reaching their decision. However it could be argued that "relevant circumstances" is a purely subjective phrase and is dependent on the interpretation given to it by courts.

In the case of *Hatton et al.* v. *United Kingdom*[287] decided by the European Court of Justice in 2001 where eight United Kingdom nationals filed a case against the United Kingdom (lodged on 6 May 1997) with the European Commission of Human Rights ("the Commission") under former Article 25 of the Convention for the Protection of Human Rights and Fundamental Freedoms ("the Convention"), the applicants charged the government of the United Kingdom with accountability and responsibility for aircraft noise over their residences emanating from aircraft landing in and taking off from London Heathrow. The Court held that:

> Heathrow airport and the aircraft which use it are not owned, controlled or operated by the Government or by any agency of the Government. The Court considers that, accordingly, the United Kingdom cannot be said to have "interfered" with the applicants' private or family life. Instead, the applicants' complaints fall to be analysed in terms of a positive duty on the State to take reasonable and appropriate measures to secure the applicants' rights under Article 8 § 1 of the Convention.[288]

In inquiring as to whether, as claimed by the applicants, their fundamental right had been breached by the aircraft noise the Court took into consideration Article 8 of the *European Commission on Human Rights and Fundamental Freedoms*, adopted in Rome in 1950 protects the right to respect for private and family life, home and correspondence, which reads as follows:

1. *Everyone has the right to respect for his private and family life, his home and his correspondence.*
2. *There shall be no interference by a public authority with the exercise of this right except such as is in accordance with the exercise of this right except such as is in accordance with the law and is necessary in a democratic society in the interests of national security, public safety or the economic well-being of the country, for the prevention of disorder or crime, for the protection of health or morals, or for the protection of the rights and freedoms of others.*

In attempting a balance between this provision and the averment of the respondent that aircraft movement at the time in question were necessary for the economic well being of the industry as well as society, the Court held that:

[287] Application No. 36022/97. See http://hudoc.echr.coe.int/sites/eng/pages/search.aspx?i=001-59686#{"itemid":["001-59686"]}.
[288] *Ibid.*

> The Court would, however, underline that in striking the required balance, States must have regard to the whole range of material considerations. Further, in the particularly sensitive field of environmental protection, mere reference to the economic well-being of the country is not sufficient to outweigh the rights of others... It considers that States are required to minimise, as far as possible, the interference with these rights, by trying to find alternative solutions and by generally seeking to achieve their aims in the least onerous way as regards human rights. In order to do that, a proper and complete investigation and study with the aim of finding the best possible solution which will, in reality, strike the right balance should precede the relevant project.[289]

The *Environmental Noise (England) Regulations 2006* (as amended), which transpose the provisions of EU Directive 2002/49/EC relating to environmental noise from transport and industry, requires that airport operators produce a Noise Action Plan which must be approved by the Government and reviewed every 5 years. Heathrow's Noise Action Plan 2010–2015 was adopted in May 2011 and London Stansted's Noise Action Plan for 2010–2015 in July 2011.[290]

A historical analysis of the *cursus curiae* in the field of nuisance as a remedy for noise pollution illustrates the fact that judges throughout the ages up to the present time have been influenced by the fundamental legal principle enunciated in the early English case of *Walter* v. *Selfe*. However, the judges in the different cases have given their own interpretation of the rationale in *Walter* v. *Selfe*. Although it could be stated that diversity in judicial opinion leads to independence in judicial thinking and that judges should not be hampered by a comprehensive definition, yet a strong argument against this type of attitude is that the law would be ambivalent when there is a multitude of judicial opinions on one point, out of which no general principle could be culled. It is this vagueness which has led to the common law of nuisance being to a large extent replaced by an array of statutory provisions which have been designed to control environmental damage.

2.5.1.6 Noise and the International Human Rights Regime

The World Health Organization (WHO) has stated that that noise pollution can cause interference with communication, sleep disturbance, increased annoyance responses, noise-induced hearing loss, learning acquisition, performance effects, and cardiovascular and psycho physiological effects. Studies have suggested that the reading age of children in schools under flight paths is below the national average.[291] WHO has said:

> Noise pollution is not believed to be a cause of mental illness, but it is assumed to accelerate and intensify the development of latent mental disorders. Noise pollution may cause or contribute to the following adverse effects: anxiety, stress, nervousness, nausea, headache, emotional instability, argumentativeness, sexual impotence, changes in mood, increase in

---

[289] *Ibid.*

[290] http://www.politics.co.uk/reference/aviation-noise

[291] *Ibid.*

> social conflicts, neurosis, hysteria, and psychosis. Population studies have suggested associations between noise and mental-health indicators, such as rating of well-being, symptom profiles, the use of psychoactive drugs and sleeping pills, and mental-hospital admission rates. Children, the elderly, and those with underlying depression may be particularly vulnerable to these effects because they may lack adequate coping mechanisms. Children in noisy environments find the noise annoying and report a diminished quality of life.[292]

In a 2001 Report issued by WHO Europe[293] it is said that the evidence from epidemiological studies on the association between exposure to road traffic and aircraft noise and hypertension and ischemic heart disease has increased during recent years. Road traffic noise has been shown to increase the risk of ischemic heart disease, including myocardial infarction. Both road traffic noise and aircraft noise increase the risk of high blood pressure. Very few studies exist regarding the cardiovascular effects of exposure to rail traffic noise.[294]

The World Medical Association (WMA), in a Statement issued in 1992 has stated that noise affects people in diverse ways pertaining to hearing, the vegetative nervous system, the psyche, spoken communication, sleep and performance. WMA states further that the stress caused by noise on humans places, an increased burden on the body leading to higher energy consumption and greater wear. WMA suspects that noise can primarily favour diseases in which stress plays a contributory role, such as cardiovascular diseases, which can then be manifested in the form of hypertension, myocardial infarction, angina pectoris, or even apoplexy, not to mention in the psychosocial aspects of human life. In its Statement WMA inter alia called upon ministers of transport and urban planners to develop alternative concepts that are capable of countering the growing level of environmental noise pollution, while advocating appropriate statutory regulations for combating environmental noise pollution.[295]

2.5.2 *Climate Change, Aviation and Human Rights*

Aircraft emit gases and particles directly into the upper troposphere and lower stratosphere where they have an impact on atmospheric composition. These gases and particles alter the concentration of atmospheric greenhouse gases, including carbon dioxide (CO_2), ozone (O_3), and methane (CH_4); trigger formation of condensation trails (contrails); and may increase cirrus cloudiness, all of which lead to climate change.[296]

---

[292] http://www.medscape.com/viewarticle/554566_3

[293] *Burden of disease from environmental noise - Quantification of healthy life years lost in Europe*, ISBN:978 92 890 0229 5.

[294] *Id. Executive Summary* at XV.

[295] http://www.wma.net/en/30publications/10policies/n3/.

[296] See IPCC Report, *infra*, note 303 (this chapter).

The effects of climate change through air pollution is already well know and addressed in depth. The effects of aviation and climate change and work carried out in that respect have also been addressed quite regularly in all its nuances.[297] However, the following discussion is on the effects of greenhouse gas emissions (which are also emitted by aircraft engines) on human rights.

On 28 March 2008 the United Nations Human Rights Council adopted Resolution 7/23 which *inter alia* decided to request the Office of the United Nations High Commissioner for Human Rights, in consultation with and taking into account the views of States, other relevant international organizations and intergovernmental bodies including the Intergovernmental Panel on Climate Change and the secretariat of the United Nations Framework Convention on Climate Change, and other stakeholders, to conduct, within existing resources, a detailed analytical study on the relationship between climate change and human rights, to be submitted to the Council prior to its tenth session. The following year, in March 2009, the Council adopted Resolution 10/4 which decided to hold a panel discussion on the relationship between climate change and human rights at its 11th session in order to contribute to the realization of the goals set out in the Bali Action Plan[298] and to invite all relevant stakeholders to participate therein.

In October 2011 the United Nations Human Rights Council adopted Resolution 18/22 which, while expressing its concern that, while implications of climate change affect individuals and communities around the world, the effects of climate change will be felt most acutely by those segments of the population that are already in vulnerable situations owing to factors such as geography, poverty, gender, age, indigenous or minority status and disability, and recognizing that climate change is a global problem requiring a global solution, and that effective international cooperation to enable the full, effective and sustained implementation of the United Nations Framework Convention on Climate Change in accordance with the provisions and principles of the Convention is important in order to support national efforts for the realization of human rights implicated by climate change-related impacts, affirmed that human rights obligations, standards and principles have the potential to inform and strengthen international and national policymaking in the area of climate change, promoting policy coherence, legitimacy and sustainable outcomes The Council reiterated its concern that climate change poses an immediate and far-reaching threat to people and communities around the world and has adverse implications for the full enjoyment of human rights and requested the Office of the United Nations High Commissioner for Human Rights: to convene, prior to the 19th session of the Human Rights Council, a seminar on addressing the adverse impacts of climate change on the full enjoyment of human rights, with a view to following up on the call for respecting human rights in all climate change-related actions and policies, and forging stronger interface and cooperation between

[297] See Abeyratne (2009). Also by the same author, Abeyratne (2011) and Abeyratne (2014).

[298] See Report of the Conference of the Parties on its Thirteenth Session, held in Bali from 3 to 15 December 2007.

the human rights and climate change communities. Resolution 18/22 also requested that States and other relevant stakeholders, including academic experts, civil society organizations and representatives of those segments of the population most vulnerable to climate change, be invited to participate actively in the seminar.

In April 2012 the Human Rights Council adopted Resolution 19/10 which decided inter alia to appoint for a period of 3 years an expert on human rights to study To study, in consultation with Governments, relevant international organizations and intergovernmental bodies, including the United Nations Environment Programme and relevant multilateral environment agreements, human rights mechanisms, local authorities, national human rights institutions, civil society organizations, including those representing indigenous peoples and other persons in vulnerable situations, the private sector and academic institutions, the human rights obligations, including non-discrimination obligations, relating to the enjoyment of a safe, clean, healthy and sustainable environment; to identify, promote and exchange views on best practices relating to the use of human rights obligations and commitments to inform, support and strengthen environmental policymaking, especially in the area of environmental protection, and, in that regard to prepare a compendium of best practices.

ICAO has produced its *Airport Air Quality Manual*[299] contains advice and practical information to assist ICAO Member States in implementing best practices with respect to airport-related air quality. Information related to State requirements, emissions from airport sources, emissions inventories and emissions allocation are addressed throughout the document. It also provides a process for States to determine the best approaches and analytical frameworks for assessing airport-related air quality and identifies best practices for different needs or scenarios. It is not intended as a basis for any regulatory action, it does not describe specific projects or actions nor does it address research-related aspects of airport air quality. The Manual contains such key recommendations on such subjects as best practices in achieving to the best possible extent clean air quality in the vicinity of airports as well as the maintenance of emissions inventories.

The Economist of January 19–25 2013 reported that, although international attempts at achieving global consensus that could result in a multilateral international treaty on cutting emissions and compensating victims are constantly ending in failure, local legislation per country is increasing. The journal claims that "according to a new study of 33 countries for GLOBE, a group of legislators from round the world, the number of new domestic environmental laws rose from 10 a year in the early 2000s to 20 in 2012. . .31 of the 33 countries now have a basic greenery law. South Korea passed legislation to set up an emissions–trading scheme in 2015. Japan introduced a carbon tax and a new law to encourage denser, more energy efficient cities". The Study concludes that half the 33 countries made significant strides in environmental legislation in 2012, prompting Sam Fankhauser of the London School of Economics, who was involved in writing the results of the

[299] *Doc 9889*, First Edition 2011.

Study, to say that the figures debunk the views of skeptics who reject the approach of countries acting alone in taking steps to counter the ill effects of climate change.

The article concludes that voters appear to favor domestic legislation to an international treaty on emissions and that this might be an indictment of years of sustained green activism for a global treaty.

So why has this activism failed? There could be many reasons. The first could be that we consider emissions as a problem that should be solved at home, and many countries are practicing self help because they can tailor the solution according to their individual problem. The problems of pollution in China and Australia may not necessarily be those of a majority of European Countries. The problems in the United States, if globally addressed, may hinder industry. Above all, the views on emissions are still polarized because emissions issues are intertwined with industry and development, human rights and health, together with and complex concepts such as Common But Differentiated Responsibilities.

Obviously, we still do not have a handle on the legal reason for establishing global principles. Most view the environment in its simplistic sense—that it is the overall summation of all things natural. In this context, as Nobel laureate Amartya Sen observes, the pervasive common view is that this "state of nature" will remain without change as long as we interfere with it as little as possible. Sen further states that this misconception can be rejected on two counts, the first being that the value of the environment does not lie in its existing state but in the opportunities it offers humankind, and the second being that it is not sufficient to ensure that the environment is passively preserved by us but there needs to be active initiatives in educating the populace of the world on the environment and the benefits that would accrue to it by our actions such as reducing the population of the world and creating employment opportunities. In this context, education is a paramount factor that would make us more environmentally conscious. I would add a third factor to Sen's *au fait* treatise, that economic instruments which are brought to bear with a view to reducing pollution, particularly in the context of global warming, would be essential to meet the reality that, no matter what we might do to avoid interference with the environment by our enforced inactivity, the growth of world trade would force us to impose economic measures to curb greenhouse gas emissions. The issue is whether such measures could be enshrined in a globally applicable regime.

Over the past two decades environmental scientists have set the climate sensitivity parameter (Cs) at a range of 1.5–4.5 °C equilibrium warming for a 100°% increase of atmospheric concentrates of carbon dioxide from pre industrial levels. The Inter Governmental Panel on Climate Change (IPCC)'s international review carried out in 2001 and its Third Assessment Report (TAR) which followed did not alter this benchmark, although the TAR indicated increased expected realized warming by the year 2100. Policy was established on the issue of climate change and global warming at the 1992 Rio Earth Summit when 150 States Parties agreed to the United Nations Framework Convention on Climate Change (UNFCC). Although the UNFCC did not set hard targets for emissions, there were two following implementing conferences in Berlin and Kyoto in 1995 and 1997 respectively. The 1997 Kyoto Protocol set quantitative emissions ceilings for developed nations while abstaining from setting such ceilings for developing nations. By March 2001, 84 nations had signed the Kyoto Protocol (with the notable exception of the United States) and by November 2003, 120 states had ratified the Protocol. The United States policy in refusing to accede to the Protocol, conveyed by President Bush's refusal to sign the Protocol in 2001, has been based on the argument that scientific reasoning that led to the Kyoto Protocol was uncertain and implementation of the Protocol's policy would be too expensive. Another reason was that the United States strongly believes that there must be commitments by developing countries towards achieving emissions ceilings. In early December 2003, Russian President Putin also indicated that he did not intend to make Russia a party to the Protocol. The policy of the European Union toward the Kyoto Protocol was initially enthusiastic, where the EU stated in October 1990 that it would control and constrain emissions in 2000 to be consistent with 1990 levels. However, in 1992 the EU indicated that its carbon tax policy will be enforced toward the above end only if all OECD States (including United States and Japan) ratified the Protocol.

The Protocol was adopted on 11 December 1997 in Kyoto, Japan, and entered into force on 16 February 2005. As of September 2011, 191 States had signed and ratified the protocol. The United States signed but did not ratify the Protocol and Canada withdrew from it in 2011.

The most compelling policy issue with regard to a rational analysis of global warming lies in attempting to discount costs and benefits of the future with a view to achieving comparable values for current policy judgements. The underlying recognition by the global community is that comparably expensive measures may have to be taken at the present time to ensure sustainability of a stable climate in the future. A known factor in this equation is that it takes at least three decades for the impact of emissions to be felt, due to ocean thermal lag and, perhaps more importantly, due to the fact that emissions are recurrent annually over a span of centuries owing to the pervasive residues of carbon dioxide in the atmosphere for long periods of time.

It must be noted that the underlying principle of the Kyoto Protocol is to have industrial and transition economies cut emissions back to 5 % below their 1990 levels and freeze them at that level, while allowing developing States the luxury of unlimited emissions. This philosophy is fundamentally flawed if one were to take into consideration the increasing gaseous emission levels from the developing countries which could seriously endanger global abatement. Another perceived inadequacy of the Kyoto Protocol is its inability to sustain consistency between its provisions on emissions trading which violates the least-cost solution of cutting emissions aimed at equating the marginal cost of cutbacks across all countries.

The issue is not what happens to Kyoto after 2012 or whether to place monetary caps and values on emissions. The IMF/World Bank reports that additional needs in developing countries to limit the rise of the global mean temperature to +2 C above pre industrial levels will continuously grow over the next decades to reach US$139 billion to $175 billion annually by 2030. Therefore, no one would doubt that mitigating the adverse effects of climate change is a costly business. It is here that the deficit of judgment occurs, where the main focus seems to be on the generation of revenue as compensation for pollution. Any such revenue collected should be based firstly on external factors such as trading on the quantity of emissions rather than internally on the quantity of consumables such as fossil fuels. Secondly, the revenue collected should be used on modernizing technology rather than on buttressing national coffers. Greener machines and engines should be developed that would lesson global engine emissions, and resources should be pumped into developing the necessary technology for the manufacture of such equipment. More sophistication is needed in the design of industry components towards this goal.

The key response to climate change should be investment, where investors accept the bulk of the scientific evidence available and invest in companies that would offer advanced technology in the field. The overall opportunity for investment is huge in that the value of low carbon energy markets would be around $500 billion by 2050 and cumulative net savings from energy efficient products by 2012 in the United States alone would be $84 billion. Investors will look at a long term model of investment, taking into account the importance of low carbon technologies; carbon prices; environmental actions by corporations and governmental environmental policies and regulations.

Internationalization of environmental law and seeking global consensus on emissions is indeed a noble approach. However, one must take in to consideration the fact that a multilateral treaty at public international law would entail State responsibility. Would such a treaty enforce penalties on States after their obligations are breached? This would be counter- productive from an environmental sense. What solution would there be after a species or ecosystem is lost? Could State responsibility be enforced in complex situations affecting global warming? Would restitution or compensation work in instances of environmental damage? How would States be held responsible for emissions of private entities? Perhaps Professor Ian Brownlie, who takes the example of the Chernobyl disaster of 1986 offers some food for thought: "Chernobyl and its aftermath casts doubts on the efficacy of the approach to environmental disasters by way of State responsibility. States clearly did not regard the legal approach as being especially relevant. The State responsibility, or liability approach is about allocation of losses and reparation. It is thus retrospective. In the case of the protection of the environment it is prospective and prevention action is called for".[300]

Have we got the whole equation upside down?

The Stern Review on the Economics of Climate Change, released in 2006 states that the ongoing proliferation of greenhouse gases will pose immense challenges to the well being of humanity, stating that "the stock of greenhouse gases could more than treble by the end of the century, giving at least a 50 % risk of exceeding 5 °C global average temperature change during the following decades. This would take

[300] *Stern Review: Economics of Climate Change - 2010*. See http://webarchive.nationalarchives.gov.uk/+/www.hm-treasury.gov.uk/sternreview_index.htm.

humans into unknown territory. An illustration of the scale of such an increase is that we are now only around 5 °C warmer than in the last ice age." The key points of the Stern Review are as follow:

All countries will be affected by climate change, but the poorest countries will suffer earliest and most.

Average temperatures could rise by 5C from pre-industrial levels if climate change goes unchecked.

Warming of 3 or 4C will result in many millions more people being flooded. By the middle of the century 200 million may be permanently displaced due to rising sea levels, heavier floods and drought.

Warming of 4C or more is likely to seriously affect global food production.

Warming of 2C could leave 15–40 % species facing extinction.

Before the industrial revolution level of greenhouse gases in the atmosphere was 280 parts per million (ppm) CO_2 equivalent (CO_2e); the current level is 430 ppm CO_2e. The level should be limited to 450–550 ppm CO_2.

Anything higher would substantially increase risks of very harmful impacts. Anything lower would impose very high adjustment costs in the near term and might not even be feasible.

Deforestation is responsible for more emissions than the transport sector.

Climate change is the greatest and widest-ranging market failure ever seen.

2.5.2.1 Recommended Actions

Three elements of policy are required for an effective response: carbon pricing, technology policy and energy efficiency.

Carbon pricing, through taxation, emissions trading or regulation, will show people the full social costs of their actions. The aim should be a global carbon price across countries and sectors.

Emissions trading schemes, like that operating across the EU, should be expanded and linked.

Technology policy should drive the large-scale development and use of a range of low-carbon and high-efficiency products.

Globally, support for energy research and development should at least double; support for the deployment of low-carbon technologies should be increased my up to five times.

International product standards could be introduced.

Large-scale international pilot programmes to explore the best ways to curb deforestation should be started very quickly.

Climate change should be fully integrated into development policy, and rich countries should honour pledges to increase support through overseas development assistance.

International funding should support improved regional information on climate change impacts.

International funding should go into researching new crop varieties that will be more resilient to drought and flood.

2.5.2.2 Economic Impacts

The benefits of strong, early action considerably outweigh the costs.

Unabated climate change could cost the world at least 5 % of GDP each year; if more dramatic predictions come to pass, the cost could be more than 20 % of GDP.

The cost of reducing emissions could be limited to around 1 % of global GDP; people could be charged more for carbon-intensive goods.

Each tonne of CO_2 we emit causes damages worth at least $85, but emissions can be cut at a cost of less than $25 a tonne.

Shifting the world onto a low-carbon path could eventually benefit the economy by $2.5 trillion a year.

By 2050, markets for low-carbon technologies could be worth at least $500 billion.

What we do now can have only a limited effect on the climate over the next 40 or 50 years, but what we do in the next 10–20 years can have a profound effect on the climate in the second half of this century.[301]

Later, in an interview given to *The Guardian,* Lord Stern, the author of the Stern Review said that he had under-estimated the danger of global warming and the dangers posed to humanity and the global economy, which is even worse than is reflected in his Review of 2006.[302] Lord Stern is reported to have told *The Guardian:* "This is potentially so dangerous that we have to act strongly. Do we want to play Russian roulette with two bullets or one? These risks for many people are existential."[303]

---

[301] As reported in *the Guardian*. See http://www.theguardian.com/politics/2006/oct/30/economy.uk.

[302] See http://www.theguardian.com/environment/2013/jan/27/nicholas-stern-climate-change-davose.

[303] *Ibid.* An Inter-Governmental Panel on Climate Change Report, on Aviation and Climate Change (IPCC) released in 1999 states that aircraft emissions were calculated to have increased NO_x at cruise altitudes in northern mid-latitudes by approximately 20 %. The uncertainty in this calculation is primarily related to uncertainties in the NO_x chemical lifetime and in the relative magnitude of the aircraft source compared to lightning, rapid vertical convection of surface NO_x, and other sources of upper tropospheric NO_x. The calculated increase was substantially smaller than the observed variability in NO_x. NO_x emissions from aircraft were calculated to have increased ozone by about 6 % in the region 30–60°N latitude and 9–13 km altitude. Calculated total ozone column changes in this latitude range are approximately 0.4 %. Calculated effects are substantially smaller outside this region. Some of the uncertainty in these calculations is captured by the range of model results. However, the models were notably deficient in coupling representations of stratospheric and tropospheric chemistry and in describing exhaust plume processes, HO_x sources, and non-methane chemistry in the upper troposphere. In addition, there was high uncertainty associated with the model description of vertical and horizontal transport in the upper troposphere/lower stratosphere. The effect of current aircraft particle and particle precursor

The Fifth Assessment Report on climate change of the Inter-Governmental Panel on Climate Change (IPCC) released in 2014 reports that:

> Each of the last three decades has been successively warmer at the Earth's surface than any preceding decade since 1850. The period from 1983 to 2012 was *likely* the warmest 30-year period of the last 1400 years in the Northern Hemisphere, where such assessment is possible (*medium confidence*). The globally averaged combined land and ocean surface temperature data as calculated by a linear trend, show a warming of 0.85 [0.65 to 1.06] °C2 over the period 1880 to 2012, when multiple independently produced datasets exist.[304]

The ICAO Assembly, at its 38th Session held in 2013 adopted Resolution A38-18 which reiterated *inter alia* that ICAO should continue to take initiatives to promote information on scientific understanding of aviation's impact and action undertaken to address aviation emissions and continue to provide the forum to facilitate discussions on solutions to address aviation emissions; and emphasis should be on those policy options that will reduce aircraft engine emissions without negatively impacting the growth of air transport especially in developing economies.[305]

References

Abeyratne RIR (1997) The use of civil aircraft and crew for military purposes. Annals Air Space Law (Annales de Droit Arien et Spatial) XXII(II)1–23

Abeyratne R (2009) Aviation and the environment. Publish America, Baltimore. ISBN 1-60813-707-4

Abeyratne R (2010) Aviation security law. Heidelberg, Springer

Abeyratne R (2011) Aviation and the carbon trade. Nova Science, New York. ISBN 978-1-61761-990-8

Abeyratne R (2014) Aviation and climate change: in search of a global market based measure. Springer, Heidelberg. ISBN:978-3-319-08442-8 (Print) 978-3-319-08443-5 (Online)

Abey-Wickrama I et al (1969) Mental hospital admissions and aircraft noise. Lancet 2 (7633):1275-1277

Akehurst M (1977) The use of force to protect nationals abroad. Int Rel 5(3)

Akehurst M (1984) Humanitarian intervention. In: Bull H (ed) Intervention in world politics, p 95

Austin R (2010) Unmanned aircraft systems, UAVs design, development and deployment. Wiley, West Sussex

Ban H (1999) Aviation Daily. Thursday, 7 October 1999

Bone E, Bolkcom C (2003) Unmanned aerial vehicles: background and issues for congress, Washington/Congressional Research Service, 25 April 2003

emissions (i.e., soot, sulfur, and water) in the stratosphere on ozone was estimated to be smaller than, and of opposite sign to, the NO_x effect. Model representations of aerosol microphysics and chemistry were, however, largely incomplete. See Aviation and the Global Atmosphere, at https://www.google.ca/#q=ipcc+report+aviation+and+the+environment.

[304] http://www.ipcc.ch/report/ar5/syr/.

[305] Resolution A38-18 (Consolidated statement of continuing ICAO policies and practices related to environmental protection—Climate change) ICAO Assembly Resolutions in Force (as at 4 October 2013) Doc 10022, at 1–71.

Brownlie I (1963) International law and the use of force by states
Butler A (2013) Italy looking to develop black MALE UAV. Aviation Week & Space Technology, 20 May 2013. Online at http://www.aviationweek.com/Article.aspx?id=/article-xml/AW_05_20_2013_p26-578980.xml
Byman D (2013) Why drones work – the case for Washington's weapon of choice. Foreign Aff 92 (4):32–43
Capaccio T (2010) Pentagon to increase stock of high-altitude drones. *Bloomberg,* 4 February 2010. Online: BLOOMBERG http://www.bloomberg.com/apps/news?pid=newsarchive&sid=aUYvb_D1LBzM
Cheng B (1962) The law of International Air Transport. Oceania Publications, London
Cohen A (1968) Noise and psychological state. National Centre for Urban and Industrial Health, United States Public Health Service, Cincinnati
Cole C, Wright J (2010a) Campaign against military drones' peace news: issue 2519, March 2010. Online at <http://peacenews.info/node/3426/campaign-against-military-drones
Cole C, Wright J (2010b) What are drones? Drone Wars UK, January 2010. Online: http://dronewars.net/aboutdrone/
Crayston J, Hupe J (1999) ICAO facing complex and evolving challenges in the environmental field. ICAO J
Cronin AK (2013) Why drones fail. Foreign Aff 92(4):44–54
Danssaert P, Finardi S (2011) The arms flyers - commercial aviation, human rights and the business of war and arms, IPIS
de Lima FX (1971) Intervention in international law. Vitgeverij Pax Nederland, Den Haag
Esler D (1999) The latest noise about noise. Bus Commer Aviat 52
Evans B (2013) Terrifying video captures moment German drone missed Afghan plane carrying 100 passengers by just two meters, 3 June 2013. Mail Online, online at http://www.dailymail.co.uk/news/article-2335122/Terrifying-video-captures-moment-German-drone-missed-Afghan-plane-carrying-100-passengers-just-metre.html
Flint P (1999) Breaking the sound barrier. Air Transp World 3
Frey AR, Kinsler LE (1982) Fundamentals of acoustics, 3rd edn. Wiley, Chichester
Friedman N (2010) Unmanned combat air systems: a new kind of carrier aviation. Naval Institute Press, Maryland
ICAO (Secretary General) (2011) Unmanned aircraft systems (UAS). Circular Cir 328 AN/190. ICAO, Montreal
Grey J (1975) Noise, Noise, Noise. The Westminster Press, Philadelphia
Harper DV (1988) Regulation of aircraft noise at major airports: past, present and future. Transp Law J 17
Harthy C (2013) France Requests $1.5B General Atomics Reaper UAV Sale, 2 July 2013. Online at http://www.suasnews.com/2013/07/23626/france-requests-1-5b-general-atomics-reaper-uav-sale/
Hyde CC (1945) International law, vol 1 (2nd Review)
International Committee of the Red Cross (1987) International Red Cross handbook, 12th edn. Martinus Nijhoff, Geneva
Joyner J, Was Afghanistan worth it? The National Interest. http://nationalinterest.org/commentary/was-afghanistan-worth-it-8182
Kahlili R (2013) Iran gives stealth drone secrets to Russia. WND, 10 February 2013. Online at http://www.wnd.com/2013/02/iran-gives-stealth-drone-secrets-to-russia/
Kryfer KD (1985) The effects of noise on man. Academic, Toronto
Laing A (2013) US drone 'shot down by al-Shabaab in Somalia'. The Telegraph, 29 May 2013. Online at http://www.telegraph.co.uk/news/worldnews/al-qaeda/10086660/US-drone-shot-down-by-al-Shabaab-in-Somalia.html
Lazarski A (2002) Lt Col USAF, legal implications of the uninhabited combat aerial vehicle. Aerospace Power J. Online: http://www.airpower.maxwell.af.mil/airchronicles/apj/apj02/sum02/lazarski.html

Lillich RB (ed) (1973) Humanitarian intervention and the United Nations. University Press of Virginia, Charlottesville
Lowe AV, Warbrick C (ed) (1993) Current developments: international law. Int Comp Law Q 42 (4)
Milde M (2008) Rendition Flights' and international law, pp 477–486. Online at www.redress.org/downloads/publications/Prof_Dr_Micheal_Milde_for_REDRESS_June_2008_2_.pdf
Milde M (2012) International air law and ICAO, 2d edn. Eleven International Publishing, Hague
Mortimer LF (1992) Ambitious programme of future work to be undertaken by CAEP. ICAO J
Nekhai O (2013) State-of-the-art UAVs to enter service in the Russian Army. The Voice of Russia, 14 June 2013. Online at http://voiceofrussia.com/2013_06_14/State-of-the-art-UAVs-to-enter-service-in-the-Russian-Army-0448/
Nelson PM (ed) (1987) Transportation in noise reference book. Butterworth, London
Newman RJ (2002) The little predator that could. Air Force Magazine, March 2002. Online at http://www.airforcemag.com/MagazineArchive/Pages/2002/March%202002/0302predator.aspx
Nichols M (2013) Italian firm to provide surveillance drone for U.N. in Congo. Reuters, 1 August 2013. Online at http://www.reuters.com/article/2013/08/01/us-congo-democratic-un-drones-idUSBRE97014720130801
O'Keeffe B (2000) The new communications, navigation, surveillance and air traffic management system guides aviation into the 21st century. J Aviat Manag 133–148
Oppenheim (1955) International law, vol 1. Lauterpacht
Peterson ME (2005) The UAV and the current and future regulatory construct for integration into the national airspace system. LL.M. Thesis, McGill University Institute of Air and Space Law
Rayleigh L (1945) The theory of sound, 2nd edn. Dover, London
Roberts K (2013) When the whole world has drones. Natl J. Online: http://www.nationaljournal.com/magazine/when-the-whole-world-has-drones-20130321
Ruwantissa I, Abeyratne R (2002) International responsibility in preventing the spread of communicable diseases through air carriage - the SARS crisis. Transp Law J 30(1):53–80
Schmitt E, Schmidt M (2012) U.S. drones patrolling its skies provoke outrage in Iraq. New York Times, 29 January 2012. Online at http://www.nytimes.com/2012/01/30/world/middleeast/iraq-is-angered-by-us-drones-patrolling-its-skies.html?pagewanted=all&_r=0
Schwarzenberger G (1971) International law and order. Stevens and Sons, London
Shachtman N (2009) CIA chief: drones 'Only Game in Town' for Stopping Al-Qaeda. Wired, 19 May 2009. Online: WIRED http://www.wired.com/dangerroom/2009/05/cia-chief-drones-only-game-in-town-for-stopping-al-qaeda/
Shaw MN (2003) International law, 5th edn. Cambridge University Press, Cambridge
Simpson G (2007) Law, war & crime, - war crimes trials and the reinvention of international law. Polity Press, Cambridge
Smith LJ (1895) Shelfer v. City of London Electric Lighting Company. 1 Ch. 287
Smith JM (2012) Drone "blowback" is real A new analysis finds five ways drone strikes in Yemen are hurting American interests. Salon, 5 September 2012. Online at http://www.salon.com/2012/09/05/drone_blowback_is_real/
Solis G (2007) Targeted killing and the law of armed conflict. Naval War Coll Rev 60(2)
Staff (2004) Commercial use of UAVs – Widespread in Japan. In Yearbook: UAVs global perspective. UVS International, Blyenburgh& CO., p 138
Starke JG (1977) An introduction to international law, 7th edn. Butterworths, London
Stein PB (1991) The price of success: mitigation and litigation in airport growth. J Air Law Commer 57
Teson FR (1956) Humanitarian intervention: an inquiry into law and morality. Transnational Publishers Inc., Dobbs Ferry, New York
The Law of the Sea (1983) Original text of the United Nations Convention on the Law of the Sea, all Annexes and Index. United Nations, New York
Vincent RJ (1974) Non-intervention and international order. Princeton University Press, New Jersey

Wagner M, Moore A, Aryel R (ed) (2006) Handbook of bio surveillance. Elsevier, New York

Walcott J, Paztor A (1987) CIA role in arming Contras seen larger than White House has acknowledged. Wall Street J

Warrick J, Finn P (2010) Amid outrage over civilian deaths in Pakistan, CIA turns to smaller missiles. Washington Post, 26 April 2010. Online at http://www.washingtonpost.com/wp-dyn/content/article/2010/04/25/AR2010042503114.html

Wehberg H (1938) La Guerre Civil et le Droit International. Hague Recueil 63

Weishaupt G (ed) (1979) Selected international agreements relating to air law

Chapter 3
Rights of the Passenger

3.1 The Disabled Passenger

3.1.1 Access to the Airport

In 1976 the World Health Organization handed down its definitions of the three words "disabilities" "impairment" and "handicap". These definitions brought to bear a distinct correlation between the words as flowing from one to the other. According to the WHO definitions 'an **impairment** is any loss or abnormality of psychological, physiological or anatomical structure or function; a **disability** is any restriction or lack (resulting from an impairment) of ability to perform an activity in the manner or within the range considered normal for a human being; and, a **handicap** is a disadvantage for a given individual, resulting from an impairment or a disability, that prevents the fulfillment of a role that is considered normal (depending on age, sex and social and cultural factors) for that individual'.[1]

[1] http://www.aarogya.com/support-groups/disability/the-un-definition-of-disability.html. However, according to activists in the disability movement, the World Health Organization has confused between the terms '**disability**' and '**impairment**'. They maintain that impairment refers to physical or cognitive limitations that an individual may have, such as the inability to walk or speak. In contrast, **disability** refers to socially imposed restrictions, that is, the system of social constraints that are imposed on those with impairments by the discriminatory practices of society. Thus, the Union of the Physically Impaired Against Segregation has defined impairment and disability as: an '**impairment** is lacking part of or all of a limb, or having a defective limb, organ or mechanism of the body'. '**Disability** is the disadvantage or restriction of activity caused by contemporary organizations which take no or little account of people who have physical impairments and thus excludes them from the mainstream of social activities'. According to the United Nations Standard Rules on the Equalization of Opportunities for Persons with disabilities, the term "**disability**" summarizes a great number of different functional limitations occurring in any population, in any country of the world. People may be disabled by physical, intellectual or sensory impairment, medical conditions or mental illness. Such impairments, conditions or illnesses may be permanent or transitory in nature.

R. Abeyratne, *Aviation and International Cooperation*,
DOI 10.1007/978-3-319-17022-0_3

Therefore, one could deduce that a passenger with disabilities in an aircraft would have a restriction or lack (resulting from an impairment) of ability to perform an activity in the manner or within the range considered normal for a human being—through a loss or inability and that such persons would be handicapped. Accordingly passengers in an aircraft in flight[2] including those: with ambulatory challenges; who are blind or mute; and who need support to consume food or drink, would be considered disabled.

The European Union defines a disabled passenger as any person whose mobility when using transport is reduced due to any physical disability (sensory or locomotors, permanent or temporary), intellectual disability or impairment, or any other cause of disability, or age, and whose situation needs appropriate attention and the adaptation to his or her particular needs of the service made available to all passengers.[3]

Provisions covering the carriage by air of the disabled passenger are contained in Annex 9 to the Chicago Convention on facilitation of air transport.[4] Regrettably, all two the provisions under Chapter 8 dedicated to the transport of disabled persons are Recommended Practices and not Standards. Recommended Practice 8.22 provides that when travelling, persons with disabilities should be provided with special assistance in order to ensure that they receive services customarily available to the general public. Assistance should be provided in a manner that respects the dignity of the individual. The following provision calls upon Contracting States to cooperate with a view to taking the necessary measures to make accessible to persons with disabilities all the elements of the chain of the person's journey, from arrival at

[2] An aircraft is considered to be in flight at any time from the moment when all its external doors are closed following embarkation until the moment when any such door is opened for disembarkation; in the case of a forced landing, the flight is deemed to continue until the competent authorities take over the responsibility for the aircraft and for persons and property on board. See *Convention on the Suppression of Unlawful Interference Relating to International Civil Aviation*, done at Beijing on 10 September 2010. This document could be accessed at http://www.icao.int/icao/en/leb/treaty.htm.

[3] Regulation (EC) No 1107/2006 of the Parliament and of the Council of 5 July 2006 concerning the rights of disabled persons and persons with reduced mobility when travelling by air. Article 2.

[4] Annex 9 defines a person with a disability as "Any person whose mobility is reduced due to a physical incapacity (sensory or locomotor), an intellectual deficiency, age, illness or any other cause of disability when using transport and whose situation needs special attention and the adaptation to the person's needs of the services made available to all passengers". The disabled passenger is not defined in Annex 9 to the Chicago Convention. The *ICAO Manual on Access to Air Transport by Persons with Disabilities* (Doc 9984) adopts the definition. The World Health Organization (WHO) regards "disabilities" as an umbrella term, covering impairments, activity limitations, and participation restrictions. WHO states that an impairment is a problem in body function or structure; an activity limitation is a difficulty encountered by an individual in executing a task or action; while a participation restriction is a problem experienced by an individual in involvement in life situations. According to WHO, disability is thus not just a health problem. It is a complex phenomenon, reflecting the interaction between features of a person's body and features of the society in which he or she lives. Overcoming the difficulties faced by people with disabilities requires interventions to remove environmental and social barriers.

the airport of departure to leaving the airport of destination. Recommended Practice 8.24 suggests that Contracting States should take the necessary steps with aircraft, airport and ground handling operators to establish and publish minimum uniform standards of accessibility with respect to transportation services for persons with disabilities, from arrival at the airport of departure to leaving the airport of destination.

Furthermore, as per Recommended Practice 8.25, Contracting States are requested to take the necessary steps with aircraft, airport and ground handling operators and travel agencies to ensure that persons with disabilities are given the information they need, in formats that are accessible to those with cognitive or sensory disabilities, and should take the necessary steps to ensure that airlines, airports and ground handling operators are in a position to give those passengers the assistance necessary for them, depending on their needs, to help them in their travel. The following provision recommends that Contracting States take all necessary steps to secure the cooperation of aircraft, airport and ground handling operators in order to establish and coordinate training programmes to ensure that trained personnel are available to assist persons with disabilities.

Standard 8.27 requires Contracting States to take the necessary steps to ensure that airport facilities and services are adapted to the needs of persons with disabilities. The following Recommended Practices—From 8.28 to 8.33 request that Contracting States ensure that lifting systems or any other appropriate devices are made available in order to facilitate the movement of persons with disabilities between the aircraft and the terminal on both arrival and departure as required where telescopic passageways are not used. Furthermore, that measures should be taken to ensure that the hearing- and vision-impaired are able to obtain flight service-related information in accessible formats. Designated points for the pick-up and drop-off of persons with disabilities at a terminal building should be located as close as possible to main entrances and/or exits. To facilitate movement within the airport, access routes should be free of obstacles and be accessible. Where access to public services is limited, every effort should be made to provide accessible and reasonably priced ground transportation services by adapting current and planned public transit systems or by providing special transport services for people with mobility needs. Adequate parking facilities should be provided for people with mobility needs and appropriate measures taken to facilitate their movement between parking areas and the terminal buildings. When assistance is provided to transfer persons with disabilities from one aircraft to another, it should be provided as efficiently as possible, with due regard for connecting flights.

3.1.2 Access to Air Services

The 2006 *United Nations Convention on the Rights of Persons with Disabilities* states *in limine* that the intent of the Convention is to promote, protect and ensure the full and equal enjoyment of all human rights and fundamental freedoms by all persons with disabilities, and to promote respect for their inherent dignity. Persons with disabilities include those who have long-term physical, mental, intellectual or sensory impairments which in interaction with various barriers may hinder their full and effective participation in society on an equal basis with others. The basic principles of the Convention are couched in Article 3 which states that Respect for inherent dignity, individual autonomy including the freedom to make one's own choices, and independence of persons; non-discrimination; full and effective participation and inclusion in society; respect for difference and acceptance of persons with disabilities as part of human diversity and humanity; equality of opportunity; accessibility; equality between men and women; and respect for the evolving capacities of children with disabilities and respect for the right of children with disabilities to preserve their identities.

Article 5 of the Convention requires all States Parties to States Parties recognize that all persons are equal before and under the law and are entitled without any discrimination to the equal protection and equal benefit of the law. States Parties are also required to prohibit all discrimination on the basis of disability and guarantee to persons with disabilities equal and effective legal protection against discrimination on all grounds. In order to promote equality and eliminate discrimination, States Parties shall take all appropriate steps to ensure that reasonable accommodation is provided. The Convention does not consider as discrimination specific measures which are necessary to accelerate or achieve de facto equality of persons with disabilities.

In the context of air transport, Article 9 of the Convention is the most relevant and applicable. It states that in order to enable persons with disabilities to live independently and participate fully in all aspects of life, States Parties are required to take appropriate measures to ensure to persons with disabilities access, on an equal basis with others, to the physical environment, to transportation, to information and communications, including information and communications technologies and systems, and to other facilities and services open or provided to the public, both in urban and in rural areas. These measures, which shall include the identification and elimination of obstacles and barriers to accessibility, shall apply to, inter alia buildings, roads, transportation and other indoor and outdoor facilities, including schools, housing, medical facilities and workplaces. Information, communications and other services, including electronic services and emergency services.

States Parties are also required to take appropriate measures to: develop, promulgate and monitor the implementation of minimum standards and guidelines for the accessibility of facilities and services open or provided to the public; ensure that private entities that offer facilities and services which are open or provided to the public take into account all aspects of accessibility for persons with disabilities; provide training for stakeholders on accessibility issues facing persons with disabilities; provide in buildings and other facilities open to the public signage in Braille and in easy to read and understand forms; provide forms of live assistance and intermediaries, including guides, readers and professional sign language interpreters, to facilitate accessibility to buildings and other facilities open to the public; promote other appropriate forms of assistance and support to persons with disabilities to ensure their access to information; promote access for persons with disabilities to new information and communications technologies and systems, including the Internet; and promote the design, development, production and distribution of accessible information and communications technologies and systems at an early stage, so that these technologies and systems become accessible at minimum cost.

Shifting the focus to Europe, Article 80 (2) of the *Treaty Establishing The European Community* provides that The European Council may, acting by a qualified majority, decide whether, to what extent and by what procedure appropriate provisions may be laid down for sea and air transport. Using this as a basis on 5 July 2006 the European Union adopted Regulation EC No. 1107/2006 concerning the rights of disabled persons and persons with reduced mobility when travelling by air.[5] The fundamental driver in the drafting of the regulation by the legislature was to ensure that disabled persons and persons with reduced mobility had access to air travel on an equal footing with passengers without mobility limitations. This objective is achieved by providing specific rights to assistance, by assigning corresponding obligations to providers and thereby protecting disabled persons and persons with reduced mobility from discrimination in exercising their rights while booking, purchasing and using air transport services. There was an overriding caveat that and fundamental purpose of flight safety for all passengers and crew members must ineluctably be observed.

This regulation recognizes that disabled persons and persons with reduced mobility[6] should be accepted for carriage and not refused transport on the grounds of their disability or lack of mobility, except for reasons which are justified on the grounds of safety and prescribed by law. Before accepting reservations from disabled persons or persons with reduced mobility, air carriers, their agents and tour operators should make all reasonable efforts to verify whether there is a reason which is justified on the grounds of safety and which would prevent such persons being accommodated on the flights concerned. In Article 3 the Regulation prescribes that an air carrier or its agent or a tour operator shall not refuse, on the grounds of disability or of reduced mobility: to accept a reservation for a flight

---

[5] Regulation (EC) No 1107/2006 concerning the rights of disabled persons and persons with reduced mobility when travelling by air ('the Regulation') was adopted by the European Parliament and the Council on 5 July 2006. In the last decade, The European Disability Forum was actively lobbying for legislative measures putting an end to this situation since several years. The European Commission has also been clear in their intention to ensure equal opportunities for air travel for disabled people. It first raised this question in 2000 in its Communication on Air Passengers in the European Union. In 2002, the European Commission published a consultation paper regarding the question on whether any legal measures should be taken on Community level or not.

[6] According to the Regulation, 'disabled person' or 'person with reduced mobility' means any person whose mobility when using transport is reduced due to any physical disability (sensory or locomotor, permanent or temporary), intellectual disability or impairment, or any other cause of disability, or age, and whose situation needs appropriate attention and the adaptation to his or her particular needs of the service made available to all passengers.

departing from or arriving at an airport to which this Regulation applies; and to embark a disabled person or a person with reduced mobility at such an airport, provided that the person concerned has a valid ticket and reservation.

Article 7 of the Regulation provides that when a disabled person or person with reduced mobility arrives at an airport for travel by air, the managing body of the airport shall be responsible for ensuring the provision of the assistance specified in Annex I in such a way that the person is able to take the flight for which he or she holds a reservation, provided that the notification of the person's particular needs for such assistance has been made to the air carrier or its agent or the tour operator concerned at least 48 h before the published time of departure of the flight. This notification shall also cover a return flight, if the outward flight and the return flight have been contracted with the same air carrier. Where use of a recognised assistance dog is required, this shall be accommodated provided that notification of the same is made to the air carrier or its agent or the tour operator in accordance with applicable national rules covering the carriage of assistance dogs on board aircraft, where such rules exist.

If no notification is made in accordance with paragraph 1, the managing body shall make all reasonable efforts to provide the assistance specified in Annex I in such a way that the person concerned is able to take the flight for which he or she holds a reservation. The above provisions of Article 7 apply on condition that: the person presents himself or herself for check-in: if no time is stipulated, not later than 1 h before the published departure time, or the person arrives at a point within the airport boundary designated in accordance with Article 5: at the time stipulated in advance and in writing (including by electronic means) by the air carrier or its agent or the tour operator, or if no time is stipulated, not later than 2 h before the published departure time.

Article 8 devolves responsibility on the airport by providing that the managing body of an airport shall be responsible for ensuring the provision of the assistance specified in Annex I without additional charge to disabled persons and persons with reduced mobility. The managing body may provide such assistance itself. Alternatively, in keeping with its responsibility, and subject always to compliance with the quality standards referred to in Article 9(1),[7] the managing body may contract with one or more other parties for the supply of the assistance. In cooperation with airport users, through the Airport Users Committee where one exists, the managing body may enter into such a contract or contracts on its own initiative or on request, including from an air carrier, and taking into account the existing services at the airport concerned. In the event that it refuses such a request, the managing body shall provide written justification. The managing body of an airport may, on a

[7] Article 9(1) states that with the exception of airports whose annual traffic is less than 150,000 commercial passenger movements, the managing body shall set quality standards for the assistance specified in Annex I and determine resource requirements for meeting them, in cooperation with airport users, through the Airport Users Committee where one exists, and organisations representing disabled passengers and passengers with reduced mobility.

non-discriminatory basis, levy a specific charge on airport users for the purpose of funding this assistance.

The charge referred to above is required to be reasonable, cost-related, transparent and established by the managing body of the airport in cooperation with airport users, through the Airport Users Committee where one exists or any other appropriate entity. It shall be shared among airport users in proportion to the total number of all passengers that each carries to and from that airport. The managing body of an airport is required by the Regulation to separate the accounts of its activities relating to the assistance provided to disabled persons and persons with reduced mobility from the accounts of its other activities, in accordance with current commercial practice.

Article 3 refers to the responsibilities of the carrier. It states that an air carrier or its agent or a tour operator shall not refuse, on the grounds of disability or of reduced mobility: to accept a reservation for a flight departing from or arriving at an airport to which this Regulation applies; to embark a disabled person or a person with reduced mobility at such an airport, provided that the person concerned has a valid ticket and reservation.

Article 10 requires that the carrier provide the assistance without additional charge to a disabled person or person with reduced mobility departing from, arriving at or transiting through an airport to which the Regulation applies provided that the person in question fulfils the conditions set out in Article 7(1), (2) and (4) as referred to above.

3.1.3 ICAO Provisions

Provisions on access to air services in Annex 9[8] are contained in Standard 8.34 and Recommended Practices 8.35–8.34 Contracting States shall take the necessary steps to ensure that persons with disabilities have equivalent access to 8.40.1. Standard 8.34 requires Contracting States to take the necessary steps to ensure that persons with disabilities have equivalent access to air services.

Recommended Practice 8.35 requests Contracting States to introduce provisions by which aircraft coming newly into service or after major refurbishment should conform, where aircraft type, size, and configuration permit, to minimum uniform standards of accessibility with respect to equipment on board aircraft which would include movable armrests, onboard wheelchairs, accessible washrooms and suitable

[8] ICAO first adopted measures to facilitate air travel by the elderly and disabled in 1968 when the Seventh Session of the Facilitation Division (FAL/7) adopted a Recommended Practice urging that assistance be rendered to invalid passengers in making a direct transfer from one aircraft to another. This recommendation now appears slightly changed as Recommended Practice 8.23 in Annex 9 (12th Edition) which provides that Contracting States should cooperate with a view to taking the necessary measures to make accessible to persons with disabilities all the elements of the chain of the person's journey, from beginning to end.

lighting and signs. This provision is followed by Recommended Practice 8.36 which recommends that disability aids required by persons with disabilities should be carried free of charge in the cabin where space, weight and safety requirements permit or should be carried free of charge and designated as priority baggage. Recommended Practice 8.37 suggests that service animals accompanying persons with disabilities should be carried free of charge in the cabin, on the floor at the person's seat, subject to the application of any relevant national or aircraft operator regulations. As per 8.38 Contracting States should notify ICAO promptly of restrictions on the transport of battery-powered devices, including mobility aids containing spillable batteries so that they can be included in Doc 9284, Technical Instructions for the Safe Transport of Dangerous Goods by Air and ensure that aircraft operators make such information publicly available and in accordance with Chapter 2, 2.5 of Annex 18. The following Recommended Practice adds that in principle, persons with disabilities should be permitted to travel without the requirement for a medical clearance. Aircraft operators should only be permitted to require persons with disabilities to obtain a medical clearance in cases of a medical condition where it is not clear that they are fit to travel and could compromise their safety or well-being or that of other passengers.

The last two Recommended Practices—8.40 and 8.40.1—add that in principle, persons with disabilities should be permitted to determine whether or not they need an assistant. If the presence of an assistant is required, Contracting States should encourage aircraft operators to offer discounts for the carriage of that assistant. Aircraft operators should require an assistant only when it is clear that the person with a disability is not self-reliant and this could pose a risk to safety or the well-being of such person or that of other passengers. Advance notice should strongly be encouraged where assistance or lifting is required.

3.1.3.1 Manual on Access to Air Transport by the Disabled Passenger

The ICAO *Manual on Access to Air Transport By the Disabled Passenger*—the first edition of which was released in 2013—begins by saying that Persons with disabilities make up a significant and growing percentage of the world's population and constitute the world's largest minority. The World Health Organization (WHO) reports that this number is increasing through population growth, medical advances and the ageing process.[9] WHO states:

[9] According to WHO, there are more than 1,000 million people with disabilities in the world, of whom between 110 million and 190 million experience significant difficulties. The total corresponds to about 15 % of the world's population and is higher than WHO's previous estimates, which date from the 1970s and suggested a figure of around 10 %. Furthermore, the prevalence of disability is growing because of ageing populations and the global increase in chronic health conditions. National patterns of disability are influenced by trends in health conditions and environmental and other factors—such as road traffic crashes, natural disasters, conflict, diet and substance abuse. Disability disproportionately affects vulnerable populations, in particular,

> Disability can occur at three levels: an impairment in body function or structure; a limitation in activity, such as the inability to read or move around; a restriction in participation, such as exclusion from school or work. As such, people with disabilities include those who are traditionally understood as disabled (for example wheelchair users, people who are blind or deaf or people with intellectual impairments), and people who experience difficulties in functioning due to a wide range of health conditions such as chronic diseases, severe mental disorders, multiple sclerosis and old age.[10]

The fundamental premise upon which ICAO has developed its Manual on disabled passengers is that all procedures forming part of an air travel journey, including reservations, check-in, immigration and customs, security clearances, transfers within airports, embarkation and disembarkation, departure, carriage and arrival should be adapted to the needs of persons with disabilities in order to facilitate the clearance and air transportation of such persons in a dignified manner.[11] This has to be accomplished by providing training for all staff in the chain of the air travel service delivery is vital for providing quality service to persons with disabilities in a consistent and respectful manner. It is essential that staff know their responsibilities and are able to perform them. Training should address the attitudinal, environmental/physical and organizational barriers that affect persons with disabilities in air transportation. Training should prepare staff to provide assistance to persons with disabilities in a manner that respects their dignity, and as a professional service to which the person is entitled, rather than as a favour or compassionate gesture.[12] For purposes of training, the categories of disability are identified in the Manual as physical, sensory, hidden and learning disabilities; mental health problems; cognitive impairments; persons who use mobility aids; persons with walking difficulties or limitations in balance, agility or coordination that affect their mobility; persons who are deaf or hard of hearing; persons who are blind or partially sighted; persons who are deaf or blind; persons with impaired speech; persons who require assistants and the roles of assistants; and persons travelling with a service animal, and the role and needs of that animal.[13]

Chapter 3 of the Manual elaborates on the assistance to be rendered to a disabled passenger prior to his/her journey, focusing on communication. The Manual recommends that States should publish, or encourage aircraft and airport operators or interested organizations to publish, general information or guidance material relating to air travel by persons with disabilities in accessible formats and technologies

---

women, older people and people that are poor. Low-income countries have a higher prevalence of disability than high-income countries. See WHO Executive Board EB 132/10, *Secretariat Report A66/12*, 132nd session, 30 November 2012, at 3.

[10] *Ibid.*

[11] See *Manual on Access to Air Transport by Persons with Disabilities* Doc 9984, First Edition: 2013 at 1.1.

[12] *Id.* 2.1.

[13] *Id.* 2.5.

appropriate to different kinds of disabilities, such as electronic, web-based, large-print and audio, in a timely manner and without additional cost.[14]

In Paragraph 3.2 the Manual recommends that aircraft and airport operators, ground handling operators, and travel agents should take necessary measures to inform those persons with disabilities planning to travel or in the course of travel of the availability of accessible services and of how to access them, whether or not the information is specifically requested. Information about the following should be made available to the public by airport and aircraft operators as applicable: (a) hours of operation; (b) location of designated parking areas; (c) location of drop-off and pick-up areas; (d) telephone numbers for accessibility information; (e) wheelchair or electric cart services; (f) location of relieving areas for service animals; (g) accessible inter-terminal transportation; (h) accessible ground transportation; (i) passes for non-travelling companions; (j) complaint resolution services; (k) advance notice requirements; (l) check-in and flight departure times; (m) requirements for the carriage of mobility aids, e.g. size; and (n) types of services available at the airport and in flight, including available boarding equipment.[15]

With regard to reservations, the Manual recommends, in paragraphs 3.6–3.10, for travel agents, tour operators and aircraft operator employees or contractors to actively enquire during booking transactions whether there will be anyone in the party who may require assistance at the airport and/or in flight. For bookings made online, there should be a place where the required assistance can be noted. If this is not possible and can only be done by telephone, a toll-free telephone number should be provided. When a reservation for a person with disabilities is being made, and such disability and the assistance required is made known to the aircraft operator, the latter should provide information to the person with disabilities on the assistance that it is required to provide, and the accessibility features of the aircraft or service limitations, whether or not such information is specifically requested. Aircraft operators should request the dimensions of any mobility aids at the time of booking to ensure that if they cannot carry a mobility aid, or if a mobility equipment requires special handling (e.g. a powered wheelchair that requires the battery to be isolated), such information is communicated as soon as possible to passengers, in order for them to make alternative arrangements, if needed. If a change in aircraft before a flight's departure results in the aircraft operator not being able to accommodate a passenger's wheelchair in the cargo hold, the aircraft operator should offer the passenger alternative transportation at a different time or refund the fare where there is no alternative. Agents should advise passengers on the limits of an aircraft operator's liability regarding the carriage of mobility aids, so that persons with disabilities can consider taking out their own insurance. Agents should also advise

---

[14] The Manual adds that web-based material and internet bookings should be accessible to persons with disabilities in accordance with international web accessibility standards found at www.w3.org/standards/webdesign/accessibility. *Id.* 3.1.1.

[15] *Id*, 3.3.

passengers of the current status of safety, security, and hazardous material restrictions, as these could impact on the carriage of medical devices or mobility aids.

Paragraphs 3.16–3.19 of the Manual are important, as the ensuing discussion on fares charged of disabled passengers with regard to accompanying persons who assist a disabled passenger throughout the flight reflects. The Manual recommends that aircraft operators should accept the self-assessment persons with disabilities as to whether they are able to travel independently and attend to their own needs on board an aircraft. However, in some circumstances, for reasons of safety, an aircraft operator may determine that it is necessary for a person with disabilities to travel with an assistant (e.g. to understand and respond to safety briefings when no alternate forms of communication can accomplish this, or to assist in an emergency evacuation). The Manual also states in this context that as aircraft operators do not provide personal care (e.g. assistance in the washroom or with eating), some persons with disabilities who require such care will need to travel with an assistant who can provide personal care during the flight. However, an aircraft operator's concern that a person with disabilities may need personal care should not be reason enough for it to impose a requirement to travel with an assistant; rather, a person's self-assessment of his or her personal care needs should be accepted. While consideration should be given to offering discounted rates or a free seat to assistants, the Manual also provides that an assistant travelling with a person with disabilities should be seated with that person.[16]

Chapter 4 of the Manual has provisions relating to facilities for disabled passengers at the airport. The Manual says that aircraft and airport operators should provide assistance to persons with disabilities in the use of airport facilities, such as assistance in getting to and from washrooms and using automated kiosks, subject to sufficient time being available and that information desks should be accessible to all persons with disabilities, for example, those who use a mobility aid such as a wheelchair, or who have a visual, auditory, dexterity, or speech impairment. Information desk staff should be provided with the training required for handling requests for information from persons with disabilities with efficiency and sensitivity As for check-in-services, airport and aircraft operators are advised to ensure that automated check-in machines or kiosks under their control are accessible and identified with the universal symbol of accessibility. If the machines and kiosks cannot be made accessible, then an equivalent level of service should be provided to those persons who are unable to use them independently. Airport and aircraft

[16] In this context, it is interesting to note that the Manual recommends that aircraft operators should have seats that are designated as accessible for persons with disabilities. Aircraft operators may choose to block these seats until close to the time of departure and should ensure that they are the last seats assigned to other passengers. Seats should be reassigned, if necessary, to ensure that persons with disabilities have appropriate seating. Aircraft operators that charge for advanced seat selection should waive the charge for a persons with disabilities in order that the latter may select the seat that best meets his or her needs. Aircraft operators are encouraged to provide persons with disabilities alternative seating, if available, where this can better meet their needs. Id, paragraphs 3.20 and 3.20.1.

operators are called upon to provide assistance to persons with disabilities at check-in counters. Check-in staff should be given appropriate training to handle requests from, and respond to the needs of, persons with disabilities, such as by assigning appropriate seating.

Paragraph 4.6 calls upon airport and aircraft operators to provide wheelchairs, upon request, at airports to persons with disabilities to ensure seamless transportation from their arrival at an airport to the time of boarding an aircraft, and from the point of disembarkation to their exit from an airport. Adequate information should be available to passengers about the availability of wheelchairs, and whether the airport or aircraft operator needs advance notification to provide wheelchair service. Passengers should be able to remain in their personal wheelchair up to the departure gate and receive it upon disembarking wherever this is possible. Airport and aircraft operators should provide assistance with baggage to persons with disabilities when requested during check-in, security, customs, immigration and boarding. Airport or aircraft operators are also advised that they should provide persons with disabilities with seating in passenger service areas under their control where there may be long waiting lines or times, including at ticket sales counters, check-in counters, security screening points, customs areas, and baggage retrieval areas. Seating should not be placed where it blocks evacuation routes. When a person using a wheelchair who is not independently mobile must await assistance with boarding or disembarking, the airport or aircraft operator staff, as appropriate, should be in frequent (e.g. every 30 min) contact with the person to advise them of the status of the request for assistance and to enquire about the person's needs.

While Chapter 5 of the Manual is dedicated to guidance on the planning of airports to suit the needs of the disabled traveller, Chapter 6 addresses issues such as immigration, customs and security operational procedures which should be adapted to the needs of persons with disabilities in order to facilitate their clearance in a dignified manner. Chapter 7 of the Manual is extremely important to a discussion to follow in this book which addresses issues of injury or death caused to a passenger while on board the aircraft or in the process of embarkation or disembarkation. The Manual calls for the assurance of assistance for persons with disabilities which should be provided by airport and aircraft operator staff trained and qualified to meet their needs, from arrival at an airport terminal until they are seated on board an aircraft, and vice versa, and with the appropriate equipment to assist them. Paragraph 7.2 in particular is important which lays down the type of assistance to be rendered which should include: assistance with boarding and disembarking an aircraft; transferring a person with disabilities from the person's own mobility aid to a wheelchair, boarding chair or other mobility aid provided by the airport or aircraft operator; transferring a person with disabilities from a wheelchair or mobility aid to his or her seat and vice versa; assistance with stowing and retrieving cabin baggage, including mobility aids and other assistive devices; assistance with proceeding to baggage retrieval areas, including the completion of immigration and customs procedures; assistance with retrieving hold baggage if the person with disabilities has difficulty doing so; and assistance with proceeding to general public areas and, if requested, to the first point of onward connection or to reach a connecting flight.

The Manual also states in paragraph 7.3 that persons with disabilities who self-identify as needing assistance or additional time should be offered the opportunity to separately pre-board (i.e. prior to all other passengers) and disembark (i.e. before or after all other passengers), as this is generally more dignified and less stressful for the person and more efficient for the aircraft operator. Assistance in getting on and off aircraft should be provided promptly to persons with disabilities. Aircraft, airport and ground handling operators should cooperate to ensure, where practicable, that persons who use wheelchairs may use their own wheelchairs to move to and from an aircraft door. However, this may not be possible with some powered wheelchairs. For example, it may not be practical to allow an individual to remain in his or her own wheelchair at an airport where wheelchairs have to be carried up and down stairs at a gate, which could potentially place staff at risk of injury. Aircraft and airport operators are called upon to provide for the timely return of wheelchairs and other mobility aids, as close as possible to the door of an aircraft so that persons with disabilities may use their own equipment to the greatest extent possible, except if there are national security or hazardous material rules to the contrary, or if the person requests otherwise.

Paragraphs 8.14 and 8.15 of the Manual are two more provisions which are contentious in the context of airline liability. As the discussion to follow would examine, courts inquire as to airlines acted prudently with regard to avoiding or coping with possible injury. The Manual states that an aircraft with accessible washroom facilities should be equipped with an on-board wheelchair. An aircraft that is not equipped with accessible washrooms should carry an on-board wheelchair when a person with disabilities requests one, subject to the aircraft having the capacity to stow and restrain such equipment. Furthermore, the Manual states that on-board wheelchairs should be designed to permit the easy transfer of an occupant and easy manoeuvring of the wheelchair. On-board wheelchairs should include footrests and armrests that are moveable or removable, an occupant restraint device, and wheel locks or other adequate means to prevent the chair from moving during transfer or turbulence.

Chapter 9 of the Manual has various provisions on how to make aircraft accessible to persons with disability. Some of the provisions broadly recommend that aircraft operators optimize the level of accessibility of new aircraft and those undergoing refurbishment by choosing design options and features that meet the needs of persons with disabilities, insofar as aircraft type, size and configuration permit, for example: aircraft should be equipped with an accessible washroom; signage should be positioned to avoid shadows and glare. Aircraft should have tactile directional signage to assist the movement of vision-impaired passengers around the cabin; and lighting on aircraft, except reading and other lighting under the control of a passenger, should be directed and controlled so as not to create shadows or glare. Lighting should not result in any sharp contrasts in intensity in the cabin.

Finally, Chapter 11 says inter alia that airport and aircraft operators should ensure that accessible ground handling services are available in landside, terminal and airside areas during operational hours, and also during extended hours when necessitated by irregular operations. Where it is not possible to provide accessible ground transportation, airport and aircraft operators should advise of alternative services available in the community and advise the public how to access these other services. Where there is no ground transportation available, the public should be informed of this fact.

3.1.3.2 Comments on Annex 9 and the Manual on Access to Air Transport by the Disabled Passenger

It must be noted that neither the *Annex* nor the *Manual* are legally binding on States, airports and airlines and therefore these three parties are not obligated to adhere to the provisions therein. To make matters worse, the Annex has only two Standards with regard to the disabled passenger which are also ineffective if a State is not able to comply with them.[17] However, in any adjudication involving a disabled passenger that brings to bear the conduct of an airline or an airport, courts may look at such guidance as persuasive of providing the court with indications of the presence or absence of prudent conduct on the part of the defendant. Both airports and airlines (and indeed other service providers as relevant) should be mindful of the weight given by courts to corporate responsibility and accountability where the question as to whether the service provider acted prudently in the interests of the claimant would be considered relevant, if not compelling.

Although legal provisions, ipso facto, and if strictly interpreted may often deprive an aggrieved person of reparation for damage. Rigid public policy driven by the letter of the law has not always been judicially acceptable and therefore Annex 9 and the Manual may not be treated by the judiciary *stricto sensu*. Justice Oliver Wendell Holmes said:

> The life of the law has not been logic: it has been experience. The felt necessities of the time, the prevalent moral and political theories, intuitions of public policy, avowed or unconscious, even the prejudices that judges share with their fellow men, have had a good

[17] Article 38 of the Chicago Convention provides that any State which finds it impracticable to comply in all respects with any international standard or procedure resulting from the provisions of the Chicago Convention (such as the Annexes), or to bring its own regulations or practices into full accord with any international standard or procedure after amendment of the latter, or which deems it necessary to adopt regulations or practices differing in any particular respect from those established by an international standard, shall give immediate notification to the International Civil Aviation Organization of the differences between its own practice and that established by the international standard. In the case of amendments to international standards, any State which does not make the appropriate amendments to its own regulations or practices shall give notice to the Council within 60 days of the adoption of the amendment to the international standard, or indicate the action which it proposes to take. In any such case, the Council shall make immediate notification to all other states of the difference which exists between one or more features of an international standard and the corresponding national practice of that State.

deal more to do than the syllogism in determining the rules by which men should be governed.[18]

This view, offered by one of the most eminent jurists of all time, was reiterated in 1994 in *Browns* v. *LumbermansMut. Cas. Co.*[19]—that public policy considerations may preclude tort liability that would otherwise be imposed.[20] This is indeed unfortunate. Pandering to public policy or prejudices should not interfere with independent judicial thinking in the dispensation of justice. It is, and remains a matter for fair, equitable, courageous and unbiased judicial interpretation.

English law has accepted the concept of *incrementalism*,[21] which rejects generalization in relation to the duty of care, in favor of a cautious development of law founded on analogies to similar fact situations, but espousing and applying *fairness and justice* to each case. This approach has its genesis in the judgment of Justice Brennan in the 1985 case of *Sutherland Shire Council* v. *Heyman*[22] handed down by the High Court of Australia where His Honour said:

It is preferable. . .that the law should develop novel categories of negligence incrementally and by analogy with existing categories, rather than by massive extension of a prima facie duty of care restrained only by the indefinable. . .considerations which ought to negative, or to reduce or limit the scope of the duty or the class of person to whom it is owed.[23]

The House of Lords found it fit to import this approach to the United Kingdom in the leading 1990 case of *Caparo Industries Plc* v. *Dickman*[24] where Lord Bridge stated:

Whilst recognizing, of course, the importance of the underlying general principles common to the whole field of negligence, I think the law has now moved in the direction of attaching greater significance to the more traditional categorization of distinct and recognizable situation, as guides to the existence, the scope and the limits of the varied duties of care which the law imposes.[25]

Therefore, there is no simple formula or touchstone to which recourse can be had in order to provide in every case a ready answer to questions as to whether the law will or will not grant recourse based on traditional rules of negligence. It would certainly be interesting if the "incrementalism" approach were to be applied along with established rules of law to instances of adjudication on determining the conduct of airlines, airports and other service providers towards persons with reduced mobility.

---

[18] Oliver Wendell Holmes, The Common Law (Mark DeWolfe ed.) 1963 at 5.

[19] 517 N.W. 2d. 432.

[20] *Id.* 439–444.

[21] See Stanton, Incremental Approaches to the Duty of Care, Chapter 2 of Mullany, *Torts in the Nineties*, Law Book Company: 1997.

[22] (1985) 157 *CLR* 424.

[23] *Id.* 481.

[24] [1990] 2 *AC* 605.

[25] *Id.* at 618.

3.1.4 Some Instances of Adjudication

In November 2008, the Federal Court of Appeal of Canada handed down its decision[26] against Air Canada that it and other carriers do not have the right to charge disabled or obese people for an extra ticket when they need an additional seat or an attendant to accompany them. The Court refused to hear an appeal from the airline, and rejected a Canadian Transportation Agency (CTA) ruling that the applicant had to satisfy the Agency both that she was disabled by reason of her obesity and that she had encountered an undue obstacle in air travel. The majority of the three-member panel dismissed the application on the preliminary ground that the applicant did not satisfy the first condition. The agency estimated the change would cost Air Canada, which reported C$10.6 billion in revenue last year, C$7.3 million ($5.7 million) annually. The ruling applies only to domestic flights and doesn't include the domestic portion of an international flight.[27]

In order to obtain the remedy which she was seeking, the applicant, who was morbidly obese and attributed her obesity to a condition known as the Stein-Leventhal syndrome, had to show that although no medical evidence establishes the cause of her obesity it was not disputed that her morbidly obese condition had been medically established. She therefore had to satisfy the Agency both that she was disabled by reason of her obesity and that she had encountered an undue obstacle in air travel.[28] The majority of the three-member panel dismissed the application on the preliminary ground that the applicant did not satisfy the first condition. The applicant appealed against a finding of the Canadian Transportation Agency which dismissed the applicant's complaint. The majority concluded that the applicant inappropriately relied on inapplicable criteria to establish her disability. The Agency equated the limitation encountered by the applicant to an obstacle, and held that considering the obstacle in determining the applicant's disability was inconsistent with the scheme of the CTA. The majority went on to find that the applicant's obesity did not constitute a disability for the purposes of Part V of the CTA and dismissed her complaint on that basis.

On January 14, 2003, the Federal Court of Appeal granted the applicant leave to appeal. The Court held that a disability cannot be determined by mere definition or in the abstract and that it should be taken in context. The Court cited with approval a dictum of Binnie J. for a unanimous Court in a case decided in 2002[29] involving workplace discrimination on account of an alleged disability:

---

[26] *McKay-Panos* v. *Air Canada* [2006] F.C.J. No. 28 2006 FCA 8 Docket A-100-03.

[27] http://www.bloomberg.com/apps/news?pid=newsarchive&sid=aA3SCZGJ9LeA.

[28] On boarding the aircraft in Calgary on August 21, the applicant discovered that the bulkhead seats were actually smaller than the other seats because the tray tables folded into the arm rests rather than the forward chair as they did in other seats. The applicant found that she could barely force herself into her seat. The passenger beside her could not access his tray table because her hips spread onto his arm rest. The flight attendants bumped into her with their serving carts.

[29] *Granovsky* v. *Canada (Minister of Employment and Immigration)*, [2000] 1 S.C.R. 703.

> A disability, unlike, for example, race or colour, may entail pertinent functional limitations. [...] An individual may suffer severe impairments that do not prevent him or her from earning a living. Beethoven was deaf when he composed some of his most enduring works. Franklin Delano Roosevelt, limited to a wheelchair as a result of polio, was the only President of the United States to be elected four times. Terry Fox, who lost a leg to cancer, inspired Canadians in his effort to complete a coast-to-coast marathon even as he raised millions of dollars for cancer research. Professor Stephen Hawking, struck by amyotrophic lateral sclerosis and unable to communicate without assistance, has nevertheless worked with well-known brilliance as a theoretical physicist. (Indeed, with perhaps bitter irony, Professor Hawking is reported to have said that his disabilities give him more time to think.) The fact they have steady work does not, of course, mean that these individuals are necessarily free of discrimination in the workplace. Nor would anyone suggest that, measured against a yardstick other than employment (access to medical care for example), they are not persons with daunting disabilities.[30]

3.1.5 International Perspectives

3.1.5.1 Action Taken by the United Nations

In 1975, the United Nations General Assembly adopted the *Declaration on the Rights of the Disabled Persons (Resolution 3447 XXX)* which *inter alia* recognized the inherent right of disabled persons to respect for their dignity and self-reliance. Paragraph 8 of the Declaration provides that disabled persons are entitled to have their special needs taken into consideration at all stages of economic and social planning. Later, in 1976, at its thirty first Session, the United Nations General Assembly adopted a Resolution which proclaimed 1981 as the International Year of the Disabled,[31] identifying in paragraph (c) as one of the objectives of the Resolution, the encouragement of study and research projects designed to facilitate the practical participation of disabled persons in daily life, for example by improving their access to public buildings and transportation systems. In 1979, at its 34th Session, the United Nations General Assembly, by *Resolution 34/154*, recalling its Resolution of 1976, further resolved that the theme of the International Year of Disabled Persons should be expanded to "full participation and equality" to promote the realization of the right of disabled persons to participate fully in the social life and development of the societies they live in. Two years later, in 1981, by *Resolution 36/77*, the General Assembly requested *inter alia*, the Secretary General to convene in 1982 a meeting of the Advisory Committee of the International Year of the Disabled Persons to finalize the draft World Programme of Action concerning disabled persons, in the light of comments by Member States, international organizations and relevant non-governmental organizations, with a view to its adoption by the General Assembly. The World Programme of Action Concerning Disabled Persons was later introduced by General Assembly

[30] *Id.* 723.

[31] *UNGA/RES/31/123* dated 16 December 1976.

Resolution 37/52 of 3 December 1982, and accepted for implementation by *Resolution 37/53* of 3 December 1982, which in paragraph 11 proclaimed the period 1983–1992 as the United Nations Decade of Disabled Persons.

3.1.5.1.1 The United Nations Convention on the Rights of Persons with Disabilities

On 13 December 2006, Member States of the United Nations adopted the Convention on the Rights of Persons with Disabilities and its Optional Protocol[32] at the United Nations Headquarters in New York. It was opened for signature on 30 March 2007. There were 82 signatories to the Convention, 44 signatories to the Optional Protocol, and 1 ratification of the Convention, which is the highest number of signatories in history to a UN Convention on its opening day. At the time of writing, there were 147 signatories and 95 ratifications. The Convention,[33] which recognizes as disabled persons those including persons who have long-term physical, mental, intellectual or sensory impairments which in interaction with various barriers may hinder their full and effective participation in society on an equal basis with others. The purpose of the Convention is to promote, protect and ensure the full and equal enjoyment of all human rights and fundamental freedoms by all persons with disabilities, and to promote respect for their inherent dignity.

The principles of the Convention are: respect for inherent dignity, individual autonomy including the freedom to make one's own choices, and independence of persons; non-discrimination; full and effective participation and inclusion in society; respect for difference and acceptance of persons with disabilities as part of human diversity and humanity; equality of opportunity; accessibility; equality between men and women; respect for the evolving capacities of children with disabilities and respect for the right of children with disabilities to preserve their identities.[34] Article 9, which directly pertains to persons with disabilities and transportation *inter alia*, provides that, in order to enable persons with disabilities to live independently and participate fully in all aspects of life, States Parties are required to take appropriate measures to ensure to persons with disabilities access, on an equal basis with others, to the physical environment, to transportation, to information and communications, including information and communications technologies and systems, and to other facilities and services open or provided to the public, both in urban and in rural areas.

---

[32] The Convention on the Rights of Persons with Disabilities and its Optional Protocol are serviced by a joint Secretariat, consisting of staff of both the United Nations Department of Economic and Social Affairs (DESA), based in New York, and the Office of the High Commissioner for Human Rights (OHCHR) in Geneva.

[33] http://www.un.org/disabilities/convention/conventionfull.shtml.

[34] Article 3.

The above measures, which include the identification and elimination of obstacles and barriers to accessibility, apply to, *inter alia*: buildings, roads, transportation and other indoor and outdoor facilities, including schools, housing, medical facilities and workplaces; and information, communications and other services, including electronic services and emergency services. The same provision requires States Parties to ensure that they develop, promulgate and monitor the implementation of minimum standards and guidelines for the accessibility of facilities and services open or provided to the public; and ensure that private entities that offer facilities and services which are open or provided to the public take into account all aspects of accessibility for persons with disabilities.

Another significant provision is Article 20 relating to personal mobility which provides that States Parties are required to take effective measures to ensure personal mobility with the greatest possible independence for persons with disabilities, including by facilitating the personal mobility of persons with disabilities in the manner and at the time of their choice, and at affordable cost; facilitating access by persons with disabilities to quality mobility aids, devices, assistive technologies and forms of live assistance and intermediaries, including by making them available at affordable cost; providing training in mobility skills to persons with disabilities and to specialist staff working with persons with disabilities; and encouraging entities that produce mobility aids, devices and assistive technologies to take into account all aspects of mobility for persons with disabilities.

3.1.5.1.2 Incorporation of the Convention in Domestic Legislation

One of the significant features of human rights treaties is that States Parties are required to give effect to their obligations within their domestic legal order. Article 4 of the Convention on the Rights of Persons with Disabilities stipulates this obligation.[35] There is a well established principle of construction at international law that when a piece of domestic legislation is adopted it means that Parliament intends to fulfil its international obligations.[36] There are two ways in which legislation gives effect to a treaty: by directly enacting the provisions of a treaty; or by resonating the provisions of a treaty within the substantive provisions of the legislation. The question at issue then is whether domestic courts should use enabling legislation as its main source or go to the treaty that the law gives effect to in applying the provisions of the law in a given instance. In the 1931 case of *Ellerman* v. *Murray*[37] the court held that if the legislation had a natural meaning it

[35] Article 4 provides that "States Parties undertake to ensure and promote the full realization of all human rights and fundamental freedoms for all persons with disabilities without discrimination of any kind on the basis of disability".

[36] *Salomon* v. *Commissioners of Customs and Excise* [1967] 2 QB 116, CA at 141 per Lord Denning M.R.

[37] [1931] AC 126 at 147 per Lord Tomlin. See also *Barras* v. *Aberdeen Steam Trawling Co. Ltd.* [1933] AC 402; *Burns Philp & Co. Ltd.* v. *Nelson and Robertson Proprietaries Ltd.* (1957–58) 98 CLR 495, HC of A.

would be improper to resort to the wording of the applicable treaty. However, later decisions[38] have held that the text of the treaty may be referred to as interpretative of an Act which gives effect to that treaty. Lord Diplock identified the necessary criteria for this approach as being twofold: when the terms of the legislation are ambiguous or may give rise to more than one interpretation; or that there is unequivocal and cogent evidence that the law concerned was intended to fulfil obligations under a particular treaty. *Brownlie* supports the latter on the basis that if it is the intention of Parliament to fulfil its international obligations, then it is logical to assume that the terms of the treaty should be important.[39]

The approach—that the intention of Parliament should be reflected in having recourse to the specific terms of a treaty—is consistent with the decision handed down in the 1997 case of *Sidhu* v. *British Airways*[40] where Lord Hope stated that "it is well established that a purposive approach should be taken to the interpretation of international conventions which have the force of law in this country".[41] In the same year, another more traditional view was expressed, that the significance of a textual interpretation of text in question should be as understood in their "ordinary meaning", seemingly implying that the primary resort should be to the ordinary meaning of statute.[42] This view is more in keeping with the *Ellerman* approach.[43] However, the prevalent current approach in Europe seems to favour the view that, as stated by Brownlie, if it is the intention of Parliament to fulfil its international obligations, then it is logical to give priority to the text in a treaty. The growing importance ascribed to treaties particularly on such subjects as human rights has impelled courts to take into account language of applicable treaties.[44]

3.1.5.2 EC Regulation No. 1107/2006

EC Regulation No. 1107/2006[45] is aimed at establishing rules for the protection of and provision of assistance to disabled persons and persons with reduced mobility travelling by air, both to protect them against discrimination and to ensure that they receive assistance. The provisions of the Regulation, which came into force on 26 July 2008, apply to disabled persons and persons with reduced mobility, using or

---

[38] *Salomon* v. *Commissioners of Customs and Excise supra*, note 36.

[39] Brownlie (1990), p. 49.

[40] [1997] 1 All E.R. 193.

[41] *Id.* at 202.

[42] *Semco Salvage* v. *Lancer Navigation*, [1997] 1 All. E.R. 502 at 512.

[43] *Supra*, note 178 (Chap. 2).

[44] See Shaw (2003), p. 142.

[45] Regulation (EC) No 1107/2006 of the European Parliament and of the Council of 5 July 2006 concerning the rights of disabled persons and persons with reduced mobility when travelling by air Text with EEA relevance. Official Journal L 204, 26/07/2006 P. 0001–0009.

intending to use commercial passenger air services on departure from, on transit through, or on arrival at an airport, when the airport is situated in the territory of a Member State[46] to which the Treaty[47] applies. Under the Regulation, an air carrier or its agent or a tour operator cannot refuse, on the grounds of disability or of reduced mobility: to accept a reservation for a flight departing from or arriving at an airport to which this Regulation applies; or to embark a disabled person or a person with reduced mobility at such an airport, provided that the person concerned has a valid ticket and reservation.[48]

Article 4 stipulates that there are two exceptions under which a carrier may refuse to carry a person with a disability: the carrier's need to meet applicable safety requirements established by international, Community or national law or in order to meet safety requirements established by the authority that issued the air operator's certificate to the air carrier concerned; and if the size of the aircraft or its doors makes the embarkation or carriage of that disabled person or person with reduced mobility physically impossible. However, if the carrier is unable to offer transportation to a person with a disability the carrier, its agent or the tour operator shall make reasonable efforts to propose an acceptable alternative to the person in question. A person with a disability or a person with reduced mobility who has been denied embarkation on the grounds of his or her disability or reduced mobility and any person accompanying such person is to be offered the right to reimbursement or re-routing as provided for in Article 8[49] of Regulation (EC) No 261/2004.

---

[46] *Id.* Article 1.

[47] Treaty establishing the European Community. Air transport in the European Community is fundamentally regulated by two treaties, i.e. the Treaty which establishes the European Coal and Steel Community (ECSC Treaty) and the Treaty which establishes the European Economic Community (EEC Treaty, now called the EC Treaty). The former, which was signed in Paris in 1951, addresses issues related to the carriage of coal and steel through the media of rail, road and inland waterways and as such is not directly relevant to aviation. The latter, on the other hand, admits of issues relating to all modes of transport in the carriage of persons and goods and is of some relevance to aviation.

The EEC Treaty, which was signed in Rome on 25 March 1957, has at its core a Common Transport Policy (CTP) concept which is calculated to achieve the fundamental purposes of the European Community. One of the most salient features of the EEC Treaty is that the tasks of the Community are set out succinctly in Article 2 of the Treaty which provides *inter alia* for the adoption of a CTP as provided for in Article 3(1) of the Treaty. This provision is linked to Article 74 which in turn provides that the objectives of the Treaty in relation to issues of transportation would be pursued by State Parties within the parameters of the CTP, which is established by the Council of Europe through secondary legislation.

[48] Supra, note 186 (Chap. 2), Article 3.

[49] Article 8 places responsibility on a managing body of an airport in which a person with disabilities is, to provide certain assistance with regard to: communicating their arrival at an airport and their request for assistance at the designated points inside and outside terminal buildings; moving from a designated point to the check-in counter; check-in and register baggage; proceeding from the check-in counter to the aircraft, with completion of emigration, customs and security procedures; boarding the aircraft, with the provision of lifts, wheelchairs or other assistance needed, as appropriate; proceeding from the aircraft door to their seats, storing and retrieving baggage on the aircraft; proceed from their seats to the aircraft door; disembarking from

The right to the option of a return flight or re-routing shall be conditional upon all safety requirements being met.

The European Civil Aviation Conference (ECAC) has its own codified guidelines[50] which follows the general principle that individual airport communities should work in partnership to review and develop the way that special assistance services for persons with reduced mobility are organised in order to support a professional and seamless service.[51]

3.1.5.3 United States Legislation

The United States has enacted 14 CFR Part 382 (Non Discrimination on the Basis of Disability in Air Travel)—effective May 13, 2009—with the purpose of implementing the Air Carrier Access Act of 1986[52] which provides that no air carrier may discriminate against any otherwise qualified individual with a disability, by reason of such disability, in the provision of air transportation. There is a general prohibition requirement in Part 382 which stipulates that no carrier is permitted to, directly or through contractual, licensing, or other arrangements: discriminate against any otherwise qualified individual with a disability, by reason of such disability, in the provision of air transportation; require a person with a disability to accept special services (including, but not limited to, pre-boarding) not requested by the passenger; exclude a qualified individual with a disability from or deny the person the benefit of any air transportation or related services that are available to other persons, even if there are separate or different services available

the aircraft, with the provision of lifts, wheelchairs or other assistance needed, as appropriate; proceeding from the aircraft to the baggage hall and retrieve baggage, with completion of immigration and customs procedures; proceeding from the baggage hall to a designated point; reaching connecting flights when in transit, with assistance on the air and land sides and within and between terminals as needed; and move to the toilet facilities if required. Where a person with a disability or person with reduced mobility is assisted by an accompanying person, this person must, if requested, be allowed to provide the necessary assistance in the airport and with embarking and disembarking. Ground handling of all necessary mobility equipment, including equipment such as electric wheelchairs subject to advance warning of 48 h and to possible limitations of space on board the aircraft, and subject to the application of relevant legislation concerning dangerous goods. Temporary replacement of damaged or lost mobility equipment, albeit not necessarily on a like-for-like basis. Ground handling of recognized assistance dogs, when relevant. The airport must also provide communication of information needed to take flights in accessible formats.

[50] Code of Good Conduct in Ground Handling for Persons with Reduced Mobility.

[51] The service must be delivered in a harmonised, transparent, non-discriminatory way and must be subject to audits and reviews in accordance with the European Ground Handling Directive. It must improve levels of customer service and safety to PRMs, through a seamless service from quality supplier/s, implemented with quality staff, equipment and a quality organisational structure, operating to meet and exceed prescribed customer service and safety standards.

[52] 49 U.S.C. 41705. Congress enacted the Air Carrier Access Act (ACAA) in 1986. The statute prohibits discrimination in airline service on the basis of disability.

for persons with a disability except when specifically permitted by another section of this part; or, take any action adverse to an individual because of the individual's assertion, on his or her own behalf or through or behalf of others, of rights protected by this part or the Air Carrier Access Act.[53]

Furthermore Part 382 requires that, unless specifically permitted by a provision of this part, a carrier cannot refuse to provide transportation to a qualified individual with a disability on the basis of his or her disability. A carrier cannot refuse to provide transportation to a qualified individual with a disability solely because the person's disability results in appearance or involuntary behaviour that may offend, annoy, or inconvenience crewmembers or other passengers. It further provides that a carrier shall not refuse to provide transportation to qualified individuals with a disability by limiting the number of such persons who are permitted to travel on a given flight. However, personnel of the carrier may refuse to provide transportation to any passenger on the basis of safety, and may refuse to provide transportation to any passenger whose carriage would violate the Federal Aviation Regulations. In exercising this authority, carrier personnel cannot discriminate against any qualified individual with a disability on the basis of disability and their actions shall not be inconsistent with the provisions of this Part. In the event that such action is inconsistent with the provisions of this Part, the carrier shall be subject to remedies provided under § 382.65.[54] When a carrier refuses to provide transportation to any person on a basis relating to the individual's disability, the carrier shall specify in writing to the person the basis for the refusal, including, where applicable, the reasonable and specific basis for the carrier's opinion that transporting the person would or might be inimical to the safety of the flight. This written explanation shall be provided within 10 calendar days of the refusal of transportation.[55]

The rule also provides that persons with a disability will undergo security screening in the same manner, and be subject to the same security requirements,

---

[53] § 382.7.

[54] § 382.65 stipulates *inter alia* that each carrier providing scheduled service shall establish and implement a complaint resolution mechanism, including designating one or more complaints resolution official(s) (CRO) to be available at each airport which the carrier serves. The carrier is required to make a CRO available to any person who complains of alleged violations of this part during all times the carrier is operating at the airport. Furthermore, this provision, *inter alia*, provides that carriers are required to categorize disability-related complaints that they receive according to the type of disability and nature of complaint. Data concerning a passenger's disability must be recorded separately in the following areas: vision impaired, hearing impaired, vision and hearing impaired, mentally impaired, communicable disease, allergies (*e.g.*, food allergies, chemical sensitivity), paraplegic, quadriplegic, other wheelchair, oxygen, stretcher, other assistive device (cane, respirator, etc.), and other disability. Data concerning the alleged discrimination or service problem related to the disability must be separately recorded in the following areas: refusal to board, refusal to board without an attendant, security issues concerning disability, aircraft not accessible, airport not accessible, advance notice dispute, seating accommodation, failure to provide adequate or timely assistance, damage to assistive device, storage and delay of assistive device, service animal problem, unsatisfactory information, and other.

[55] § 382.31.

as other passengers.[56] Possession by a such a person of an aid used for independent travel would not subject the person or the aid to special screening procedures if the person using the aid clears the security system without activating it. However no security personnel can be precluded from examining a mobility aid or assistive device which, in their judgment, may conceal a weapon or other prohibited item. Security searches of qualified individuals with a disability whose aids activate the security system will be conducted in the same manner as for other passengers. Private security screenings will not be required for persons with a disability to a greater extent, or for any different reason, than for other passengers, except that if a person with a disability requests a private screening in a timely manner, the carrier will provide it in time for the passenger to enplane.

The rule applies to all U.S. carriers, regardless of where their operations take place, except where otherwise provided in the rule. With respect to foreign carriers, the application of the rule is more limited. Only flights[57] of foreign carriers that begin or end at a U.S. airport, and aircraft used in these operations, are covered. Notably, a foreign carrier is not covered under the rule with respect to an operation between two foreign points, even if, under a code-sharing arrangement with a U.S. carrier, the foreign carrier transports passengers flying under the U.S. carrier's code. The U.S. carrier, however, is covered under the rule with respect to the passengers travelling under its code on such a flight, such that if there is a violation of the Part 382 rights of a passenger travelling under the U.S. carrier's code, the Department would hold the U.S. carrier, not the foreign carrier, responsible. Finally, a charter flight on a foreign carrier from a foreign airport to a U.S. airport and back would not be covered if the carrier did not pick up any passengers in the U.S.

The issue as to what measures should be taken to alleviate the lot of the person with a disability in an aircraft has already been addressed.[58] A commentator[59] has wisely expressed the view that there must be harmonization of efforts, particularly between the United States and European States on this subject. There is no room for doubt that, if a meaningful approach is to be taken, there must be global harmony on the measures that need to be taken. However, such harmonization should not be relegated to mere philosophy but should ensure applicability by giving effect to global principles enunciated through incorporation. It is evident that global harmonization has not worked so far, judging by the few Recommended Practices that are in Annex 9 to the Chicago Convention, which cannot be even upgraded to the level of Standards. The other disconcerting fact is that it is not sufficient to harmonize rules and regulations between the United States and Europe. All 190 ICAO member States (or at least a significant proportion of membership) must reach an agreement on enforceable and harmonized Standards.

---

[56] § 382.49.

[57] A flight means a continuous journey of a passenger in the same aircraft or using the same flight number. The rule provides several examples of what constitutes a "flight" and what does not.

[58] Abeyratne (1995), pp. 52–59.

[59] O'Keefe (2006), pp. 408–421.

From a policy perspective, it must not be forgotten that ICAO serves the people of the world in meeting their needs for safe, regular, efficient and economical air transport.[60] The incontrovertible fact is that the aviation industry is about one core product—air transport—and the primary client is the passenger. Therefore, member States must ascribe to ICAO its proper *raison d'etre*. One way to approach global harmonization is to obtain agreement from States on Annex 9 provisions as Standards (and not as Recommended Practices) and subject them to the current ICAO safety audit system. The basis for this would be that a disabled passenger would compromise his safety if he is not accorded the facilities expected or recognized by domestic laws or he is not allowed to travel unless he pays for a needed attendant. Harmonized ICAO Standards would also ensure the recognition of global ethics as propounded by international treaty, thus preventing fragmentation of policy on a nation to nation basis.

States Parties to the Convention on the Rights of Persons with Disabilities should not have an objection to this approach as they are required to incorporate the Convention in their national legislation, as discussed above. All ICAO member States could recognize ICAO Standards in Annex 9 as their minimum practice standards and should be able to enact more favourable legislation for persons with disabilities if they so wish. This approach could help the estimated 500 million persons with disabilities who are potential air travellers.[61]

3.2 Racial Profiling

Racial profiling is any action taken by a person of authority in police or defence that is based upon racial or ethnic stereotypes and that has the effect of treating minority races differently than non-minority races.[62] Amnesty International USA defines racial profiling as "the targeting of individuals and groups by law enforcement even partially on the basis of race, ethnicity, national origin, or religion, except where there is trustworthy information, relevant to the locality and timeframe, that links persons belonging to one of the aforementioned groups to an identified criminal incident or scheme". The backlash to the events that unfolded on 11 September 2001 when thousands of persons were killed by terrorist activity on a single day in the United States by terrorists of a particular ethnic origin (Arab Muslim) reflected

---

[60] Chicago Convention, *supra*, Article 44 d). All ICAO's aims and objectives as contained in Article 44 addresses the primeval importance of the air transport product and the need to ensure that it serves the passenger and cargo consignor well.

[61] *Supra*.

[62] See Verniero and Zoubek (1999), p. 5 for an analogical definition. See also, *White* v. *Williams*, United States District Court for the District of New Jersey, 179 F. Supp. 2d 405, 1/9/02.

instances where racial profiling might have been exercised in air transport. Chandrashekhar writes:

> The terrorist attacks of September 11, 2001 ("9/11") permanently transformed the American civil liberties landscape. After nineteen Arab Muslim men hijacked and crashed commercial aircraft into the World Trade Center towers and the Pentagon, people of South Asian,' Arab, and Middle Eastern descent have become targets of hundreds of hate crimes and incidents of racial profiling across the country[63]

Amnesty International chronicles the case of 8 year old Omar Rana who was pulled out of line at the airport to the absolute horror of his mother Dr. Sandra Rana (A Pakistani national who no longer wears the *Hijab* at airports). She says:

> Imagine how I felt when my eight-year-old son was pulled from the line because of his name and I could not go with him. Imagine how he felt when they started to take apart his Boy Scout pinewood derby car in the Boy Scout box. . . . It is now routine for my son, for Omar Rana, to get extra security checks at the airport. He knows it's going to happen, and he expects it. . . . But how do I tell my . . . son that it's okay? He is now ten. He is learning about civil liberties and civil rights. What meaning do they have for him. . .?[64]

Another case cited is that of Herb Boyd, a Dutch journalist who often wears Muslim garb. He recalls and instance where, during a flight to Detroit in December 2001, he noticed that his ticket had a red dot on it. He had not paid any attention to that until he noticed he was standing with several others, all of them identifiably Arabs or [of] Arab descent. This group of persons, including Mr. Boyd, had all been stripped before entering the aircraft. As it turned out two passengers in the group had later advised Mr. Boyd that they had been subjected to such searches on every flight they travelled in.[65]

In October 2001, a passenger of Indian origin was refused boarding a Northwest Airlines flight. In the case of *Chowdhury* v. *Northwest Airlines Corporation*,[66] Arshad Chowdhury claimed that Northwest Airlines refused to permit him to board a Northwest flight because of his race and national origin. The defence of the airline was based on preemption of liability for purposes of enforcement of security. The court noted that in *Woodson v. U.S. Airways, Inc.*,[67] for example, African-American passengers sued for racial discrimination for being denied boarding. The defendant argued that the state civil rights claims were preempted by the express preemption provision of the ADA. The court tentatively concluded that the claims were not preempted, but denied the defendants' motion without prejudice.

One of the most ludicrous and unacceptable instances of racial profiling occurred post 9/11 as reflected in the case of *Dasrath* v. *Continental Airlines*.[68]

---

[63] Chandrashekhar (2003), pp. 215–252 at 215.

[64] *Threat and Humiliation – Racial Profiling, Domestic Security and Human Rights in the United States*, Amnesty International, 2nd Printing, October 2004 at 8.

[65] *Ibid.*

[66] *No. C 02-02665 CR*, in the District Court of Northern California. See 238 *F.Supp.2d* 1153 (2002).

[67] 67 *F.Supp.2d* 554 (M.D.N.C.1999).

[68] 228 *F. Supp.* 2d 531.

On December 31, 2001 Michael Dasrath and Edward Cureg, two passengers on a Newark-to-Tampa flight, were discriminated against and removed from the aircraft they were in. Dasrath and Cureg did not know one another, and were both flying to Tampa to be with family for New Year's Eve. While on board awaiting take off, Cureg, a Philippino national, spoke with his mathematics professor, an Indian national who was travelling e on the same flight. Dasrath, who is an American citizen originally from South America, sat in his assigned seat one row behind them and did not converse with them. It was reported that a female passenger walked up the aisle to first class on a few occasions and had stared at the three men who seemingly formed an unusual assortment of nationalities. Consequently she called upon the captain and informed him that "three brown-skinned men are behaving suspiciously." The pilot had neglected request the woman to elaborate on those behaviors. Instead, he had left the cabin area. Thereafter, a flight supervisor had come aboard the plane and asked Dasrath, Cureg and Cureg's professor to leave the plane. After their removal, the flight supervisor had explained that the pilot had informed that a passenger was uncomfortable with their presence. Dasrath, Cureg, and Cureg's professor had to wait hours for a later flight, which flew to Orlando as there were no more flights to Tampa. The three passengers were placed on the later flight without any additional security checks.

The District Court of the District of New Jersey noted that Federal law expressly provided that an "air carrier or foreign air carrier may not subject a person in air transportation to discrimination on the basis of race, color, national origin, religion, sex, or ancestry." 49 U.S.C. § 40127(a). Nevertheless, since September 11, reports of air carriers discriminating against passengers perceived to be Arab or Middle Eastern have risen dramatically. The Court held that the defendant airline had Defendant was liable for the unlawful acts of its agents and employees directly and/or under the doctrine of *respondeat superior*.

Furthermore, the defendant had engaged in intentional discrimination based on plaintiff's perceived race, color, ethnicity, and/or alienage in removing plaintiff from Continental Flight #1218 on December 31, 2001. In doing so, defendant had discriminated against plaintiff in the making and enforcement of his contract with defendant, namely his ticket to travel on Continental Flight #1218 on December 31, 2001. Consequently, defendant had caused the plaintiff to suffer deprivation of his right to make and enforce contracts as enjoyed by white citizens under 42 U.S.C. § 1981. The Court concluded that the defendant's actions were intentional, malicious, wilful, wanton, callous, and showed reckless disregard for plaintiff's civil rights, and have directly and proximately caused plaintiff financial injury and humiliation, mental pain, and suffering.

It is an incontrovertible fact that profiling is a useful tool in the pursuit of the science of criminology. Profiling is also a key instrument in a sociological context and therefore remains a sustained social science constructed through a contrived process of accumulation of single assumptions and propositions that flow to an eventual empirical conclusion. However, profiling raises well reasoned latent fears when based on a racial platform. Jonathan Turley, Professor of Constitutional Law at George Washington University, in his testimony before a United States House of

Representatives Committee on Airport Security regarding the use of racial profiling to identify potentially dangerous travelers observed:

> [R]acial profiling is to the science of profiling as forced confessions are to the art of interrogation. Like forced confessions, racial profiling achieves only the appearance of effective police work. Racial profiling uses the concept of profiling to shield or obscure a racist and unscientific bias against a particular class or group. It is the antithesis of profiling in that it elevates stereotypes over statistics in law enforcement.

Notwithstanding this telling analogy, and the apprehensions one might have against racial profiling, it would be imprudent to conclude that racial profiling is *per se* undesirable and unduly discriminatory, particularly in relation to profiling at airports which should essentially include some considerations of ethnic and national criteria. This article will examine the necessary elements that would go into effective and expedient airport profiling of potential undesirable passengers. It will also discuss legal issues concerned with the rights of the individual with regard to customs and immigration procedures. The rights of such persons are increasingly relevant from the perspective of ensuring air transport security and refusing carriage to embarking passengers who might show profiles of criminality and unruly behavior on board.

3.2.1 Airport Profiling

A legitimate profiling process should be based on statistically established indicators of criminality which are identified through a contrived aggregation of reliable factors. The application of this criterion to airport profiling would immediately bring to bear the need to apply nationality and ethnic factors to passenger profiles. Although one might validly argue that racial profiling would entail considerable social and political costs for any nation, while at the same time establishing and entrenching criminal stereotypes in a society, such an argument would be destitute of effect when applied to airport security which integrally involves trans boundary travel of persons of disparate ethnic and national origins. This by no means implies that racial profiling is a desirable practice. On the contrary, it is a demeaning experience to the person subjected to the process and a *de facto* travel restriction and barrier. It is also a drain on law enforcement resources that effectively preclude the use of proven and conventional uses of enforcement.

The sensitive conflict of interests between racial profiling *per se*, which at best is undesirable in a socio-political context, and airport profiling, raises interesting legal and practical distinctions between the two. Among these the most important distinction is that airport profiling is very serious business that may concern lives of hundreds if not thousands in any given instance or event. Profiling should therefore be considered justifiable if all its aspects are used in screening passengers at airports. Nationality and ethnicity are valid baseline indicators of suspect travelers together with other indicators which may raise a ‘flag’ such as the type of ticket a passenger holds (one way instead of return) and a passenger who travels without any luggage.

Racial profiling, if used at airports, must not be assumptive or subjective. It must be used in an objective and non discriminatory manner alongside random examinations of non-targeted passengers. All aspects of profiling, including racial and criminal profiling, should as a matter of course be included in the Computer Assisted Passenger Screening System (CAPS) without isolating one from the other. In this context the now popular system of compliance examination (COMPEX) is a non threatening, non discriminatory process which transcends the threshold debate on "profiling" by ensuring a balanced and proper use of profiling in all its aspects by examining "non targeted" passengers as well as on a random basis.

Another critical distinction to be drawn between discriminatory and subjective racial profiling on the one hand and prudent airport profiling on the other is the blatant difference between racism and racial profiling. The former is built upon the notion that there is a causal link between inherent physical traits and certain traits of personality, intellect or culture and, combined with it, the idea that some races are inherently superior to others. The latter is the use of statistics and scientific reasoning that identify a set of characteristics based on historical and empirical data. This brings to bear the clear difference between "hard profiling", which uses race as the only factor in assessing criminal suspiciousness and "soft profiling" which uses race as just one factor among others in gauging criminal suspiciousness.

Article 13 of the Chicago Convention provides that the laws and regulations of a Contracting State as to the admission to and departure from its territory of passengers, crew or cargo of aircraft, such as regulations relating to entry, clearance, immigration, passports, customs and quarantine shall be complied with by or on behalf of such passengers, crew or cargo upon entrance into or departure from, or while within the territory of that State. This provision ensures that a Contracting State has the right to prescribe its own internal laws with regard to passenger clearance and leaves room for a State to enact laws, rules and regulations to ensure the security of that State and its people at the airport. However, this absolute right is qualified to preclude unfettered and arbitrary power of a State, by Article 22 which makes each Contracting State agree to adopt all practicable measures, through the issuance of special regulations or otherwise, to facilitate and expedite navigation of aircraft between the territories of Contracting States, and to prevent unnecessary delays to aircraft, crews, passengers and cargo, especially in the administration of the laws relating to immigration, quarantine, customs and clearance. Article 23 follows this trend to its conclusion by providing that each Contracting State undertakes, to the extend practicable, to establish customs and immigration procedures affecting air navigation in accordance with the practices which may be established or recommended from time to time pursuant to the Convention.

Annex 9 to the Chicago Convention (Facilitation) in Standard 3.2, recognizes that, in developing procedures aimed at the efficient application of border controls on passengers and crew, Contracting States shall take into account the application of aviation security, border integrity, narcotics control and immigration control measures, where appropriate. This Standard gives States the flexibility of enacting procedures, rules and regulations to ensure the security and integrity of their borders and view passengers with necessary caution.

The aims and objectives of ICAO, as contained in Article 44 of the Chicago Convention are, *inter alia*, to ensure the safe and orderly growth of international civil aviation; meet the needs of the peoples of the world for safe, regular, efficient and economical air transport; and avoid discrimination between Contracting States. Annex 17—*Security* to the Chicago Convention, in paragraph 2.1.1, identifies the main aim of aviation security as being the safeguarding of international civil aviation operations against acts of unlawful interference. These aims have been established by the international community with a view to ensuring that security and safety of civil aviation apply not only in the context of individual States, but also of their peoples, in accordance with the distinction made in the Preamble of the Chicago Convention.

The above-mentioned principles and aims make it abundantly clear that the role of ICAO and the international community is to consider the offence of unlawful interference with civil aviation in its entirety, as a generic term encompassing a wide range of offensive activity on the part of the perpetrators. This does not admit of the offence being restricted to a species or other category of offensive activity. It is for the above reason that the ICAO Assembly, at its 33rd Session, held subsequent to the events of 11 September 2001, adopted Resolution A33-2, strongly condemning all acts of unlawful interference against civil aviation wherever and by whomsoever and for whatever reason they are perpetrated. This all-encompassing approach to the offence of unlawful interference with civil aviation effectively precludes parochial assumptions that the offence would be recognizable as such only if it is perpetrated pursuant to or as an act which can only or mostly be perpetrated by a certain type of individual of a certain race, nationality or religious persuasion.

The first issue, that of encompassing aviation security under the umbrella of threats or considering aviation security as being affected only by certain acts, is therefore a critical consideration if a balanced and focussed approach to remedial action on aviation security were to be seriously addressed. Broadly considered, unlawful interference with civil aviation is an offence, which is any wrong that affects the security or well-being of civil aviation, in the context of persons and property, and creates an interest in the public in its suppression. The offence itself should therefore not be confused with a type of conduct or activity.

The second issue, within the purview of the security guidelines of the Chicago Convention, pertains to stereotyping and racial profiling of individuals in the pursuit of ensuring aviation security. In essence, racial profiling is intersectional in nature and may consist of multiple grounds of institutionalized discrimination such as nationality, race, age, gender, socio-economic status, disability, health status, descent, language, class, culture and religion. At the World Conference against Racism, Racial Discrimination, Xenophobia and Related Intolerance, held in Durban, South Africa from 31 August to 7 September 2001, the Conference, referring to the International Convention on the Elimination of All Forms of Racial

Discrimination, urged States to implement or strengthen legislation and administrative measures prohibiting racial discrimination and related intolerance.

In its Declaration, the Conference urged States "to design, implement and enforce effective measures to eliminate the phenomenon popularly known as 'racial profiling' and comprising the practice of police and other law enforcement officers relying, to any degree, on race, colour, descent or national or ethnic origin as the basis for subjecting persons to investigatory activities or for determining whether an individual is engaged in criminal activity."

It becomes important to draw attention to practices of racial profiling involving discrimination during civil aviation operations, at airports at departure and arrival, that have re-emerged, especially after the events of 11 September 2001. Such unacceptable practices, which are diametrically opposed to international tenets and norms of human rights, only succeed in causing insult and injury to individuals discriminated upon, creating rancour, and being totally inconsistent with the fundamental principles of civil aviation as enunciated in the Chicago Convention.

3.2.2 Profiling and the Right of Privacy

Profiling of airline passengers could be carried out primarily through examination of the passenger's passport. At its 33rd Session, held in September/October 2001, the ICAO Assembly, while acknowledging that new measures should be taken to enhance security, observed that such measures should not impede ICAO's ongoing work in improving border control systems at airports and ensuring the smooth flow of passengers and cargo. Consequently, the Assembly stressed that ICAO's work on these issues should continue on an urgent basis. The machine readable travel document was among specific areas mentioned by the Assembly as requiring urgent continuing work, in keeping with UN Security Council Resolution 1373 of 23 September 2001, which re-affirmed the need for continuing work to ensure the integrity and security of passports and other travel documents. In this context the Assembly agreed that all contracting States should be urgently encouraged to issue their travel documents in machine readable format and enhance their security in accordance with ICAO specifications, while introducing automated travel document reading systems at their international airports.

These measures of the ICAO Assembly bring to bear the essential link between aviation security and facilitation and the fact that one cannot be ignored while the other is given some prominence, as is the case with aviation security at the present time.

The data subject, like any other person, has an inherent right to his privacy. The subject of privacy has been identified as an intriguing and emotive one. The right to privacy is inherent in the right to liberty, and is the most comprehensive of rights and the right most valued by civilized man. This right is susceptible to being eroded, as modern technology is capable of easily recording and storing dossiers on every man, woman and child in the world. The data subject's right to privacy, when applied to the context of the machine readable travel document (MRTD) is brought into focus by Alan Westin who says:

> Privacy is the claim of individuals, groups or institutions to determine for themselves when, how, and to what extent information is communicated to others.

There are three rights of privacy relating to the storage and use of personal data:

> 1. the right of an individual to determine what information about oneself to share with others, and to control the disclosure of personal data;
> 2. the right of an individual to know what data is disclosed, and what data is collected and where such is stored when the data in question pertains to that individual; the right to dispute incomplete or inaccurate data; and
> 3. the right of people who have a legitimate right to know in order to maintain the health and safety of society and to monitor and evaluate the activities of government.

It is incontrovertible therefore that the data subject has a right to decide what information about oneself to share with others and more importantly, to know what data is collected about him. This right is balanced by the right of a society to collect data about individuals that belong to it so that the orderly running of government is ensured.

The role played by technology in modern day commercial transactions has affected a large number of activities pertaining to human interaction. The emergence of the information superhighway and the concomitant evolution of automation have inevitably transformed the social and personal life styles and value systems of individuals, created unexpected business opportunities, reduced operating costs, accelerated transaction times, facilitated accessibility to communications, shortened distances, and removed bureaucratic formalities. Progress notwithstanding, technology has bestowed on humanity its corollaries in the nature of automated mechanisms, devices, features, and procedures which intrude into personal lives of individuals. For instance, when a credit card is used, it is possible to track purchases, discovering numerous aspects about that particular individual, including, food inclination, leisure activities, and consumer credit behaviour. In similar vein, computer records of an air carrier's reservation system may give out details of the passenger's travel preferences, *inter alia*, seat selection, destination fondness, ticket purchasing dossier, lodging keenness, temporary address and telephone contacts, attendance at theatres and sport activities, and whether the passenger travels alone or with someone else. This scheme of things may well give the outward perception of surveillance attributable to computer devices monitoring individuals' most intimate activities and preferences, leading to the formation of a genuine "traceable society".

A good airport profiling system must originate from a repository of research based on the characteristics of a person evoking criminal suspiciousness. These characteristics must match the automated passenger information of airlines including the API technology generated prior to a flight. In addition, they must also be consistent with information on travel documents used by passengers. One way of accomplishing this objective is to use the profiles of known or suspected criminals and terrorist categories. Additionally, a diligent and energetic State instrumentality must be established for the purpose of constantly monitoring and ensuring that airport profiling does not discriminate between categories of persons on a subjective basis and that a balanced system of compliance examination is in place.

In the ultimate analysis, the socio-legal relevance of non discrimination in the profiling process lies in the importance of respecting and safeguarding human rights and the rights of certain identified ethnic groups which may form a minority in a particular jurisdiction. Due recognition and active protection of a minority's rights pertaining to racial, cultural and religious issues is the first characteristic of a prudent and balanced profiling process. This feature guarantees freedom from discrimination based on race, language, nationality and national origin or religion. Western democracies, particularly after World War II and the Nuremberg trials which ensured punishment for those responsible for the organized murder of thousands of innocent persons by the commission of atrocities during the war, have been particularly sensitive to the need to ensure human rights. This has led to a gradual evolution where focus on collective rights of national minorities has replaced earlier emphasis on individual rights.

The protection of human rights is the most significant and important task for a modern State, particularly since multi ethnic States are the norm in today's world. Globalization and increased migration across borders is gradually putting an end to the concept of the nation State, although resistance to reality can be still seen in instances where majority or dominant cultures impose their identity and interests on groups with whom they share a territory. In such instances, minorities frequently intensify their efforts to preserve and protect their identity, in order to avoid marginalization. Polarization between the opposite forces of assimilation on the one hand and protection of minority identity on the other inevitably causes increased intolerance and eventual armed ethnic conflict. In such a scenario, the first duty of governance is to ensure that the rights of a minority society are protected.

Racial profiling is an issue related to minority rights and must not be ignored In essence, racial profiling is intersectional in nature and may consist of multiple grounds of institutionalized discrimination such as nationality, race, age, gender, socio-economic status, disability, health status, descent, language, class, culture and religion. At the World Conference against Racism, Racial Discrimination, Xenophobia and Related Intolerance, held in Durban, South Africa from 31 August to 7 September 2001, the Conference, referring to the International Convention on the Elimination of All Forms of Racial Discrimination, urged States to implement or strengthen legislation and administrative measures prohibiting racial discrimination and related intolerance.

Truth and justice are unhappily mutually exclusive. While in legal terms, legislative parameters will define acts and qualitize their reprehensibility, in truth, speech and conduct that ingratiate themselves to a society have to be addressed politically. This is the dilemma that legislators will face in dealing with racial discrimination. Racial discrimination primarily erode ethical boundaries and convey an unequivocal message of contempt and degradation. The operative question then becomes ethical, as to whether societal mores would abnegate their vigil and tolerate some members of society inciting their fellow citizens to degrade, demean and cause indignity to other members of the very same society, with the ultimate aim of harming them. Conversely, the question arises as to whether there is any obligation on a society to actively protect all its members from indignity and physical harm caused by hatred. The answer to both these questions lies in the fundamental issue of restrictions on racial hatred and discrimination and the indignity that one would suffer in living in a society that might tolerate such discrimination. Obviously, a society committed to protecting principles of social and political equality cannot look by and passively endorse such atrocities, and much would depend on the efficacy of a State's coercive mechanisms. These mechanisms must not only be punitive, but should also be sufficiently compelling to ensure that members of a society not only respect a particular law but also internalize the effects of their proscribed acts.

The intrinsic value to a society and perhaps to the whole world, of eschewing racial and national discrimination is portrayed in the aftermath of the Holocaust—the defining event of this century. Human rights in our lifetime cannot be comprehended without touching our own conceptual proximity to this and other recent events which marred the dignity of human civilization. The result of the Holocaust was the adoption by the United Nations of the Universal Declaration of Human Rights which has now stood its ground over the past 50 plus years. The Universal Declaration, which has flourished both internationally and nationally, has been supplemented by the International Covenant on Economic, Social and Cultural Rights adopted in 1966. Both the Universal Declaration and the International Covenant have committees established to oversee their implementation. The Universal Declaration of Human Rights is composed of 30 articles which asserts a human being's just rights to civilized and dignified living. In this context, we are traversing a thin line between genuine, acceptable processes of profiling possible criminal elements and the danger of racial and national discrimination.

3.3 Delayed or Denied Carriage

3.3.1 ICAO

Article 22 of the Chicago Convention Article 22 of which provides, *inter alia*, that each Contracting State agrees to adopt all practical measures, to facilitate and expedite navigation by aircraft ... and to prevent unnecessary delays to aircraft, crews, passengers and cargo. Annex 9 to the Chicago Convention, on facilitation, is the operative document providing guidelines to States on alleviating the lot of the passenger. Annex 9 derives its authority from Article 37 of the Chicago Convention, which vests authority in ICAO to adopt and amend from time to time international standards and recommended practices dealing with facilitation. Numerous provisions of the Annex require States to provide services and facilities that would obviate delays and other procedural impediments that would otherwise act to the detriment of the passenger who undertakes a journey by air.

Chapter 3 of the Annex commences with the general requirement in Standard 3.1 requiring regulations and procedures applied to persons travelling by air not to be less favourable than those applied to persons travelling by other means of transport. This provision establishes, *in limine*, a certain parity of status between air transport and surface transport, particularly in relation to passenger clearance at border crossings and entry points. Standard 3.4 requires that Contracting States shall not require from visitors travelling by air any other document of identity than a valid passport. Although this provision could be perceived as unduly restrictive on the sovereign prerogative of a State to require specific entry documentation from persons, it is not intended to discourage States, who wish to be more liberal, from accepting official documents of identity other than passports in lieu of valid documents of identity. The Annex also contains provisions calculated to simplify travel documents; open access to facilities to obtain a passport; improve quality of passports, visas and other travel documents; make easier the visa obtaining process; expedite entry to and exit from States visited; obviate difficulties of transit through States not intended to be visited; and ensure the rapid return of baggage which becomes separated from the owners in the transportation process.

The utility of air travel is often diluted by the apprehension an airline passenger feels when he approaches the check-in counter to obtain his seat allocation. On the other side of the counter are the check-in staff of the airline, with their computers, waiting to punch the passenger's name into their machines, take over his baggage for the flight and issue him with his boarding card. However much a passenger reconfirms his seat long before the flight, he could, in some circumstances be "bumped" (denied boarding) on the flight he will not get on his flight and would have to wait for the next flight…or the next…or the next…whenever that may be. Worse still, most airlines include a condition in their passenger tickets and baggage checks that they do not guarantee dates and times of departure and arrival that are indicated in the documents of carriage.

One of the unpleasant side effects of being deprived an opportunity to board an aircraft that one believes holds a confirmed seat for him, is the ensuing disagreement between the passenger and airline staff. Usually, instances of denied boarding are unpleasant both to the passenger and the airline check-in-staff. Some jurisdictions such as the United States have envisaged circumstances wherein a passenger could arbitrarily be denied boarding, leading to the imputation of discrimination on the part of the airline. These jurisdictions have incorporated statutory provisions for remedies in instances of discriminatory "bumping".

Admittedly, denied boarding is not a regular feature in commercial aviation any more. Airline reservation systems are now more streamlined and booking profiles of most airlines are conservative and prudent. Unfortunately however, there are still instances where airlines overbook (accept reservations for more passengers than a flight could actually take) or oversell, on the expectation that a certain percentage of passengers would be "no shows" (fail to turn up for the flight). As one commentator observed:

> Although most airlines deliberately overbook, it is often not done with the intent to defraud passengers or to deny advertised and contracted services, but rather for the purpose of limiting losses due to the failure of passengers to use their reservations. The effect on passengers of overbooking and overselling may be similar, in that harassment, mental anguish, distress, discrimination and material damages are likely to arise. However, the expectations of the parties to the contract and the legal implications are in the end clearly different.
>
> Booking is generally defined as the process by which a reservation is made for the passenger until the passenger confirms at a later date that reservation with a contract of carriage. The carrier confirms the reservation and guarantees the seat only when the passenger pays the price and is given the ticket, that is, when the seat on the aircraft has been sold to the passenger. An airline reservation for a seat on a flight, therefore, merely amounts to a promise of future performance, on the condition that the number of bookings and sales is not greater than the seating capacity of the aircraft. However, in a widely-adopted international airline practice, the number of bookings often exceeds seating capacity, with the promise of future performance becoming void upon entry into the contract, that is, at the time when the passenger discovers that his or her reservation cannot be confirmed. From the outset, therefore, the intent of the carrier may seem less than well-meaning. Overbooking, as well as the logical consequence of possible overselling, implies that a passenger, who believed he or she was going to fly, will necessarily be bumped.[69]

In the 1976 case of *Nader* v. *Allegheny Airlines Inc.*,[70] Justice Powell, addressing the problem of overbooking observed:

> Such overbooking is a common industry practice, designed to ensure that each flight leaves with as few empty seats as possible despite the large number of "no-shows" – reservation-holding passengers who do not appear at flight time. By the use of statistical studies of no-show patterns on specific flights, the airlines attempt to predict the appropriate number of reservations necessary to fill each flight. In this way, they attempt to ensure the most efficient use of aircraft while preserving a flexible booking system that permits passengers to cancel and change reservations without notice or penalty. At times the practice of overbooking results in oversales, which occur when more reservation-holding passengers than can be accommodated actually appear to board the flight. When this occurs, some passengers must be denied boarding, that is, they are "bumped". The chance that any particular passenger will be bumped is so negligible that few prospective passengers aware of the possibility would give it a second thought.[71]

A last minute aircraft change from a larger aircraft to a smaller one (mostly due to unforeseen technical problems or aircraft scheduling difficulties) is another reason for a passenger to be denied boarding. The good news however, is that "bumping" is often a "democratic" process where a busy airline executive, an even busier lawyer, and a footloose tourist may all get "bumped" together from the same flight.

Of course, delinquent airlines do their best to make amends. Some even offer cash gifts and plush hotel accommodation (with full board) on the house. Although this mode of "conciliation" sometimes proves to be an unexpected "bonus" to a

---

[69] Guerreri (1989), p. 192.

[70] 426 US 290 (1976); 14 Av LR 17, 148.

[71] *Id.* at 294.

tourist who would rather leave his holiday resort the next day (or better still 3 days later), the airline executive and the lawyer, who are both tightly scheduled for their overseas meetings the same day would lose immensely if they do not find accommodation on the flight that they thought they had obtained "confirmed" reservations well in advance. Would these hapless passengers have a legal remedy against the errant airline? what damages could they claim?

The ICAO *Manual on the Regulation of International Air Transport*[72] has several statements on the subject of delays or cancellation of flights, one of which is that, the passenger is informed that when (due to circumstances beyond their control) airlines cancel or delay a flight or cause a passenger to miss a connection, most airlines accept the obligation to refund or provide alternative flight arrangements at no further cost to the passenger, but do not generally accept any further liability for any damages incurred as a result (for example, lost vacation or work time). Advice that where major financial or other losses are incurred as a consequence of delay, passengers may wish to ascertain their rights at law.

In case of delay to passengers, most airlines customarily provide certain amenities, such as the free provision or reimbursement of the cost of communications, meals, refreshments, transit taxes and, usually in circumstances of prolonged delay during the night hours (at points which are not origin, destination or stopover points), hotel accommodation and ground transportation. In this context, the passenger is provided with advice that in the case of delayed baggage, airlines may, in certain circumstances, provide or reimburse passengers for the purchase of personal necessities.

On reconfirmation of flights, many airlines require reconfirmation of an onward reservation wherever a journey is broken for more than 72 h, and failure to comply may result in cancellation of the reservation in question, together with those for any subsequent flights in the journey. This is a crucial issue for most passengers, particularly regular travellers who may neglect reconfirmation of flight.

Advice as to complaint procedures and avenues for action, including information on differing time periods within which formal claims must be made is also a key piece of information recommended by ICAO that States disseminate to passengers.

An essential pre requisite for the passenger is his awareness of the conditions of carriage and ICAO strongly recommends that passengers (and air carriers or charterers) are made to understand that they are bound by conditions appearing on the ticket, by tariff conditions, by carrier or charterer regulations and, outside Canada and the United States, by conditions of carriage, all of which should be readily available from the carriers or charterers for reference. Also of importance is the fact that all these conditions may vary for different segments of a journey, if provided by more than one airline.

On the issue of liability, a brief summary of limits of liability as provided for by international treaty (supplemented in some instances by other arrangements) and available for reference in summary on the ticket coupon and in detail in air carrier

[72] *Doc 9626*, Second Edition 2004.

conditions of carriage (air carrier tariffs in Canada and the United States) are considered essential information for the passenger. ICAO also recognizes the need for, and recommends strongly that States concerned make available to the passenger a glossary of descriptions commonly used air carrier terms and abbreviations.

3.3.2 Europe

The European Commission, in early 2000, issued a comprehensive consultative document on consumer protection in Air Transport.[73] This statement introduced a common air transport policy and raised a number of issues such as the contractual rights of passengers, tariffs, and comfort and health. This paper maps out an inventory of passenger rights containing the following:

Information about Flights and Reservation of Tickets—Any passenger has the right to neutral and accurate information; *Check-in and boarding*—European Community rules require that passengers must receive fair treatment and proper compensation when they are denied boarding; *Liability in Case of an Accident*; Passengers travelling with an European Community airline will receive full compensation in the case of an accident.

Data protection; Passengers reserving their tickets in the EC have the right to know what personal details about them are being stored on the computerized reservation systems; *Air Travel as Part of a Package Holiday*—Air passengers travelling as part of a package tour or holiday bought in the EC must receive clear and precise information from the organizer about their trip. They also enjoy clear rights concerning the performance of the contract; *Enforcing Passenger Rights;* The passenger rights set out above are laid down either directly in EC law or in national laws which have been made to implement EC Directives. Therefore, Airlines, travel agents, tour operators and all other business involved in the provision of air transport services must observe them. In turn, passengers should stand firm in demanding that their rights are respected and complain when they are not.

Regulation (EC) No. 261/2004[74] of the European Parliament and Council contains the basic principle that denied boarding and cancellation or long delay of flights cause serious trouble and inconvenience to passengers and that action by the Community in the field of air transport should aim, among other things, at ensuring a high level of protection for passengers. Moreover, full account should be

[73] Air Passenger Rights in the European Union, European Commission, Regulatory Affairs Review, January–March 2000, p. 47.

[74] On 17 February 2005, EC Regulation 261/2004 establishing common rules on compensation and assistance to passengers in the event of denied boarding and of cancellation or long delay of flights, entered into force across the European Community. Member States are required to ensure and supervise general compliance by their air carriers with this legislation and to lay down rules on sanctions applicable to infringements that are effective, proportionate and dissuasive.

taken of the requirements of consumer protection in general. This regulation has its genesis in Council Regulation (EEC) No 295/91 of 4 February 1991 which established common rules for a denied boarding compensation system in scheduled air transport. Regulation 261/2004 goes on to say that the trouble and inconvenience to passengers caused by cancellation of flights should also be reduced. This should be achieved by inducing carriers to inform passengers of cancellations before the scheduled time of departure and in addition to offer them reasonable re-routing, so that the passengers can make other arrangements. Air carriers should compensate passengers if they fail to do this, except when the cancellation occurs in extraordinary circumstances which could not have been avoided even if all reasonable measures had been taken. A salient principle in the regulation is that passengers whose flights are cancelled should be able either to obtain reimbursement of their tickets or to obtain re-routing under satisfactory conditions, and should be adequately cared for while awaiting a later flight.

The Regulation cites the principle embodied in the Montreal Convention of 1999 (which will be discussed in some length in the next chapter) in that as under the Montreal Convention, obligations on operating air carriers should be limited or excluded in cases where an event has been caused by extraordinary circumstances which could not have been avoided even if all reasonable measures had been taken. Such circumstances may, in particular, occur in cases of political instability, meteorological conditions incompatible with the operation of the flight concerned, security risks, unexpected flight safety shortcomings and strikes that affect the operation of an operating air carrier.

The Regulation in Whereas clause 15 states that extraordinary circumstances should be deemed to exist where the impact of an air traffic management decision in relation to a particular aircraft on a particular day gives rise to a long delay, an overnight delay, or the cancellation of one or more flights by that aircraft, even though all reasonable measures had been taken by the air carrier concerned to avoid the delays or cancellations.

Whereas clause 17 requires that passengers whose flights are delayed for a specified time should be adequately cared for and should be able to cancel their flights with reimbursement of their tickets or to continue them under satisfactory conditions. The next provision states that care for passengers awaiting an alternative or a delayed flight may be limited or declined if the provision of the care would itself cause further delay. Whereas clauses that follow go on to say that operating air carriers should meet the special needs of persons with reduced mobility and any persons accompanying them and that passengers should be fully informed of their rights in the event of denied boarding and of cancellation or long delay of flights, so that they can effectively exercise their rights. Member States are called upon to lay down rules on sanctions applicable to infringements of the provisions of the Regulation and ensure that these sanctions are applied. The sanctions should be effective, proportionate and dissuasive in addition to ensuring and supervising general compliance by their air carriers with the Regulation and designate an appropriate body to carry out such enforcement tasks. The supervision should not affect the rights of passengers and air carriers to seek legal redress from courts under procedures of national law.

Article 4 of the Regulation lays down the principle that when an operating air carrier reasonably expects to deny boarding on a flight, it shall first call for volunteers to surrender their reservations in exchange for benefits under conditions to be agreed between the passenger concerned and the operating air carrier. Volunteers shall be assisted in accordance with Article 8,[75] such assistance being additional to the benefits mentioned in the paragraph.

If an insufficient number of volunteers comes forward to allow the remaining passengers with reservations to board the flight, the operating air carrier may then deny boarding to passengers against their will. If boarding is denied to passengers against their will, the operating air carrier shall immediately compensate them in accordance with Article 7 and assist them in accordance with Articles 8 and 9.

Article 6 of the Regulation addresses the issue of how to complain against delay. It states that when an operating air carrier reasonably expects a flight to be delayed beyond its scheduled time of departure: for 2 h or more in the case of flights of 1,500 km or less; or for 3 h or more in the case of all intra-Community flights of more than 1,500 km and of all other flights between 1,500 and 3,500 km; or for 4 h or more in the case of all flights not falling under the above two criteria. In these instances passengers shall be offered by the operating air carrier: free of charge: meals and refreshments in a reasonable relation to the waiting time; hotel accommodation in cases where a stay of one or more nights becomes necessary, or where a stay additional to that intended by the passenger becomes necessary; transport between the airport and place of accommodation (hotel or other) In addition, passengers shall be offered free of charge two telephone calls, telex or fax messages, or e-mails. Article 6 also states that when the reasonably expected time of departure is at least the day after the time of departure previously announced, hotel accommodation in case where a stay of one or more nights becomes necessary, o where a stay additional to that intended by the passenger becomes necessary.

Article 7, which pertains to the quantum of compensation states that where reference is made to this Article, passengers shall receive compensation amounting to: EUR 250 for all flights of 1,500 km or less; EUR 400 for all intra-Community flights of more than 1,500 km, and for all other flights between 1,500 and 3,500 km; EUR 600 for all flights not falling under the above two criteria. In determining the

[75] Article 8 provides that passengers shall be offered the choice between: reimbursement within 8 days, by the means provided for in Article 7, of the full cost of the ticket at the price at which it was bought, for the part or parts of the journey not made, and for the part or parts already made if the flight is no longer serving any purpose in relation to the passenger's original travel plan, together with, when relevant; a return flight to the first point of departure, at the earliest opportunity; re-routing, under comparable transport conditions, to their final destination at the earliest opportunity; or re-routing, under comparable transport conditions, to their final destination at a later date at the passenger's convenience, subject to availability of seats.

distance, the basis shall be the last destination at which the denial of boarding or cancellation will delay the passenger's arrival after the scheduled time. When passengers are offered re-routing to their final destination on an alternative flight pursuant to Article 8, the arrival time of which does not exceed the scheduled arrival time of the flight originally Booked by 2 h, in respect of all flights of 1,500 km or less; or by 3 h, in respect of all intra-Community flights of more than 1,500 km and for all other flights between 1,500 and 3,500 km; or by 4 h, in respect of all flights not falling under the above. The operating air carrier may reduce the compensation provided for in paragraph 1 by 50 %. The compensation is required to be paid in cash, by electronic bank transfer, bank orders or bank cheques or, with the signed agreement of the passenger, in travel vouchers and/or other services.

Article 14 is also an important provision pertaining to passenger information. It requires the carrier to ensure that at check-in a clearly legible notice containing the following text is displayed in a manner clearly visible to passengers: "If you are denied boarding or if your flight is cancelled or delayed for at least 2 h, ask at the check-in counter or boarding gate for the text stating your rights, particularly with regard to compensation and assistance". Article 14 also states that the carrier denying boarding or cancelling a flight shall provide each affected passenger with a written notice setting out the rules for compensation and assistance in line with the Regulation. It is also required to provide each passenger affected by a delay of at least 2 h with an equivalent notice. There is also the requirement that contact details of the national designated body should also be given to the passenger in written form. Article 14 will be applied using appropriate alternative means in respect of blind and visually impaired persons.

In September 2014 The European Court of Justice (ECJ) issued a ruling on arrival at the destination which said that Regulation (EC) No 261/2004 of the European Parliament and of the Council, must be interpreted as meaning that the concept of 'arrival time', which is used to determine the length of the delay to which passengers on a flight have been subject, refers to the time at which at least one of the doors of the aircraft is opened, the assumption being that, at that moment, the passengers are permitted to leave the aircraft. In an earlier ruling the ECJ in November 2009 stated Regulation (EC) 261/2004 involved cash compensation similar to flight cancellations if the delay is 3 h or longer at the destination. However, this unless the airline can prove that the delay was caused by 'extraordinary' circumstances. Before the ruling cash compensation was only rewarded for cancelled flights but not for delays. Flight delays prior to the ruling gave the passenger the right to assistance free of charge as mentioned above. The crux of the ECJ's judgment was that Passengers on flights starting or ending in the 27-country European Union are entitled to between 250 euros (£203) and 600 euros for delayed or cancelled flights under EU rules. They also apply to any flight outside the EU which takes place on an airline company based in the Union. These rights still apply to the air passengers when faced with a delay. This Regulation is applicable to all worldwide airlines when departure takes place within the EU and, in the case of flights from outside the EU to a destination within the EU, only to airlines licensed in a Member State of the EU.

3.3.3 United States

The general philosophy pertaining to air travel in the United States is that passengers should be aware, when purchasing their tickets that airlines don't guarantee their schedules. Each airline has its own policies about what it will do for delayed passengers waiting at the airport; there are no federal requirements. Airlines are also generally exempt from liability for delay in instances of hard to predict bad weather, air traffic delays, and mechanical repairs, are hard to predict and beyond the airlines' control. However, liability may be ascribed to an airline or airport for predicted delays particularly in instances of bad weather where no adequate preparations or precautionary measures are taken. In the case of *Vumbaca* v. *Terminal One Group Association L.P.*[76] decided in April 2012 by the United States District Court, E.D. New York. Vivian Vumbaca—the Plaintiff—an Italian citizen who was a permanent resident of the United States who arrived in New York during the snow storm of 26–27 December 2010 from Rome, alleged that she was kept locked in an aircraft on the ground without food, water, or adequate sanitary facilities for 7 h, suffering mental distress. She sued Terminal One Group Association, L.P. (TOGA), which operates Terminal One, and sought to represent similarly situated passengers claiming emotional harms resulting from negligence, false imprisonment, and intentional infliction of emotional distress under her contract of carriage on the ground that Terminal One Group Association did not afford her the facility of disembarking at her destination and kept her on board the aircraft for 7 h causing her mental distress.[77]

In terms of jurisdictional competence of the court, the court recognized Article 33 of the Montreal Convention[78] which stipulates that an action for damages must be brought, at the option of the plaintiff, in the territory of one of the States Parties, either before the court of the domicile of the carrier or of its principal place of business, or where it has a place of business through which the contract has been made or before the court at the place of destination. The Convention goes on to say that In respect of damage resulting from the death or injury of a passenger, an action

[76] 859 F Supp. 2d. 353 (E.D.N.Y).

[77] From December 26th to 27th, 2010, during the height of the holiday travel season, the New York metropolitan area was—somewhat unexpectedly—blanketed with over a foot of snow. John F. Kennedy International Airport (JFK) was closed to air traffic for the worst of the storm. When it reopened, there were continuing problems. Passengers on arriving flights were forced to endure substantial waits after landing before they were able to disembark. Difficulties appear to have been particularly severe at terminals serving international flights.

[78] The Convention for the Unification of Certain Rules for International Carriage by Air which will be discussed in some detail in the next chapter.

may be brought before one of the courts mentioned in paragraph 1 of Article 33, or in the territory of a State Party in which at the time of the accident the passenger has his or her principal and permanent residence and to or from which the carrier operates services for the carriage of passengers by air, either on its own aircraft or on another carrier's aircraft pursuant to a commercial agreement, and in which that carrier conducts its business of carriage of passengers by air from premises leased or owned by the carrier itself or by another carrier with which it has a commercial agreement.

The court held that the Montreal Convention is an international treaty to which the United States is a party, and which establishes a uniform system of liability for international air carriers.

The plaintiff averred that her claim was based on Article 30 of the Montreal Convention which stipulates that if an action is brought against a servant or agent of the carrier arising out of damage to which the Convention relates, such servant or agent, if they prove that they acted within the scope of their employment, shall be entitled to avail themselves of the conditions and limits of liability which the carrier itself is entitled to invoke under this Convention. It was the contention of the plaintiff that Terminal One Group Association, L.P. (TOGA) was an agent of the carrier and therefore could be held liable under the Montreal Convention.

The court held *inter alia* that The Convention does not define "agent" and that the Supreme Court has provided no guidance. It also observed that the Court of Appeals for the Second Circuit had held that the airline employees are agents covered by the Convention[79] and that the Convention's "basic principle" required that air carriers be "protected from having to pay out more than a fixed and definite sum for passenger injuries sustained in international air disasters."

The Court held:

> As a general rule, whenever one person is by circumstances placed in such a position with regard to another that every one of ordinary sense who did think would at once recognize that if he did not use ordinary care and skill in his own conduct with regard to the circumstances he would cause danger of injury to the person or property of the other, a duty arises to use ordinary care and skill to avoid such danger.[80] The court's power to modify this rule "is reserved for very limited situations. ...In determining whether a defendant had a duty of care toward the plaintiff, courts look to morality, logic, and the social consequences of imposing the duty."[81]

The Court found that the defendant TOGA had acted as an agent of the carrier. Despite the fact that TOGA was a terminal operator, its services were vital to the performance by the carrier of the contract of carriage with the plaintiff and therefore formed an integral part thereof and that TOGA had a common law duty to ensure that passengers on arriving flights had safe and prompt access to Terminal One. It provided a necessary link in the chain of transportation, facilitating the common

[79] Quoting *Reed* v. *Wiser*, 555 F.2d. at 1089–93.

[80] *Havas v. Victory Paper Stock Co.*, 49 N.Y.2d. 381 (1980).

[81] *Eiseman v. State of New York*, 70 NY.2d. 175 (1987).

carrier airline's service of its passengers. The defendant's duty to passengers extended beyond merely ensuring that stairs are not slippery, or that gates are properly maintained. It must ensure that those stairs and gates are made available in a timely manner when needed for use by the passengers, and that adequate ground handling staff is present to facilitate access. The Court further observed that TOGA should have foreseen that a breach of this duty would cause passengers to remain trapped on their aircraft in cramped and increasingly unpleasant or dangerous conditions. Imposing on defendant this duty was neither novel nor undesirable as a matter of public policy.

The court concluded:

> In the instant case, the defendant was in a favorable position to prevent harm to the plaintiff. TOGA—arguably, unlike the air carrier, Alitalia—was fully aware of the situation on the ground and its inability to handle incoming flights. It should have taken a robust position, ordering Alitalia not to let its flights to JFK take off since no disembarkation facilities were available. Instead, it made a soft, timid suggestion that it might not be able to handle incoming flights adequately, creating no strong sense of the serious dangers in letting planes take off for New York.[82]

The Court also held that TOGA had sole responsibility for managing the gates by which passengers moved between the terminal and an airplane. This included the duty and discretion to select a ground handling provider to move planes to and from the gates.

The recognition that an airport operator could be considered an agent of the airline establishes an inextricable link between the two—a link hitherto not recognized—places airports in a new light and widens the scope of their liability and responsibility.

In May 2001 at the 107th Congress a Bill was presented introducing the Airline Passenger Bill of Rights Act, which, inter alia contained a provision that an air carrier shall ensure access to necessary services and conditions, including food, water, restroom facilities, and the ability to deplane in the event of a weather or other emergency, for all passengers boarded on a flight segment of the air carrier in air transportation. The proposed legislation also carried a provision that an announcement by an air carrier of (A) a delay or cancellation of a flight segment, or (B) a diversion of a flight segment to an airport other than the airport at which the flight segment is scheduled to land, shall include an explanation of the reason or reasons for the delay, cancellation, or diversion.[83]

In April 2009 the Department of Transport (DOT) proposed a comprehensive and final Bill titled *Enhancing Airline Passenger Protection*[84] which *inter alia* required Under the proposed rule, a certificated or commuter air carrier 1 that operates domestic scheduled passenger service or public charter service using any aircraft with 30 or more passenger seats 2 would be required to develop and

[82] *Supra*, note 76 (this chapter) at 371.

[83] https://www.govtrack.us/congress/bills/107/hr1734/text.

[84] 14 CFR Parts 234, 253, 259, and 399 [Docket No. DOT-OST-2007-0022] RIN No. 2105-AD72.

implement a contingency plan for lengthy tarmac delays. As proposed, it would apply to all of a covered U.S. carrier's flights, both domestic and international, including those involving aircraft with fewer than 30 seats if a carrier operates any aircraft with 30 or more passenger seats.

The Bill also envisioned that protection of passengers on international flights on board US carriers was as important as those on domestic flights and recommended that passengers on international flights are also afforded protection from unreasonably lengthy tarmac delays. Therefore, the proposed Bill applied the requirement to develop and implement a contingency plan for lengthy tarmac delays to both the domestic and international flights of each U.S. carrier operating any aircraft with 30 or more passenger seats. This requirement was calculated to apply to U.S. carriers even if they operated only international scheduled or charter service.

In April 2011 US Transportation Secretary announced new rules pertaining to consumer protection regarding airline passengers. According to these rules the amount of compensation that passengers are eligible to receive for denied carriage (equal to a ticket price of $400), if the airline is able to get them to their destination within a short period of time (i.e., within 1–2 h of their originally scheduled arrival time for domestic flights and 1–4 h of their originally scheduled arrival time for international flights), was doubled. Passengers denied boarding due to overselling by the airline were therefore made eligible for compensation up to $800 for a ticket for which they paid $400, if they are delayed for a lengthy period of time (i.e., over 2 h after their originally scheduled arrival time for domestic flights and over 4 h after their originally scheduled arrival time for international flights). Under the new rule, passengers who are denied boarding subject to short delays receive compensation equal to double the price of their tickets up to $650, while those subject to longer delays would receive payments of four times the value of their tickets, up to $1,300. Inflation adjustments will be made to those compensation limits every 2 years.

In terms of delays on the tarmac, the new rule extended the existing ban on lengthy tarmac delays that only covered US carriers to apply to foreign airlines' operations at U.S. airports and establishes a 4 h hard time limit on tarmac delays for international flights of U.S. and foreign airlines, with exceptions allowed only for safety, security or air traffic control-related reasons. A fundamental right recognized by the Rule in such instances is that carriers are required to ensure that passengers delayed on the tarmac are provided adequate food and water after 2 h, as well as working lavatories and any necessary medical treatment.[85] The DOT Rule discussed above accomplishes its goal of requiring airlines to allow reservations to be held at the quoted fare without payment, or cancelled without penalty, for at least 24 h after the reservation is made, if the reservation is made one week or more prior

[85] The *Vumbaca* case, referred to earlier in the text of this chapter was one of the reasons as well as the extended tarmac delays experienced by passengers on international flights operated by foreign carriers at New York's JFK Airport during the December 2010 blizzard were important factors in the DOT's decision to extend the tarmac delay provisions to foreign air carriers and establish a four hour tarmac delay limit for international flights. *Supra*, note 71 (this chapter).

to a flight's departure date. It also requires airlines to promptly notify consumers of delays of over 30 min, as well as cancellations and diversions. This notification must take place in the boarding gate area, on a carrier's telephone reservation system and on its website.

The FAA Modernization and Reform Act of 2012 (P.L. 112-95), signed into law by the President on February 14, 2012, contained numerous provisions regarding the rights of airline passengers and introduced a more robust legislative basis for the rules adopted by DOT in 2009 and 2011. Section 406 of the Act directs the Inspector General (IG) of DOT to review and assess air carrier flight delays, cancellations, and their causes. Congress also required the IG to update its 2000 report, "Audit of Air Carrier Flight Delays and Cancellations" and submit the review results to the House and the Senate within 1 year after enactment.

3.3.4 International Treaty Law

3.3.4.1 The Warsaw Convention of 1929

Article 19 of the Warsaw Convention[86] provides that the carrier shall be liable for damage occasioned by delay in the transportation by air of passengers, baggage and goods. Legally, the notion of delay implies a discrepancy between the time when one party is entitled to expect the performance of the carrier's duty and the time when the duties are actually performed.[87] Thus when a passenger is denied boarding on a flight he has contracted with the airline to board, the airline becomes *prima facie* liable for the damage caused by the delay, provided the passenger was subsequently carried by that airline after some delay. Therefore, in an instance where an airline fails to carry a passenger at all it is not envisaged that Article 19 of the Warsaw Convention would apply. One of the perceived inadequacies of the Warsaw Convention is the absence of a definition of "delay". The lapse is aggravated by the words of Article 18(2) which implicitly provide for a definition of "delay" in relation to the carriage by air of baggage and goods. The "delay" referred to is one envisaged to be caused when the baggage or goods are in the carrier's charge. Therefore, even if Article 18(2) is to be applied analogically to passengers it would be regarded only to cover cases where a passenger had already been accepted

[86] Convention for the Unification of Certain Rules Relating to International Transportation by Air, 12 October 1929, 137 L.N.T.S. 11 49 Stat. 3000, TS No 876, ICAO Doc 7838. The Warsaw Convention was later amended by several protocols, none of which changed the substantive provisions of the Convention. The Warsaw Convention establishes a regime of liability where the liability of the carrier is presumed, once a plaintiff claims that he has suffered damage in the hands of the carrier. The carrier has to rebut this presumption. To balance this deviation from normal evidential rules, the Convention has set up limits of the carrier's liability in terms of monetary compensation payable to the plaintiff if the carrier is unable to rebut the presumption of liability.

[87] Miller (1977), p. 154.

for carriage. Most instances of denied boarding, which cause a delay in the passenger's transportation is caused when the passenger is not checked into the flight at all. The latter instance is not alluded to by the Warsaw Convention. Even Article 19 can directly be applied only to instances where a passenger is accepted for a flight and delayed in the process of being carried to his destination. It may, however, still be arguable that liability as referred to in Article 19 can be imputed to the carrier in cases of denied boarding, since anyway, the act of denial of boarding results in a delay in the carriage of a passenger.

In a 1956 decision,[88] which discussed the carriage of goods by air, the courts interpreted Article 19 to mean that the carrier was bound to deliver goods quickly to shippers, since shippers resort to carriage by air only because it allows quick delivery. By analogy therefore, the decision would show early tendencies of the courts to be supportive of enforcing passengers' rights under Article 19 to the speed inherent in air transport. In a later case[89] the courts articulated this attitude by laying down the law to be stringent in the requirement that airlines cannot be exonerated when they cause long delays, flight cancellations or postponement of flights.[90]

Early common law attitudes were seemingly compatible with the early civil law attitude discussed above. In the case of *Bart* v. *British West India Airways Ltd.*[91] the court held that the carrier cannot be left to deliver passengers and goods as he pleases and had to ensure delivery within a reasonable time having regard to all the circumstances of a particular case.

The Warsaw Convention, as could be expected, is silent on the question of damages caused by the delay in the carriage of passengers. The determination of damages therefore devolves upon the *lex fori*—the court seized with jurisdiction to hear the case. Accordingly, it is evident that damages in each case would be assessed in accordance with the internal laws of the country in which a case is heard. The carrier can exonerate himself in certain circumstances, where he shows that he took all the necessary precautions to avoid the delay,[92] or that the contributory negligence of the plaintiff was a causative factor of the delayed carriage.[93] The court is empowered by the Warsaw Convention to exonerate the carrier totally or partially from liability. Article 25 of the Warsaw Convention precludes the carrier from availing himself of the limitation of liability provisions of the Convention if the damage is caused by the carrier's wilful misconduct or by such default on his part as, in accordance with the law of the court to which the case is submitted, is considered to be equivalent to wilful misconduct. Similarly, the carrier is not able to seek the exclusions and limitations of liability under the Warsaw Convention if the damage is caused under the same circumstances by any agent of the carrier

[88] *Ste General Air fret c. Ste TWA Trans World Airlines* (1956) 10 RFDA 324.

[89] *Sonillac* v. *Air France* (1965) 28 R.G.A.E. 15.

[90] *Id.* at 17.

[91] (1967) 1 Lloyds Rep. 239 (Guyana Ct. App. 1966).

[92] *Warsaw Convention*, *supra*, note 86 (this chapter), Article 20.

[93] *Id.* Article 21.

acting within the scope of his employment. Any action grounded on damages arising from delay in the carriage of passengers, however founded, can only be brought subject to the conditions and limits set out in the Convention.

Whenever an airline issues a ticket to a passenger in return for the payment of money, the airline undertakes to carry the passenger to the destination given in his ticket with reasonable care and within a reasonable time.[94] Courts have accordingly followed the principle that the airline is professionally bound to carry a passenger who has been issued with a passenger ticket[95] and that the airline is bound by the following legal parameters:

The airline is bound to carry any person, without the imposition of any unreasonable conditions, if he is not in an unfit condition, and has paid his fare; the airline has to carry the passenger safely; the passenger has to be treated with due care throughout his flight; and, the airline has to ensure that the passenger is given the benefit of the speed inherent in air transport.

Of course, the parties to a contract of carriage of passengers may generally incorporate in the contractual document (the airline passenger ticket) any terms and conditions upon which they agree, as long as such conditions are not contrary to the provisions of the Warsaw Convention.[96]

The terms and conditions in an airline ticket are not binding upon the passenger unless he consents to them either expressly or impliedly.[97] Although generally, if a passenger accepts his ticket without objection, he will be taken to have accepted the conditions in his ticket, a court would be very reluctant to accept strictly a condition in a ticket that totally exonerates an airline where it fails to carry a passenger with reasonable dispatch. Also, as the *Lisi* case held, the passenger cannot be taken to have assented to conditions of carriage incorporated in his ticket if he has no opportunity to examine his ticket or dissent from the provisions thereof,[98] although the airline could, in its defence, prove that the passenger knew of the condition in the ticket even though he did not read the provisions incorporated in his ticket or that the passenger was given notice of the existence of the conditions in the ticket.[99]

Clearly, "bumping" of passengers by airlines is therefore not an eventuality that the Warsaw Convention has considered *strict sensu*. It would be largely left to the

---

[94] Corrigan (1978), p. 25.

[95] *Clark* v. *West Ham Corp*., (1909) 2 K.B. 858, *Readhead* v. *Midland Railway Co*., (1869) L.R. 4 Q.B. 382, *Overseas National Airways* v. *C.A.B*., 307 F. 2d. 634, *U.S.* v. *Stephen Bros. Lines*, 384 F. 2d. 118, *S.M.T. Ltd.* v. *Ruch*, 50 C.R.T.C. 369, *Roussel* v. *Aumais* 18 Que. S.C. 474, *Thibault* v. *Garneau* (1959) Que. P.R. 377.

[96] *Warsaw Convention, op.cit.* Article 32.

[97] *Watkins* v. *Rymill* (1883) 10 Q.B. 178.

[98] See also generally the common law decision of *Thornton* v. *Shoe lane Parking* (1971) 2 Q.B. 163, which upheld the contractual principle that for a contractual clause to be considered applicable the parties to the contract should have the opportunity to read the clause and to dissent from it.

[99] *Montreal Trust and Stampleman* v. *CP Air* (1976) 72 D.L.R. (3d) 282.

courts to interpret the Convention on this point with the help of Article 19, the analogous case law, and the general principles of the law of contract.

Courts in the United States have in recent times had occasion to view denied boarding in its various contexts. In March 1995, in the case of *Azubuko* v. *Varig Airline*,[100] the trial court had to determine an instance where transportation was refused to a validly ticketed passenger because of an erroneous belief by the carrier's employee that the passenger was required to have a visa to his destination. The air carrier sought dismissal of the plaintiff's claim for breach of contract amounting to $35,000,000 in damages for alleged loss of business as a result of the denied boarding on the ground that the claim was based upon a refusal to transport and was pre-empted by the *Airline Deregulation Act* of 1978.[101] The court held that the plaintiff was seeking only to enforce a private agreement for transportation and that the case did not involve any enlargement or enhancement of the agreement based on state laws or policies external to the terms of the contract of transportation. The reasoning in this case arguably has its genesis in the rationale of the PANAM adjudication relating to the Lockerbie case, where the court held that the Warsaw Convention preempts all State law causes of action falling within the purview of the Warsaw Convention.[102] One court has distinguished between claim arising as a result of a delay in transportation and a claim arising from total non-performance of the contract of transportation where, for instance, after being denied boarding, the passenger does not travel at all. In the latter instance, the court held that the claim is not covered by the Warsaw Convention and the passenger may pursue a remedy based on local law rather than under the Convention.[103] A clear distinction has to be made however, between non-performance and subsequent transportation of the passenger, where in the latter case, the delay in transportation would clearly come under the Warsaw Convention's Article 19, thus preempting local law on the subject in the United States.[104]

In the case of *Lathigra* v. *British Airways PLC*[105] it was alleged that the issuing carrier British Airways neglected to inform the passenger that one of the reconfirmed connecting flights operated by another carrier had been discontinued. As a result, the passengers were stranded *en route* for 5 days. British Airways applied to remove the plaintiff's claim to a federal court on the ground that it fell within the purview of the Warsaw Convention. The plaintiff applied to the Court of Appeal on the ground that the claim did not arise under the Warsaw Convention. In addressing the question of whether the claims arose under Article 19 of the Warsaw

---

[100] An account of this case appears in *Lloyd's Aviation Law*, Vol. 15, No. 9 May 11 1995 at 2.

[101] 49 U.S. App. S. 1305 (1994).

[102] See *In re Air Disaster at Lockerbie, Scotland*, 928 F.2d. 1267, at 1273 (2d Cir. 1991).

[103] See *Wogel* v. *Mexicana Airlines* 821 F.2d 442 (7th Cir. 1987).

[104] *Malik* v. *Butta*, unreported. See *Lloyd's Aviation Law*, Vol. 12, No. 21, November 1 1993 at 1 for a discussion of this case.

[105] Unreported at the time of writing. This case is discussed in *Lloyd's Aviation Law*, Vol. 13, No. 22, November 15, 1994 at 3–4.

Convention, the Court of Appeals held that they did not, because the damages claimed did not arise from the delay in transportation by air for purposes of Article 19. The court accepted the argument of the plaintiff that the alleged negligence of British Airways did not occur during performance of the contract of carriage but rather days before, when the airline mistakenly reconfirmed the reservations of a non-existent flight. The court also recognised that British Airways, in confirming a non-existent flight of another carrier was acting as the latter's agent. The court concluded:

> ...We emphasise that BA's liability here is based upon its negligent act of reconfirming appellant's reservation days before departure in a situation where the "carrier" no longer served the route in question. Once the passenger presents herself to the carrier of its agents as ready to begin the air journey, the Convention generally governs liability for delays in the carrier's performance, and its provision apply until completion of disembarkation at the destination airport.[106]

In an interesting case indirectly arising out of the 1990 Gulf War, which concerned a British Airways flight which arrived in Kuwait on August 2, 1990, shortly after the War had broken out, resulting in some of its passengers being detained by the Iraqi Government, an action for damages was brought against British Airways by one of the passengers in a Scottish court. The court, addressing the question of "delay" which the passenger experienced while being detained by the Iraqi government, was of the view that any loss experienced by the passenger detained by the Iraqi government after arrival in Kuwait could not in any reasonable sense be said to have been caused by any "delay" in the flight in question.[107]

In *Jamil* v. *Kuwait Airways Corporation*[108] an airline passenger sued the airline under the Warsaw Convention, claiming that a 4-day delay in transportation resulted in the passenger's failure to finalise crucial documents to secure funding for a government project with the Pakistani government. The United States District Court for the District of Columbia held that under the Warsaw Convention, a 4-day delay in transportation suffered by the passenger was not a proximate cause of his failure to finalise crucial documents related to his contract as claimed. The court further held that the airline could not reasonably have foreseen that a 4-day delay would result in an intervening government coup that would in turn affect that passenger's chances of finalising his contract. Sporkin D. J. held:

> The chain of events which has led to Mr. Jamil's alleged loss is as long as it is unlikely. Tracing the claim of this disappointed traveller takes this court over three continents, through the policies of at least two foreign governments and one election, and involves a time period spanning over a year...Mr. Jamil would have the court trace this entire causal chain and land the blame for his lost project squarely at the feet of KAC (sic). I decline,

[106] *Id.* at 4.

[107] See *Abnet* v. *British Airways PLC*, cited in *Lloyd's Aviation Law*, Vol. 3, No. 2, 15 January 1992 at 4–5.

[108] 773 F Supp. 482 (D.D.C. 1991).

however, plaintiff's invitation to find causation in the air. Such a determination would fly in the face of common sense and legal notions of proximate cause.[109]

Mahaney v. *Air France*[110] which was decided in 1979, is an interesting case where a claim for delay in transportation caused by denied boarding was combined with the claim that the plaintiff was subjected to unjust discrimination or was unduly or unreasonably placed at a disadvantage by the airline.[111] Pierce D.J. citing precedent[112] observed that "bumping" which often occurs as a result of overselling by the airline, may be considered in certain circumstances as an economic necessity for an airline in view of inevitable cancellations or no shows. His honour observed that however, when a flight is thus oversold, the airline must fill the plane in a reasonable and just manner:

To establish that a preference or discrimination has occurred, it suffices that the passenger is able to prove that he possessed a confirmed reservation and a resultant right to a seat, and that this priority was not honoured.[113]

In this instance, court recognised that applicability of Section 404(b) of the *Federal Aviation Act*[114] which provided that a passenger is entitled to recover compensatory and punitive damages under the Act for bumping in a discriminatory manner from the airline concerned.

The provision relating to delay in the carriage of passengers by air, as imputed to Article 19 of the Warsaw Convention, is rendered ineffective in the face of the Convention's two year statute of limitations as stipulated in Article 29. While Article 24 of the Warsaw Convention stipulates that in the cases covered by Articles 18 and 19 of the Convention, any action for damages, however founded, can only be brought subject to the conditions and limits set out in the Convention, Article 29 (1) of the Convention provides:

The right to damages shall be extinguished if an action is not brought within two years, reckoned from the date of arrival at the destination, or from the date on which the aircraft ought to have arrived, or from the date on which the transportation stopped.

The *Mahaney* decision brought to bear an interesting fact relating to judicial interpretation of this provision. Since the plaintiff had brought two concurrent actions one for delay in transportation, and the other for discriminatory bumping the court held that had the plaintiff's action been only for the delay in transportation, Article 29 (1) of the Warsaw Convention may have applied. However, the court recognised the claim against discriminatory bumping as coming under the purview of section 404 (b) of the *Federal Aviation Act*, thereby excluding the purview of Article 29(1) of the

[109] *Id*. 484–485.
[110] 474 F. Supp. 532 (1979).
[111] *Id*. 534.
[112] *Archibald* v. *Pan American World Airways Inc*., 460 F.2d 14, 16 (9th Cir. 1972).
[113] *Id*. 17.
[114] 49 U.S.C.A. sec. 1374(b).

Warsaw Convention from this case.[115] Eight years after the *Mahaney* decision, a similar case *Wolgel* v. *Mexicana Airlines*[116] considered the dual question of delay and discriminatory bumping and concluded that if a plaintiff claimed damages for bumping itself rather than incidental damages due to delay in carriage the Warsaw Convention would not apply, as the Convention's purview was limited in Article 19 to instances of delay in carriage. On the question of whether Article 19 extended to cases of discriminatory bumping, Flaum Circuit Judge held:

> Treaties are construed more liberally than private agreements, and to ascertain their meaning we may look beyond written words to the history of the treaty, the negotiations, and the practical construction adopted by the parties...The history of the Warsaw Convention indicates that the word "delay" in Article 19 to extend to claims, such as the Wolgels', that arise from the total non-performance of a contract.[117]

In the 1991 decision of *Sassouni* v. *Olympic Airways*[118] the United States District Court of New York had no difficulty in identifying a rather uncomplicated instance of denied boarding as coming within the purview of Article 29(1) of the Warsaw Convention. The plaintiff, who had a round-trip ticket from New York to Athens and return through Tel Aviv, had been denied boarding at Athens, on his way to Tel Aviv. The issue was one of delay in transportation and Haight D.J. held:

> Very few courts have confronted the issue of the application of Article 19 to being "bumped" from an airline flight. However, those that have, hold uniformly that damages arising from a delay in transportation caused by being bumped are governed by Article 19.[119]

The court, while endorsing the *Wolgel* decision, reiterated that Article 19 does not cover claims for damages arising from actually being bumped, such as discriminatory bumping or intentional infliction of emotional distress.

3.3.4.2 The Montreal Convention of 1999

The Montreal Convention,[120] which replaced the Warsaw Convention, provides in Article 19 that the carrier is liable for damage occasioned by delay in the carriage by air of passengers, baggage or cargo. Nevertheless, the carrier shall not be liable for damage occasioned by delay if it proves that it and its servants and agents took all measures that could reasonably be required to avoid the damage or that it was impossible for it or them to take such measures. Article 22 quantifies the

---

[115] *Mahaney* v. *Air France, supra,* note 110 (this chapter) at 534.

[116] 821 F. 2d. 442 (7th Cir. 1987).

[117] *Id.* 444.

[118] 769 F. Supp. 537 (S.D.N.Y. 1991).

[119] *Id.* 540.

[120] *Convention for the Unification of Certain rules for International Carriage by Air* done at Montreal on 28 may 1999. The Convention entered into force on 4 November 2003 109 States Parties have ratified the Convention.

compensation for delay caused by providing that In the case of damage caused by delay as specified in Article 19 in the carriage of persons, the liability of the carrier for each passenger is limited to 4,150 Special Drawing Rights.[121]

A perceived inconsistency arose between interpretations of compensation awarded by Regulation 261/2004 of the European Commission and the compensation awarded under the Montreal Convention's Article 22. Clearly the former was a regulation of a group of sovereign States forming a regional body and the latter was a multilateral treaty adopted by sovereign States. This dichotomy was addressed by the European Court of Justice in the *Sturgeon* Case[122] which was filed by the plaintiffs on the basis of cancellation of their flight owing to an inordinate delay in carriage. The Sturgeons booked return tickets with Condor from Frankfurt am Main (Germany) to Toronto (Canada).

The plaintiffs" return flight from Toronto to Frankfurt was due to depart at 16.20 on 9 July 2005. Following check-in, passengers on that flight were informed that the flight was cancelled, as was indicated on the airport departures board. Their luggage was returned to them and they were then driven to a hotel where they spent the night. The following day, the passengers were checked in at another airline's counter for a flight with the same number as that on their booking. Condor did not schedule another flight with the same number for the day concerned. The passengers were given different seats from those they had been allocated on the previous day. The booking was not converted into a booking for a flight scheduled by another airline. The flight concerned arrived in Frankfurt at around 07.00 on 11 July 2005, some 25 h after its scheduled arrival time. The plaintiffs averred that the 25 h delay must be construed as a cancellation of the flight rather than a delay.

The aircraft was delayed due to a technical fault.

The court had to determine whether a flight delay must be regarded as a flight cancellation for the purposes of Articles 2(l) and 5 of Regulation No 261/2004 where the delay is long and whether Articles 5, 6 and 7 of Regulation No 261/2004 must be interpreted as meaning that passengers whose flights are delayed may, for the purpose of the application of the right to compensation laid down in Article 7 of that regulation, be treated as passengers whose flights are cancelled. Another consideration before the court was whether a technical problem in an aircraft is covered by the concept of 'extraordinary circumstances' within the meaning of Article 5(3) of Regulation No 261/2004.

The court held that:

> passengers whose flights are delayed may be treated, for the purposes of the application of the right to compensation, as passengers whose flights are cancelled and they may thus rely

[121] The Special Drawing Right (SDR) is an international reserve asset, created by the International Monetary Fund (IMF) in 1969 to supplement its member countries' official reserves. Its value is based on a basket of four key international currencies, and SDRs can be exchanged for freely usable currencies. With a general SDR allocation that took effect on August 28, 2009 and a special allocation on September 9, 2009, the amount of SDRs increased from SDR 21.4 billion to around SDR 204 billion (equivalent to about $309 billion), converted using the rate of September 4, 2014.

[122] In Joined Cases C-402/07 and C-32/07, Fourth Chamber, 19 November 2009.

> on the right to compensation laid down in Article 7 of the regulation where they suffer, on account of a flight delay, a loss of time equal to or in excess of three hours, that is, where they reach their final destination three hours or more after the arrival time originally scheduled by the air carrier. Such a delay does not, however, entitle passengers to compensation if the air carrier can prove that the long delay was caused by extraordinary circumstances which could not have been avoided even if all reasonable measures had been taken, namely circumstances beyond the actual control of the air carrier.[123]

As to whether the airline could exonerate itself on the basis that the technical problem faced which caused the delay constituted "extraordinary circumstances" the court held that Article 5(3) of Regulation No 261/2004 must be interpreted as meaning that a technical problem in an aircraft which leads to the cancellation of a flight is not covered by the concept of 'extraordinary circumstances' within the meaning of that provision, unless that problem stems from events which, by their nature or origin, are not inherent in the normal exercise of the activity of the air carrier concerned and are beyond its actual control. Thus the court concluded that a technical problem in an aircraft which leads to the cancellation or delay of a flight is not covered by the concept of 'extraordinary circumstances' within the meaning of that provision, unless that problem stems from events which, by their nature or origin, are not inherent in the normal exercise of the activity of the air carrier concerned and are beyond its actual control.

This gives rise to the now entrenched European position that the notion of extraordinary circumstances must be interpreted strictly when Regulation 261/2004, as clarified by the 2008 decision in *Friederike Wallentin-Hermann* v. *Alitalia*[124] where the ECJ held that political instability or meteorological conditions incompatible with the operation of the flight are relevant only if they create an unexpected risk, but are not directly an exemption. As an example, a technical problem in an aircraft should be considered "extraordinary" only if it is the result of an event that is not normal to the activity of the aircraft.

In Air France v. Folkers[125]—a case with circumstances to those of *Sturgeon*, the court held that since [the relevant] inconvenience materializes, with regard to delayed flights, on arrival at the final destination, a delay must be assessed, for the purposes of the compensation provided for in Article 7 of Regulation No 261/2004, "in relation to the scheduled arrival time at that destination".

In April 2004 The International Air Transport Association (IATA) filed an application for judicial review at the UK High Court to challenge Regulation 261/2004 on the ground that the Regulation seriously endangered the interests of the consumer and that airlines, although acknowledging that they should compensate passengers for denied boarding, they cannot be expected to pay compensation for areas beyond their control. IATA claimed *inter alia* that delays or cancellations

---

[123] Id. Chapter 69. See http://curia.europa.eu/juris/document/document.jsf?text=&docid=73703&pageIndex=0&doclang=en&mode=lst&dir=&occ=first&part=1&cid=845141.

[124] ECJ in case C-549/07.

[125] Case C-11/11, see http://eutopialaw.com/2013/03/18/sturgeon-revisited-yet-again-case-c-1111-air-france-v-folkerts/.

due to inclement weather, air traffic control strikes or government security requirements are not within airline control, yet the new ruling penalizes the airline for all situations equally. *A fortiori*, IATA claimed:

> The EU is a signatory to the Montreal Convention which limits a carrier's liability for delays and cancellations in extraordinary circumstances like inclement weather. How can airlines run a business when the regulator gives us two conflicting guidelines? The wisdom of the internationally agreed Montreal convention should form the basis of the EU's thinking as the accepted global standard[126]

The formal European view[127] distinguishes between Regulation 261/2004 and the Montreal Convention with regard to the argument that in the face of an international treaty which many States' members of the European Union have ratified, the Regulation is destitute of effect in terms of its compensatory quantifications. Although the EU accepts that there are two uniform laws in the EU regarding air carrier liability whose interpretation falls to the ECJ—*i.e.* Regulation (EC) No 261/2004 and the Montreal Convention), which forms part of EU Law, by Council Decision 2001/539/EC of 5 April 2001 the two instruments should not give rise to a dichotomy that ascribes ambivalence and confusion to the issue. The European opinion is that the Montreal Convention is exclusively concerned with delays, whereas Regulation 261/2004 does not create any compensation rights in cases of flight delays, but is applicable only to cases of denial of boarding as referred to in Article 4, and cancellation of flights as addressed in Article 5 which in turn creates a compensable right of a lump sum payment for an amount determined under Article 7. On the other hand, it is contended that the Montreal Convention sets in Article 19 its applicability in cases of flight delays, with a liability cap of SDR 4,150 as stated in Article 22.

In support of this argument the European submission cites the decision in the *Sturgeon* case as its *cursus curiae*. It goes on to say that there is a delay in the case where none of the elements of the trip but the times of departure and arrival are altered. If the number of the flights changes or new boarding passes are issued, we therefore face a cancellation.

The European view acknowledges that prior to the *Sturgeon* case, the Montreal Convention was the right instrument to obtain compensation in case of flight delays, and the Regulation was the right instrument to obtain compensation in case of flight cancellations or denials of boarding. Now, both norms are in conflict.

---

[126] http://www.iata.org/pressroom/pr/Pages/2004-04-21-01.aspx.

[127] See http://blogs.law.nyu.edu/transnational/2011/10/the-capacity-of-the-court-of-justice-of-the-european-union-to-promote-homogeneous-application-of-uniform-laws-the-case-for-air-carrier-liability-for-flight-delays-and-cancellations/.

References

Abeyratne RIR (1995) Proposals and guidelines for the carriage of elderly and disabled persons by air. J Travel Res XXXIII(3):52–59

Brownlie I (1990) Principles of public international law, 4th edn. Clarendon, Oxford

Chandrashekhar CA (2003) Flying while brown: federal civil rights remedies to post-9/11 airline racial profiling of south Asians. Asian Law J 10(2):215–252

Corrigan JM (1978) The right of the air carrier to refuse carriage. Ann Air Space Law III:25 (McGill)

Guerreri G (1989) Overbooking, overselling and denial of boarding. Ann Air Space Law XVI: 192 (McGill)

Miller G (1977) Liability in international air transport. Kluwer, Deventer, p 154

O'Keefe C (2006) Air travel for the disabled: isn't it time for a harmonized approach? Air Space Law XXXI(6):408–421

Shaw MN (2003) International law, 9th edn. Cambridge University Press, Cambridge, p 142

Verniero P, Zoubek P (1999) Interim report of the state police review team regarding allegations of racial profiling, April 20

Chapter 4
Injury or Death to Passengers

4.1 The Warsaw and Montreal Conventions

As shown in the aftermath of the accident of Air Asia flight QZ 8501 which crashed into the Java sea on 28 December 2014 while operating from Surabaya to Singapore, killing all 162 passengers on board, and an earlier Malaysian Airlines Flight MH 370 which vanished in March while operating from Kuala Lumpur to Beijing consumer rights extend to dependants of the deceased.

The Warsaw Convention of 1929[1] and the Montreal Convention of 1999[2] are successive treaties which address compensation for death or injury to passengers caused during their carriage by air. The latter replaced the former and its attendant supplements and protocols as well as subsequent treaties that carried provisions of private air carrier liability. In essence both treaties carry the same or similar provisions with relevant changes when necessary. Some of these changes which were introduced by the Montreal Convention are a two tier liability for death or injury with strict liability in the first instance up to 100,000 SDRs. If this limit were to be exceeded the second tier of liability would be applicable but this could be obviated by the carrier if it proves that the damage in question was not due to the negligence

[1] Convention for the Unification of Certain Rules Relating to International Transportation by Air, 12 October 1929, 137 L.N.T.S. 11 49 Stat. 3000, TS No 876, ICAO Doc 7838. The Warsaw Convention was later amended by several protocols, none of which changed the substantive provisions of the Convention. The Warsaw Convention establishes a regime of liability where the liability of the carrier is presumed, once a plaintiff claims that he has suffered damage in the hands of the carrier. The carrier has to rebut this presumption. To balance this deviation from normal evidential rules, the Convention has set up limits of the carrier's liability in terms of monetary compensation payable to the plaintiff if the carrier is unable to rebut the presumption of liability.

[2] Convention for the Unification of Certain rules for International Carriage by Air done at Montreal on 28 may 1999. The Convention entered into force on 4 November 2003 109 States Parties have ratified the Convention.

R. Abeyratne, *Aviation and International Cooperation*,
DOI 10.1007/978-3-319-17022-0_4

or other wrongful act or omission of the carrier or its servants or agents, or such damage was solely due to the negligence or other wrongful act or omission of a third party, as prescribed in Article 21 of the Convention. There is also Article 28 which states that in the case of aircraft accidents resulting in death or injury of passengers, the carrier is liable, if required by its national law, to make advance payments without delay to a natural person or persons who are entitled to claim compensation in order to meet the immediate economic needs of such persons. Such payments are not presumptive of the carrier's liability and could be offset against any subsequent compensatory amounts paid by the carrier.

Another new provision in the Montreal Convention is the fifth jurisdiction in which an action can be brought. This is the domicile or permanent residence of the passenger. Article 33 of the Montreal Convention gives the five jurisdictions in which an action against the carrier can be brought. They are: the territory of one of the contracting parties that the flight involved, either before the court of the domicile of the carrier; its principle place of business or where the carrier has a place of business through which the contract was made or the court where the destination of the flight was to be. These are the same jurisdictions cited in the Warsaw Convention to which the Montreal Convention has added the domicile or permanent residence of the passenger.

Another new feature of the Montreal Convention which was absent in the Warsaw Convention is reflected in Article 50 which calls for the carrier to obtain adequate insurance. The State into which the carrier operates the flight is entitled to require the carrier to provide evidence to the fact of existence of adequate coverage.

4.1.1 General Principles

Generally in law an accusation has to be proved by the person who alleges it. Therefore, a presumption of innocence applies to an accused person until he is proven guilty. However in the instance of carriage by air of passengers the airline is presumed liable if a passenger alleges personal injury or if his dependants allege his death as having been caused by the airline.[3] Of course the airline can show in its defence that it had taken all necessary measures to avoid the damage[4] or that there was contributory negligence[5] and obviate or vitiate its liability. This curious anomaly of the law imposing on the airline a presumption of liability is contained in the Warsaw Convention, Article 17 of which states:

> The carrier shall be liable for damage sustained in the event of the death or wounding of a passenger, if the accident which caused the damage so sustained took place on board the aircraft or in the course of any of the operations of embarking or disembarking.

[3] Shawcross and Beaumont *Air Law* (4ed. 1 Re-Issue 1988) VII (152).

[4] *Id.* VII (116).

[5] *Id.* VII (117).

In order to control the floodgates of litigation and discourage spurious claimants the Convention admits of certain defences the airline may invoke and above all limits the liability of the airline to passengers and dependents of deceased passengers in monetary terms. The Warsaw System therefore presents to the lawyer an interesting and different area of the law which is worthy of discussion. Article 17 of the Warsaw Convention needs analysis in some detail in order that the circumstances in which a claim may be sustained against an airline for passenger injury or death be clearly identified. Further, the defences available to the airline and the monetary limits of liability need also to be discussed.

4.1.2 Defences Available to the Airlines

There are two key factors which govern the civil liability of airlines. They are, the presumption of liability that is imposed upon the airline and the liability limits that apply to the protected the airline from unlimited liability and spurious claimants. There are two other factors which operate as adjuncts to the initial concepts. They are that the airline may show certain facts in its defence to rebut the presumption and that if the airline is found to be guilty of wilful misconduct it is precluded from invoking the liability limits under the Warsaw Convention. Viewed at a glance, the said four concepts seem to the grouped into two sets of balancing measures. The end result is that whilst on the one hand the airline is subject to stringent standards of liability, on the other, it is protected by two provisions which limit its liability in monetary terms and allows a complete or partial defence in rebuttal of the presumption.

Article 20(1) of the Warsaw Convention provides that the airline shall not be liable if it proves that the airline and its agents had taken all necessary measures to avoid the damage or that it was impossible for the airline and its agent to take such measures. *Shawcross and Beaumont* are of the view that the phrase "all necessary measures" is an unhappy one in that the mere happening of the passenger injury or death presupposes the fact that the airline or its agents had not in fact taken all necessary measures to prevent the occurrence.[6] The airline usually takes such precautions as making regular announcements to passengers on the status of a flight starting with instructions on security and safety measures that are available in the aircraft. These measures are taken by the airline to conform to the requirements of the Warsaw Convention that the airline has to take all necessary measures to prevent an accident in order that the presumption of liability is rebutted. Thus in a case decided in 1963[7] it was held that a passenger who leaves her seat when the aircraft goes through turbulent atmosphere is barred from claiming under the

[6] Shawcross & Beaumont, *Air Law*, Op Cit, VII (116).

[7] *Chisholm v. British European Airways* (1963) 1 Lloyds Rep. 626. Also *Grein v. Imperial Airways Ltd.* (1937) 1 KB 50 CA at 69–71 per Greer L.J.

Warsaw Convention for personal injury. Here it was held that an admonition of the airline that the passengers were to remain seated with their seat belts fastened during the time in question was proof of the airline having taken the necessary measures as envisaged in the Warsaw Convention.[8] This case also established the fact that all necessary measures was too wide in scope and that a proper interpretation of the intention of the Warsaw Convention would be to consider the airline to require taking all reasonably necessary measures. In a more recent case Chapman J imputed objectivity to the phrase reasonably necessary measures by declaring that such measures should be considered necessary by the reasonable man.[9] A similar approach was taken in a subsequent case where the court held that the airline should show more than the fact that it was not negligent in order to invoke Article 20 (1) of the Warsaw Convention.[10] The United States also follows this approach of objectivity. In *Manufacturers Hanover Trust Co. v. Alitalia Airlines*[11] it was emphasised that the airline must show that all reasonable measures had been taken from an objective standpoint in order that the benefit of the defence be accrued to the airline. Some French decisions have also approached this defence on similar lines and required a stringent test of generality in order that the criteria for allowing the defence by approved.[12]

The airline which has the burden of proof cannot seek refuge in showing that normal precautions were taken. For example, normal precautions in attending to the safety of the passengers prior to a flight is not sufficient. If therefore the airline cannot adduce a reasonable explanation as to why the accident occurred despite the reasonably necessary precautions being taken it is unlikely to succeed in its defence.[13] Insofar as the requirement of impossibility to take precautions is concerned, the courts have required clear evidence of the difficulties faced by the airline in avoiding the disaster. In one case of a crash landing the court required that it was insufficient for the airline to show that the aircraft was in perfect condition and that the pilot took all steps to effect a good landing. The airline had to show that the weather conditions were so bad that the aircraft could not land in another airport.[14] In *Haddad v. Cie Air France,*[15] where an airline had to accept suspicious passengers who later perpetrated a hijacking, the court held that the airline could not deny boarding to the passengers who later proved to be hijackers. In that instance the airline had found it impossible to take all necessary precautions and

---

[8] See *Chisholm v. British European Airways, Id.* at 629.

[9] *Goldman v. Thai Airways International Ltd.* (1981) 125 Sol Jo 413 (High Ct). Also in (1983) 1 All E.R. 693.

[10] (1986) 2 All E.R. 188.

[11] 429 F Supp. 964 (SDNY 1977).

[12] *Preyvel v. Cie Air France* (1973) 27 RFDA 198. Also *Riviere-Girret v. Ste-Aer-Inter* (1979) Uniform L.R. 173.

[13] *Panalpina International Transport Ltd. v. Densil Underwear Ltd.* (1981) 1 Lloyds Rep. 187.

[14] *Mandreoli v. Cie Belge d'Assurance Aviation*, Milan 1972 (1974) Dir Mar 157.

[15] (1982) 36 RFDA 342.

was considered sound in defence under Article 20 (1). A similar approach was taken in the case of *Barboni v. Cie Air-France*[16] where the court held that when an airline receives a bomb threat whilst in flight and performs an emergency evacuation, a passenger who is injured by evacuation through the escape chute cannot claim liability of the airline since it would have been impossible for the airline to take any other measure.

If the airline proves that the damage was caused by or contributed to by the negligence of the injured person the court may, in accordance with the provisions of its own law, exonerate the carrier wholly or partly from his liability.[17] Contributory negligence under the Warsaw Convention has been treated subjectively as and when cases are adjudicated. The courts have not set an objective standard as in the earlier defence. For instance in *Goldman v. Thai Airways International Ltd.*[18] it was held that a passenger is not guilty of contributory negligence if he keeps his seat belt unfastened through the flight and suffers injury when there is no sign given by the aircraft control panel to keep the seat belt on. However, if a passenger removes a bandage or brace that he is required to keep on for an existing injury and he suffers injury in flight due to the removal of the support he would be found to have contributed to the negligence resulting in his injuries.[19]

Article 25(1) of the Warsaw Convention states that the airline shall not be entitled to avail itself of the provisions of the Warsaw Convention which excludes or limits its liability, if the damage is caused by the wilful misconduct or by such default on the part of the airline as, in accordance with the law of the court to which the case is submitted, is considered to be equivalent to wilful misconduct. Article 25 (1) extends this liability to acts of the agent of the airline acting within the scope of his employment and attributes such wilful misconduct to the airline. Such action as the failure of the technical crew of the aircraft to monitor weather conditions and the failure to execute a proper approach on adverse weather conditions are examples of wilful misconduct of the airline[20] Similarly the failure of a crew which is going off duty to inform the incoming drew of a defect in the aircraft or any such relevant issue which would affect the safety of the aircraft could be construed as an act of wilful misconduct on the part of the airline.[21]

The effect of Article 25 is that the plaintiff becomes entitled to lift the limit of liability of the airline as prescribed in Article 22 of the Warsaw Convention if he proves that the airline was guilty of wilful misconduct. Thus the burden of proof falls on the plaintiff and if he succeeds he may claim an amount over and above the prescribed limits of airline liability.

---

[16] (1982) 36 RFDA 355.

[17] Warsaw Convention *Op. cit.* Article 21.

[18] (1983) 3 All E.R. 693.

[19] *Bradfield v. Trans World Airlines Inc.* 152 Cal. Rptr 172 (Ca. CA 1972).

[20] *Butler v. Aeromexico* 774 F. 2d. 499. (11th Cir. 1985).

[21] *Piano Remittance Corp. v. Varig Brazilian Airlines Inc.* 18 Av. Cas (CCH). 18, 381 (SDNY 1984).

The limitation of liability of the carrier that the Warsaw Convention imposes could be circumvented by the plaintiff proving that the carrier was guilty of wilful misconduct in causing the injury. Wilful misconduct as an exception to the limitation of liability rule appears in all three air law conventions that admit of liability limitations.[22] The original French text of the Warsaw Convention states that if the carrier causes the damage intentionally or wrongfully or by such fault as, in accordance with the court seized of the case, is equivalent thereto, he shall not be entitled to claim the limitation of liability.[23] Drion[24] maintains that the English translation inaccurately states that the liability limitations of a carrier will be obviated if the damage is caused by his *wilful misconduct* or by such *default.*[25] The contentious issue in this question is what kind of misconduct is required? Drion is of the opinion that by approaching the issue in terms of conflicting concepts, the question whether *faute lourde* as proposed originally in the French text and for which there was an English equivalent of gross negligence was in fact more appropriate than the word *dol* which now occupies the document and for which no accurate English translation exists, has emerged as to what standards may be used in extrapolating the words dol or wilful misconduct.[26] Miller[27] takes a similar view when she states that the evils of conceptualistic thinking that had pervaded the drafting of Article 25 which rendered it destitute of coherence, has now been rectified by the Hague Convention which has introduced the words "done with intent to cause damage or recklessly and with knowledge that the damage would probably result."[28]

This confusion was really the precursor to diverse interpretations and approaches to the concept of wilful misconduct under Article 25 of the Warsaw Convention. The French Government took steps by its *Air Carrier Act* of 1957 to rectify ambiguities in this area by interpreting *dol* in the Convention as *faute inexcusable*, or deliberate fault which implies knowledge of the probability of damage and its reckless acceptance without valid reason,[29] making a strong analogy with the Hague Protocol's contents. This interpretation, needless to say, brought out the question whether such reckless acceptance would be viewed subjectively or objectively.

The Belgian decision of *Tondriau* v. *Air India*[30] considered the issue of Article 25 of the Convention and the Hague interpretation. The facts of the case were usual,

---

[22] The Convention for the Unification of Certain Rules Relating to the Assistance and Salvage of Aircraft at Sea, Brussels, 1938, The Rome Convention 1933, and the Warsaw Convention 1929.

[23] Article 25.

[24] Drion (1954), p. 195.

[25] *Ibid.*

[26] *Supra*, note 24 at p. 200.

[27] Miller (1977), p. 200.

[28] *Hague Protocol* 1955, Article X111.

[29] Miller (1977), p. 202.

[30] *Revue Francaise de droit arien* (R.F.D.A.) 1977 at 193.

involving the death of a passenger and a consequent claim under the Convention by his dependents. The significance of the case lay, however, in the fact that the Belgian court followed the decision of *Emery and others* v. *SABENA*[31] and held that, in the consideration of the pilot's negligence under Article 25, an objective test would apply, and the normal behaviour of a good pilot would be the applicable criterion. The court held:

> Whereas the plaintiffs need not prove, apart from the wrongful act, that the pilot of the aircraft personally had knowledge that damage would probably result from it; it is sufficient that they prove that a reasonably prudent pilot ought to have had this knowledge.[32]

The court rationalised that a good pilot ought in the circumstances to have known the existence of a risk and no pilot of an aircraft engaged in air transport ought to take any risk needlessly. The Brussels Court of Appeal however, reversed this judgment and applied a subjective test, asserting that the Hague protocol called for "effective knowledge." Professor Bin Cheng seems to prefer the objective test in the interpretation of "wilful misconduct" in Article 25, on the grounds that a subjective test would defeat the spirit of the Convention and that judges would be "flying in the face of justice in search of absolute equity in individual cases".[33]

Peter Martin, analysing the Court of Appeal decision in *Goldman* v. *Thai Airways International Ltd.*,[34] agrees with Bin Cheng and criticizes the lower court decision which awarded Mr. Goldman substantial damages for injuring his hip as a result of being thrown around in his seat in turbulence, in an instance where the captain had not switched on the "fasten seat belt" sign.[35] Martin maintains that Mr. Goldman failed to prove that the pilot knew that damage would probably result from his act, as envisaged in the Hague Protocol principle. Being an aviation insurance lawyer, Martin is concerned that, while the English courts have a proclivity towards deciding Article 25 issues subjectively, insurance underwriters could view the breach of the limits stringently. Both on the count of the need for objectivity and on the count of the adverse effects on insurance, it is difficult to disagree with Cheng and Martin.

The question of air carrier liability and the approach taken in its context by the Warsaw Convention has seen the emergence of the scholarly analysis of two issues: Should liability of the carrier be based on fault and consequently on the principles of negligence and limited liability or should liability be based on strict liability? Drion, in his 1954 treatise on liability[36] inquires into the various rationales and scenarios that may come up in an intellectual extrapolation of the subject. He examines the fact that an insurance system for liability, which would inextricably

---

[31] 5 December 1967; R.F.D.A. 184.

[32] Transcript of judgment, p. 4.

[33] Cheng (1977a), p. 55 at 99.

[34] (1983) *Law Society's Gazette*, 8 June 1983 at 1485.

[35] Martin (1983), pp. 145–149.

[36] Drion (1954), p. 7.

be linked to a strict liability concept, would be desirable, as a plaintiff would be able to claim compensation from an impecunious defendant through the latter's insurer, on the deep pocket theory,[37] and that insurance underwriters may, in their own interest, be impelled to formulate aviation accident preventive schemes, strengthening the effects of accident prevention.[38] Drion also puts forward 9 rationales for the rebuttable limitation of liability presumption that appears in Article 17, quantified by Article 22 of the Convention. These are: maritime principles carry a limitation policy; the protection of the financially weak aviation industry; the risks should be borne by aviation alone; the existence of back-up insurance; the possibility of the claimants obtaining insurance; limitation of liability being imposed on a *quid pro quo* basis on both the carrier and operator; the possibility of quick settlement under a liability limitation n regime; and the ability to unify the law regarding damages.[39]

These rationales, and whatever else that form considerations of policy in the assessment whether a liability system should be based on negligence or strict liability should be addressed with the conscious awareness that while the Convention imposes a rebuttable presumption of limited liability on the carrier, the contributory negligence of the plaintiff can exculpate the carrier and obviate or apportion compensation. More importantly, wilful misconduct of the carrier, transcends liability limits and makes the liability of the carrier unlimited. Strict liability on the other hand, as proposed in the Montreal Protocols 3 and 4, does not admit of breaking liability limits, sets a maximum limit of compensation that the carrier has to pay, and makes this limit unbreakable by such extraneous factors as the carrier's wilful misconduct.

The ultimate question therefore is, does one keep the Warsaw-Hague concept of fault and limited liability, or does one embrace a system of strict liability which assures the aggrieved party of pecuniary or repersecutory recompense, while obviating the need for lengthy determinations of who was at fault after the fact. In other words, does one point a finger at the carrier in the first instance, then limit his liability and again break the limit if he is at fault? or make the carrier pay a sum of money, the maximum limits of which have been set, with the assurance that such limits would not shoot up unconscionably if the carrier was negligent?

The Convention unified legal principles relating to air carrier liability, thus precluding the application of scores of differing domestic laws.[40] It however, did not succeed in presenting to the world unequivocally objective and quantified rules of liability. This precludes a plaintiff from knowing that he would be, as a rule, compensated if he is injured in an air accident, since the Convention admits of challenge on the grounds of the plaintiff's conduct before, during or after the accident. The strict liability principle introduced by the Guatemala City Protocol

---

[37] *Id.* at 8.

[38] *Loq. cit.*

[39] Drion (1954), pp. 12–13.

[40] *Reed v. Wiser* 555 F. 2d. 1079 (2nd Cir) at 1090.

and carried through by the Montreal protocols on the other hand has been applauded on the grounds that:

> First, it gets money into the hands of the passengers much more quickly. Second, it saves transaction expenses which includes legal fees and other substantial litigation costs. Third, it provides compensation to passengers in those factual situations where no responsible party is at fault, such as in an act of terrorism.[41]

Alexander Tobolewski points out very validly that actual aviation practice in terms of aviation insurance by the airlines has nothing to do with limitation of liability and claims, since airlines insure their fleets and liabilities for colossal amounts in the insurance market.[42] He suggests therefore, the harmonisation of the law and actual practice (presumably by infusing more specific quantum in damages) and simplification of the system of recovery *inter alia*, both of which strongly suggests a regime such as the one envisaged in the Montreal Protocols.[43] Werner Guldimann concludes:

> The most important and urgent matter in the present decade is the continuation of the efforts undertaken by ICAO to re-establish the former universality and uniformity of the Warsaw system by having the Montreal Protocols No 3 and 4 rapidly ratified by the greatest possible number of contracting States.[44]

Although Professor Bin Cheng holds the view that the Montreal Protocols are: heavily weighted towards the carrier; the limits therein are inadequate; and that the unbreakability of the limit of the SDR value is undesirable,[45] the view that strict liability should be embraced seems more sensible, in view of the inconceivable number of passengers carried every year by air, the possible eradication of legal contingency fees, and above all, giving teeth to the meaning and purpose of law—that it should be an instrument of solace, not an opportunity for debate. In an evaluation of the Warsaw System[46] it has been said in 1979 that during the first 25 years of the existence of the Warsaw Convention, it had served the aviation community satisfactorily.[47] Peter Martin bases this observation on the argument that when the Hague Protocol was being drafted in 1955, it was recorded that only 55 Warsaw cases had been adjudicated, and that is a very small number of cases for an instrument of the stature of the Warsaw Convention.[48] The unifying process of

---

[41] Matte (1983), p. 151 at 164.

[42] Tobolewski (1978), p. 261 at 263.

[43] *Id.* 266.

[44] Guldimann (1991), p. 93 at 104.

[45] Cheng (1989), p. 220 at 232.

[46] The Warsaw Convention of 1929 was amended by: The Hague Protocol 1955, the Guadalajara Convention 1961, The Guatemala City Protocol, 1971, and the Montreal Protocols 1, 2, 3, 4, of varying dates. It should also be noted that the Montreal Agreement of 1966, a private arrangement between air carriers, also purported to amend the Warsaw Convention. Hereafter, joint references to all these instruments shall be referred to as the Warsaw system.

[47] Martin (1979), p. 233 at 234.

[48] *Ibid.*

the liability of an air carrier, started by the Warsaw Convention, dealt with liability concepts, quantum of compensation, exceptions on liability, jurisdictional issues and prescription of action. It is sad however, that together with the original Warsaw Convention, there are now 7 other international agreements, few of which have ever seen the light of day. This means that the unification process started by the Warsaw Convention had been criticised and found wanting at various stages of its chequered history. The original document has been excoriated so many times, prompting Professor Cheng to call it the "Warsaw shambles"[49] although it remained, when these comments were made of it, the most widely implemented private international law convention.[50]

Ex facie, from a strictly practical standpoint, it would appear that many facets of unification of the Warsaw Convention have come under interpretation by different philosophies, presumably due to the lack of specificity of the principles of unification and *a fortiori*, the language used. For instance, the delivery of the passenger ticket and the attendant carrier liability came under a series of confounding judicial thought processes, where in two cases[51] the courts decided that the ticket had to be delivered in such a manner as to afford the passenger a reasonable opportunity to take measures to protect against liability insurance, only to decide in *Chan* v. *Korean Airlines*[52] that the only requirement of Article 3 of the Convention was that a ticket be delivered. *Goldman* v. *Thai Airways International Limited*[53] was another case where two confusing issues were decided upon. The first involved the question whether the concept of 'wilful misconduct' as reflected in Article 25 of the Convention was to be interpreted objectively or subjectively. The second issue concerned compensatory limits which were so confusing to both the courts and the parties to litigation that an outside settlement was effected on a mutually acceptable basis.[54] The issue regarding compensatory limits for death or personal injury has had a consistent evolution, starting from the Warsaw Convention at Approximately 8,300/US dollars, increased twofold by the Hague Protocol 1955, increased again by the Guatemala City Protocol to 100,000 special drawing rights (SDR) (about 130,000 US dollars) with the Montreal Protocols going even higher. The currency conversion to gold value has been anther contention of many parties to litigation and the case of *Franklin Mint* v. *TWA*[55] left the situation in fiscal anarchy by deciding that in the United States, the Poincare gold franc has to converted to the last official price of gold before the US left the gold market, and not the free market

49 Cheng (1977b), p. 55. Rene Mankiewicz also uses the word 'shambles' when he describes the Warsaw Convention. See, Mankiewicz (1989), p. 26.

50 Martin (1979), p. 239.

51 *Warren v. Flying Tiger Line Inc.* 352 F. 2d. 494 (CA9 1965), *Mertens v. Flying Tiger Line Inc.*, 341 F. 2d. 841 (CA2 1965).

52 21 Avi 18,228, (1989).

53 1983 3 All. E. R. 693.

54 Kilbride (1989), p. 183 at 187.

55 18 Avi 17,778, 1984.

price of gold. This not only made the overall American attitude towards seeking enhanced compensation turn 360°, but also awarded unrealistically low compensation to the plaintiff. Further, a case in Australia has given a new interpretation to the notion of carrier negligence in the carriage of cargo[56] and a New Zealand case has decided that any interested party can now claim compensation under a cargo claim.[57]

The Montreal Agreement of 1966—a private agreement between carriers plying the United States was also the result of failure by contracting States to reach an international solution to the problem of unifying principles of liability, particularly insofar as the quantum of damages was concerned. The Montreal Agreement amply demonstrates, as an ICAO document[58] points out, that a private agreement between air carriers, sponsored by IATA can unhinge and question the credibility of a multilateral international treaty between sovereign States. Mankiewicz attributes this chaotic state of disagreement to the stand taken by the United States when he states:

> Indeed, there is real irony in the history of the Warsaw Convention. For more than thirty years, the United States of America have steadily and successfully fought for, and obtained and signed six protocols to amend the Warsaw Convention as well as a "Convention Supplementary to the Warsaw Convention." But they have ratified not one of these Warsaw instruments. In spite of the huge amounts of time and money spent all these years by ICAO and it's member States, the US judiciary is still saddled with the awkward task of applying, construing constructively or destructively, misinterpreting and circumventing a convention which is now 60 years old...[59]

There is only one viable alternative towards rectifying this anomaly and preserving the unification efforts of the Warsaw Convention, and that comes in the nature of ratifying the Montreal Protocols 3 and 4. As Michael Milde States:

> There is hardly any viable alternative to a determined effort to bring the Montreal Protocols Nos. 3 and 4 into force. If that aim is not accomplished in the very near future, we may witness a trend to denunciation of the Warsaw System by several States with the ensuing chaotic conflicts of laws, conflicts of jurisdiction, unpredictably high compensation claims and skyrocketing increase in insurance premiums.[60]

The civil liability of an airline for the causing of death or injury to passengers has been established by international treaty and entrenched in law by judicial interpretation. The courts have attempted to balance the interest of both the airline and the passenger as indeed has been the perceived intention of the Warsaw Convention. The predominant feature of this area of civil liability is that air transport in terms of the commercial transportation of passengers is incontrovertibly the mode of

---

[56] *SS Pharmaceutical Co. Ltd., v. Qantas Airways Limited* 1988 1 Lloyds Law Reports 319.

[57] *Tasman Pulp and Paper Co. Ltd., v. Pan American World Airways Inc.* and others., see X1 Annals of Air and Space L. 1987 323 for a detailed account.

[58] Ref. LE 3/27, 3/28–91/3 at 5.

[59] Mankiewicz (1989), p. 259.

[60] Milde (1989), p. 193 at 206.

transport that involves the highest levels of technology. Therefore, courts may find difficulty in ascertaining negligence, wilful misconduct and the overall liability of the airline in the face of complex technical arguments and defence. However, this reason alone should not justify obviating the tortious element that has so carefully been entrenched in the Warsaw Convention by its founders and used by Courts over the last 66 years. As the foregoing discussions reflect, liability issues under the Warsaw Convention has been consistently addressed by the Courts on the basis of their interpretation of negligence, wilful misconduct and contributory negligence, all of which are exclusively issues involving principles of tort law.

4.1.3 Relevance of Accident to the Illness of the Passenger

It is clear that the conventional interpretation of the term accident in tort liability has been extended in aviation cases under Article 17 of the Warsaw Convention where the Courts have imputed intention to the carrier in certain instances. To this extent, as the *Seguritan*[61] case which addressed the issue of the carrier's liability in not being able to regular medical assistance when necessary and the *O'Leary* case[62] more liquor than he could consume in flight perforce proves that the courts have interpreted the Warsaw Convention to enforce liability of the carrier on the principles of intention. Wilful misconduct, therefore has played an important role in establishing that, in certain circumstances at least, it would be justified in considering that the extent of the carrier's fault is a valid consideration in the award of damages.

Fault liability as enforced by the Warsaw Convention may also be adequately reflected in intentional negligence, where the carrier intentionally breaches the duty of care he owes the passenger. Determination of a breach of a duty or care as a distinct evidentiary tool by the Courts would act towards accident prevention in that instances of carrier liability which emerge from accident investigations could then be used as admissible evidence.

The new trend introduced by the Montreal Convention of 1999 in doing away with fault liability and introducing a system of liability that may apply irrespective of fault but aligned to monetary compensation based upon subjective assessments of jurisdictional liability has its genesis in the decade between 1960 and 1970. During this period civil law liability in tort entered a new phase, effectively superannuating the existing system of liability and replacing it with a system of liability insurance. Tortious liability was no longer considered cost effective, and was no match for less expensive insurance. Jurists thought it more equitable, and, above all, practical to embrace a legal system that espoused loss distribution, which acted as the national precursor to liability insurance. This system of liability was

[61] *Infra* note 93 (this chapter).

[62] *Infra* note 97 (this chapter).

assisted along the way by three reasons which militated against fault liability and acted as catalysts towards the successful launch of liability insurance. Firstly, a tort system based upon fault was expensive to administer, when compared with any system of insurance; secondly, litigation was fraught with delay, which a plaintiff could often ill-afford; thirdly, the unpredictability of the result of cases based upon fault liability often put plaintiffs under pressure to settle their claims for amounts less than they would receive if their claims went successfully to trial.

4.1.4 Wilful Misconduct of the Carrier

Of the two instances in which the Warsaw Convention provides that the carrier's liability is unlimited, one relates to the absence of documentation (absence of the passenger ticket and baggage check or air waybill) on the grounds that the document of carriage evidences the special regime of limited liability as prescribed in the Warsaw Convention. The other, which has turned out to be contentious, deals with instances where the damage is caused by the carrier's wilful misconduct, or such default on his part as, in accordance with the law of the court which exercises jurisdiction in the case, is considered to be the equivalent of wilful misconduct. Article 25 of the Warsaw Convention provides:

> The carrier shall not be entitled to avail himself of the provisions of this Convention or exclude or limit his liability, if the damage is caused by his wilful misconduct or by such default on his part as, in accordance with the law of the Court seized of the case, is considered to be equivalent to wilful misconduct.[63]

The provision further stipulates that the carrier shall not be entitled to avail himself of the above provisions, if the damage is caused as aforesaid by any agent of the carrier acting within the scope of his employment.[64]

The primary significance of Article 25 is that it addresses both wilful misconduct and the equivalent (my emphasis) of wilful misconduct. There is a palpable inconsistency between English translation of the original text and the original text itself in that the French word dol personifies the intention to inflict an injury on a person, whereas the English words wilful misconduct requires the defendant carrier to be aware of both his conduct and the reasonable and probable consequences of his conduct, in the nature of the damage which may ensue from the carriers act. Wilful misconduct, therefore, may not necessarily involve the intention of the carrier, his servants or agents and remains wider in scope as a ground of liability.

Most civil law jurisdictions have considered dol to be consistent with gross negligence. Drion[65] dismisses the element of intention by citing examples such as

---

[63] See Convention for the Unification of Certain Rules Relating to International Carriage by Air, signed at Warsaw on 12 October 1929, reproduced in *Annals Air and Space Law*, Vol. XVIII, 1993, Part II, p. 323 at 339.

[64] *Id.*, Article 25, 2.

[65] Drion (1954), para 181 at p. 212.

the theft or pilferage of goods or baggage (which are more frequent in occurrence than aircraft accidents) which may not necessarily always occur with the concurrence or knowledge of the carrier and cites a list of possible instances where gross negligence would form more justification for the invocation of Article 25. Notable examples are assault or indecent behaviour by personnel of carrier; accidents caused by conduct of personnel; serving bad food; bumpy rides causing passenger injury; and failure to instruct passengers of rough weather etc.[66] Drion also makes the valid point of citing delay in carriage as having many dimensions which may be accommodated within the purview of Article 25 without warranting the consideration of intention.[67]

Common law jurisdictions on the other hand have separated wilful misconduct from negligence and insisted that the conduct of the carrier has to be wilful or intentional for a successful case to be grounded on Article 25 of the Warsaw Convention. This approach is consistent with the original contention of the British delegate to the Warsaw Conference, who claimed that wilful misconduct should pertain to acts committed deliberately or acts of carelessness without any regard for the consequences.[68] In the 1952 British case of *Horabin* v. *British Overseas Airways Corporation* the Court held:

> To be guilty of wilful misconduct the person concerned must appreciate that he is acting wrongfully, or is wrongfully omitting to act and yet persists in so acting or omitting to act regardless of the consequences, or acts or omits to act with reckless indifference as to what the result may be.[69]

In the same year, in the United States, the New York Supreme Court appellate Division held that wilful misconduct was:

> . . . dependent upon the facts of a particular case, but in order that acts may be characterized as wilful there must be on the part of the person or persons sought to be charged, a conscious intent to do or to omit doing the act from which harm results to another, or an intentional omission of a manifest duty. There must be a realization of the probability of injury from the conduct and a disregard of the probable consequences of such conduct.[70]

The above approach has been followed by subsequent American decisions which have classified wilful misconduct as requiring conscious intent to do or omit doing an act from which harm results to another[71] and wilful performance of an act that is likely to result in damage or wilful action with a reckless disregard of the probable consequences.[72]

As to the second limb of Article 25.1 which provides that the equivalent of wilful misconduct would suffice to impose liability, the Convention leaves the scope of

---

[66] *Id.* para 181.2 at p. 213.

[67] *Id.*, para 181.4 and 5 at p. 213.

[68] Horner and Legrez (1975), p. 42.

[69] (1952) 2 All.E.R. 1016 at 1022.

[70] *Goepp* v. *American Overseas Airlines*, New York Supreme Court, Appellate Division (1st Dep) December 16, 1952; [1952] US Av R 486; IATA ACLR, No. 12.

[71] *Grey* v *American Airline Inc.* 4 Avi. 17, 811 (2d Cir. 1955).

[72] *Wing Hang Bank Ltd* v. *Japan Air Lines Co.* 12 Avi. 17,884 (S.D.N.Y. 1973).

the provision wide open to include an instance of the carrier knowingly providing small seats and not advising the passenger of the inherent dangers related thereto.

4.1.4.1 Judicial Decisions on Wilful Misconduct

Arguably the watershed decision on the notion of wilful misconduct in recent times was contained in the case *In re Korean Airlines Disaster of September 1, 1983*[73] where the trial court considered wilful misconduct to be:

> The performance of an act with knowledge that the act will probably result in an injury or damage, or in some manner as to imply reckless disregard for the consequences of its performance.

The above pronouncement was used by the American Courts, in the 1994 decision of *Pasinato* v. *American Airlines Inc.*[74] who concluded that the act in question of a flight attendant did not constitute wilful misconduct within the purview of Article 25.2 of the Warsaw Convention. In the *Pasinato* case, a passenger of an American Airlines flight which was bound for Chicago from Italy was struck on the head when a heavy totebag fell from an overhead bin in the cabin. The incident was the outcome of an initial request by the passenger for a pillow immediately after take off, where the flight attendant, in a bid to open the overhead bin above the passenger to retrieve the pillow, was unable to prevent a totebag falling from the bin onto the passenger's head. The passenger and her husband sued American Airlines under Article 25 on the grounds of wilful misconduct. The trial court was of the view:

> there is no dispute that the flight attendant opened the overhead bin to get a pillow for another passenger. The flight attendant's disposition indicates that she opened the bin with one hand, in her customary manner, with the other hand placed defensively above her head near the bin to prevent an object from falling upon her or a passenger sitting below. Further, the flight attendant stated that she tried to catch the totebag that fell from the bin (and may have touched it as it fell), but that it fell too quickly.[75]

The court took cognizance of the contention of American Airlines that the technical and cabin crews give reported warnings to passengers of the dangers of opening overhead bins, both over the public address system of the aircraft and by personal messages. The evidence of the flight attendant that incidents of objects falling from overhead bins were infrequent and generally harmless based on her experience, was also considered relevant. The Court found difficulty in applying the criterion of the *Korean Airlines Disaster* case[76] in that it was difficult for the Court, if not impossible, to envision how the flight attendant's actions could amount to

---

[73] 932 F.2d 1475, 1479 (D.C. Cis), cert denied, 1125.ct.616 (1991).

[74] No. 93 C 1510, 1994 Westlaw 17 1522 (N.D. Ill. May 2, 1994).

[75] *Ibid.*

[76] *Supra*, note 73 (this chapter).

wilful misconduct. It was of the view that the pivotal criterion for determining the existence of wilful misconduct knowledge that the act would probably result in an injury or damage was absent. *A fortiori*, the Court observed that the other criterion established in the *Korean Airlines* case that of an act which is performed in a manner indicating reckless disregard for the consequences was also missing in the *Pasinato* case.

In the 1994 case of *Baba* v. *Compagnie Nationale Air France*,[77] involving damage to cargo, a Federal trial court in Washington found for the plaintiff and awarded damages against the act of the defendant carrier for improperly packing and storing hand-woven Persian carpets, as a result of which some of the carpets were damaged owing to the seepage of rain water when the carpets were kept outside by the carrier pending their loading onto the aircraft. the Court in this instance followed the Bench in *Pasinato* by reiterating the criteria for the proof of wilful misconduct as established by the *Korean Air* litigation. A compelling piece of evidence which enabled the court to arrive at its conclusion in the *Saba* case was the fact that the air carrier had disregarded its own cargo handling regulations in storing the carpets outdoors, in the rain. In its findings the court held:

> In short, through a series of acts, the performance of which were intentional, the carrier has demonstrated a reckless disregard of the consequences of its performance, This disregard is emphasized by the fact that no damage report was ever produced.[78]

The court, while waiving the liability limits of the Warsaw Convention in the *Saba* case noted that a combination of facts can, taken together, amount to wilful misconduct. It was sufficient, in the Court's view for an act to be intended, and not necessary for the resulting injury or wrongfulness of the act to reflect intention or knowledge. It was also significant that the Court further observed that a finding of wilful misconduct was appropriate when the act or omission constituted a violation of a rule or regulation of the defendant carrier itself.

Courts in the United States have been cautious to determine the parameters of scope of employment as envisaged in Article 25.2 of the Warsaw convention, which imputes liability to the carrier with regard to acts of its employers acting within the scope of their employment. In the 1995 case of *Uzochukwu* v. *Air Express International Ltd.*[79] where a New York Federal trial court had to decide on a case of theft by two airline employees of cargo of the two carriers, it was held that the fact that the employers had used forged documents to perpetrate the offence of theft was sufficient to conclude that the act was outside the scope of employment and that the carrier could not be held liable under Article 25.2. It is arguable that the conclusion of the court was based on the fact that generally, in the United States, wilful misconduct is regarded as the intentional performance of an act with knowledge that the performance of that act would probably result in injury or damage or that

---

[77] 866 F. Supp. 588 (D.D.C. 1994)

[78] *Ibid.*

[79] 1995 Westlaw 151 793 (E.D.N.Y. March 27, 1995).

intentional performance of an act in such a manner as to imply reckless disregard of the probably consequences.

In *Robinson* v *Northwest Airlines Inc.*,[80] a case decided in March 1996 and involving circumstances similar to the *Pasinato* case, the United States Court of Appeals dismissed the appeal of the plaintiff who had lost judgment in the trial court against the carrier. The trial court had allowed a motion of the carrier that the plaintiff's claim in relation to her being injured by a piece of hand luggage falling from an overhead bin while the plane was taxiing, and additional injuries caused to her by a passenger striking her on the head with the latter's baggage were valid at law.

The Court of Appeals in affirming the dismissal of the action of the plaintiff noted that while a common carrier (a carrier who opens itself to the world to conduct business in the carriage by air of passengers, baggage and goods) owes a high degree of care to its passengers, it cannot be considered an insurer of the passenger's safety. The court found that the plaintiff failed to raise an issue of fact regarding the carrier's breach of duty towards her. the court was of the view:

> Short of physical constraint of each passenger until each is individually escorted off the plane, we fail to see what Northwest could have done to prevent this accident. At best, that is precisely what the plaintiff has established; the fact that an accident occurred. However, as noted above, common carriers are not insurers of their passenger safety.[81]

A similar approach can be seen in the contemporaneous case of *Bell* v *Swiss Air Transport Co. Ltd*[82] where an Intermediate Appellate court in New York State refused to allow the plaintiff's claim that the loss of his laptop computer during a security check of the airline was due to the airline's wilful misconduct. In the court's view, the plaintiff had failed to prove that the airline intentionally mishandled his baggage with knowledge or reckless disregard for the probable consequences of his conduct. The court also noted that it was the local police, and not the airline, who had required the carrying out of the security check.

The case of *Singh* v. *Pan American World Airways*[83] decided in May 1996 offers a helpful insight into the rationale for determination of wilful misconduct. In wrongful death and personal injury actions arising out of the 1995 hijacking of a Pan Am flight between Bombay and New York, the jury concluded that the carrier had been guilty of wilful misconduct on the reasoning that the management of the carrier knew, or ought to have known of serious lapses in its security programme. In fact, there had been representations made by the carrier's staff to the management on several occasions prior to the hijacking. Furthermore, the jury was influenced in its conclusion by the fact that the carrier was aware of terrorist activity at European, Middle Eastern and Asian high risk airports and that very little had been done by the carrier to provide enhanced security at these airports.

---

[80] No. 94-2392 (6 cir. Mar 15, 1996).

[81] *Ibid.*

[82] 25 Avi. Cas (CCH) 17, 259 (Sup. Ct. App. Tm. N.Y. 1st Dep't. 1996)

[83] 920 F. Supp. 408 S.P.N.Y. (1996).

In the case of the crash of *Thai Airways Flight TG-311 near Katmandu, Nepal in July 1992,*[84] the question at issue was whether the air crew had been guilty of wilful misconduct in flying into terrain. The fatal crash occurred during approach to Kathmandu airport an airport known to be one of the most difficult in the world to land.[85] Evidence had revealed that the captain had given the bearings of the aircraft to the control tower shortly before the crash, and that such were inconsistent with instruction previously given by the tower to the crew in the cockpit of the aircraft. The court concluded that the plane had veered towards terrain surrounding the airport due to the crew's conscious failure to monitor their navigational instruments. The court held:

> the captain and the first officer knew or should have known that failing to perform their duty to continuously monitor the aircraft's navigational instruments would create a grave danger under the circumstances . . . both the captain and the first officer were well aware that their duty to consciously monitor navigational instruments was an act *necessary for safety* . . . their duty to perform this crucial act was so obvious under the circumstances that failing to perform it was *reckless in the extreme* . . .[86]

The Thai Airways case therefore marks an instance where the elements of wilful misconduct were imputed to the crew on the basis that due to their expertise, they knew or ought to have known the reasonable and probable consequences of their act.

A further dimension to the notion of wilful misconduct was added in the *Northwest Airlines Air Crash Case*[87] of August 1996, where the Court of Appeals of the Sixth Circuit added that a finding of wilful misconduct may be based upon consideration of a series of actions or inactions. The court was of the view that since many complex safety systems interact during an air plane flight, an air disaster would usually require multiple acts. In other words, the court held that it was permissible for a jury to consider an airline's individual errors or a series of errors and not restrict itself to the only act which seemingly caused an accident.

If one were to analyse the rationale of wilful misconduct in the light of the *cursus curiae* so far discussed, one would conclude that wilful misconduct hinges itself on knowledge of the perpetrator that damage would result or reckless disregard for consequences of an act on the part of the perpetrator. The question which then arises is whether an instance of the carrier knowingly providing small seats and not advising the passengers of the dangers of prolonged air travel in confined spaces or as would subscribe to the notion of wilful misconduct as it is perceived at the present time.

---

[84] See *Koirola* v. *thai Airways International*, 1996, Westlaw 402403 (N.D. Calif. Jan. 26, 1996).

[85] See Article 25 Thai Airways found guilty of Wilful Misconduct in 1992 Kathmandu Crash Litigation, *Lloyd's Aviation Law*. Vol 15, No 6, March 15, 1996 at 1.

[86] *Id*. 2–3.

[87] 86 F. 3d 498 (6th Cir. 1996).

4.1.5 Accident in Air Law

In commercial aviation, the word accident is sometimes given as broad a definition as those just considered. The Chicago Convention of 1944 defines accident as "occurrence associated with the operation of an aircraft".[88] The Montreal Convention of 1999, in Article 16, refers to the accident which caused the death or injury and the Warsaw Convention in Article 17 speaks of the the accident which caused the damage—both Conventions reducing the accident to the cause rather than to the death or injury.[89] The United States Supreme Court has held that *in limine* an accident must be unexpected and external to the passenger.[90] It is not sufficient that the plaintiff suffers injury as a result of his own internal reaction to the usual, normal and expected operation of the aircraft.[91] Such incidents as hijackings, terrorist attacks and bomb threats, have been considered to be accidents, together with aircraft crashes.[92] An accident could even involve such lesser incidents as tyre failure on take-off[93] and the supply of infected food causing food poisoning of passengers.[94]

The allegation against the airline was that as a result of the failure of In 1982 a passenger traveling from New York to Manila suffered a massive coronary seizure in flight. The employees of the airline to render medical assistance the patient's condition suffered irreparable deterioration resulting in death. Responsibility devolved upon the court to fit this incident to that of an accident within the meaning of the Warsaw Convention. The court readily did this by deeming that the word accident in air law in this instance was not the heart attack itself but the failure on the part of the airline to render medical assistance in flight. The Court said:

> After all, it is no different from an airline's liability in a hijacking incident where the accident is not the acts of the hijackers but the alleged failure on the part of the carrier to provide adequate security.[95]

The airline was accordingly found liable for damage so sustained by the deceased passenger.

In a contemporaneous case, a passenger brought action in the US District Court of Puerto Rico for a hernia sustained by the lifting of a heavy suitcase from the

---

[88] Chicago Convention, *op.cit*, Annex 13.

[89] Shawcross and Beaumont, *Air Law* (4ed. Reissue 1988) Op cit at VII(153).

[90] *Air France v. Saks* 105 S Ct 1338 (1985).

[91] *Ibid.*

[92] *Husserl v. Swiss Air Transport Co. Ltd.* 485 F.2d. 1240 (2nd Circ. 1975) *Day v. Trans World Airlines Inc.* 528 F2d 31 (2nd Circ 1975) *Evangelinos v. Trans World Airlines Inc.* 550 F2d. 152 (3rd Circ. 1976). See also *Salerno v. Pan American World Airways* 19 Avi cas 17,705. (SDNY 1985).

[93] *Arkin v. Trans International Airlines Inc.* 19 Avi Cas 18, 311 (EDNY 1985).

[94] *Abdulrahman Al-Zamil v. British Airways Inc.* 770 F2d. 3 (2nd Circ. 1985).

[95] *Seguritan v. Northwest Airlines Inc.* 86 A.D. 2d. 658 (2d Dept. 1982). See also *Lloyds Aviation Law* Vol. 1 No. 4 August 1 (1982) 1.

conveyor belt. A baggage handler of the defendant airline had refused to carry the suitcase and the plaintiff had solicited aid from of her relatives who were not allowed to enter the baggage area by a guard on duty. The action against the airline was dismissed by the court primarily on the grounds that the plaintiff did not suffer an unexpected injury as she had previously undergone a gall bladder operation and would have known her condition to be delicate.[96]

In 1983, a medical practitioner suffering from a head cold and respiratory infection boarded an aircraft. He disembarked completely deaf. The plaintiff averred that he suffered discomfort in his ears at descent probably due to sudden pressure changes that may have occurred. He alleged that the airline knew or ought to have known that passengers suffering from head colds would risk losing their hearing. In addition, it was alleged that the airline owed a duty to warn the passenger that it was dangerous to travel with a head cold. The airline denied the existence of such a duty. The U.S. District Court for the Southern District of New York reasoned that it would be incongruous to impose a duty on an airline to envisage all possible human afflictions and assess their effect on air travel and warn passengers accordingly. In any event, the airline was in this instance not aware that the passenger was suffering from a head cold. In this decision the court clearly indicated that the presumption of liability imposed by the Warsaw Convention on airlines and the highest-degree-of-care doctrine applicable thereto should not be taken advantage of by plaintiffs.[97] Similarly there would be no cause of action against an airline where a passenger's ill health is aggravated due to acceleration at take off or deceleration at landing.[98]

In April 1984 an intermediate Appellate Court in New York was faced with the task of deciding whether an airline can be held liable for the death of a passenger who chokes to death owing to his own intoxication. The decision was in the affirmative and the court in enforcing judgment against the airline drew the analogy between a dispensing druggist and an airline. The airline serves its passengers with drink and thus undertakes the responsibility not to serve in excess and to exercise reasonable care for the safety of passengers. In addition, in the event of excessive intoxication of a passenger, the airline is under a legal duty to render such medical assistance as is necessary to revive the passenger or in any event to keep him out of danger. In the light of this principle, the airline has a further duty to protect others from a drunken passenger who gets out of control.[99]

In *Air France v. Saks*[100] the United States Supreme Court interpreted the word accident in the context of the Warsaw Convention to mean an occurrence whereby a

---

[96] *Vincenty v. Eastern Airlines* 528 F. Supp. 171 (D.P.R. 1982). Also *Lloyds Aviation Law* Vol. 1. No. 3 July 15, 1982 at 2.

[97] *Sprayregen v. American Airlines Inc.* 570 F. Supp. 16 (SDNY 1983). See also *Warshaw v. Trans World Airlines Inc.* 443 F. Supp. 400 (ED Pa. 1977). *Pironneau v. Cie Air-Inter* (Pan CA 03 July 1986). Cf *De Marines v. KLM Royal Dutch Airlines* 586 F 2d 1193 (3rd Circ. 1978).

[98] See *Warshaw v. Trans World Airlines Inc. Id.* at 408.

[99] *O'Leary v. American Airlines* 475 N.Y.S. 2d. 285 (A.D. 2d Dept. 1984).

[100] 105 S. Ct. 1338 (1985).

passenger is injured owing to an unexpected or unusual event or happening that is external to the passenger,[101] and that where the injury results from the passengers own internal reaction to the normal exigencies of air travel such injury would not be construed as having resulted from an accident. In this case, the plaintiff was a passenger on an Air France flight from Paris to Los Angeles. During the descent the plaintiff suffered severe pain in her left ear which was aggravated thereafter. The plaintiff who consulted a doctor after the plane landed was informed that she was rendered completely deaf in her left ear. The plaintiff brought an action in a California State court on the grounds that her hearing loss was due to the negligent maintenance by the airline of the pressurization system of the aircraft which transported her. Air France moved that the allegation of the plaintiff cannot be sustained within the meaning of the word accident of Article 17 of the Warsaw Convention was meant to be an unusual and unexpected happening. Further, the airline alleged that at all times the pressurization system of the aircraft had been normal. The District Court granted summary judgment to the plaintiff on the basis that accident in Article 17 was meant to be an unusual and unexpected happening. The Supreme Court rejected the rationale adopted by the lower court on the ground that Article 17 refers to an accident which causes an injury and therefore it is the cause and not the effect that is the determinant. Accordingly, the Supreme Court held that air carriers would be liable only if an accident caused the passenger injury. Thus an injury that was in itself an accident was insufficient to satisfy the requirements of Article 17 of the Warsaw Convention.

There will be no accident if in a normal flight free of turbulence a passenger suffers discomfort from a condition he suffers from such as a hiatus hernia[102] or thrombophlebitis.[103] In *Abramson v. Japan Airlines* an airline passenger suffered an aggravation of a pre-existing hiatal hernia shortly after take-off from Anchorage on a flight to Tokyo. The passenger, who was under medication for his condition for 6 years, had not informed the carrier prior to boarding. The passenger, however, claimed that had he been given occupation of a few empty seats he could have massaged his stomach to normalcy. The airline had claimed that there were no empty seats in flight, contrary to the passenger's claim that there were in fact nine empty seats in flight in the first class section of the aircraft. The passenger claimed that his hernia attack constituted an accident within the provisions of Article 17 of the Convention. The court rejected this claim and held that the plaintiff's difficulty was not in any way related to his transportation by air and accordingly, there was no accident under Article 17.

It would have been interesting if the court applied the principle of the *Seguritan* case[104] where failure to render medical assistance by the airline was construed as falling within the purview of the word accident. After all, the airline did not make

[101] *Id.* 1345.

[102] *Abramson V. Japan Airlines Company Ltd.* 739 F. 2d. 130 (3rd Circ. 1984).

[103] *Scherer v. Pan American World Airways Inc.* 387 NYS 2d. 581 (1976).

[104] *Supra*, note 95 (this chapter).

any attempt at rendering assistance to the passenger in the *Abramson* case. The court's reasoning in the latter case contradicts the earlier decision and leads to a logical absurdity. The intention of the Convention was seemingly to provide a uniform system of compensation for passengers bringing claims against airlines operating international air services. To suggest that the failure of an airline to render required assistance is excusable under the Convention is completely at odds with earlier decisions and also arguably with the intention and purpose of the Convention itself.

Insofar as the word wounding of a passenger in Article 17 is concerned, courts have initially held that such would only be in instances of bodily injury and consequently would be palpably conspicuous physical injury.[105] This excluded mental injury. However, a later decision[106] held that types of injuries enumerated should be construed expansively to encompass as many types of injury as are colourably within the ambit of the enumerated types including mental and psycho-somatic injuries. This decision has been followed consistently in a strong line of cases.[107] In the United States mental injury is now entrenched in most jurisdictions as an independently compensable head of damages.[108] As indeed Burnett C.J. said in *Medlin v. Allied Investment Co*[109]*:*

> Memory and empathy tells us that hurt perceived through sensory media other than that of touch may be just as painful if not more than the hurt perceived by the tactile sense. Moreover, physicians tell us that the consequences of invasion of the person accomplished through the perceptory media of sight and sound may be also as damaging if not more damaging than invasions of the persons accomplished through the sense of touch.[110]

Indeed, therefore, mental anguish or injury would now be recognized by most jurisdictions, as falling within the purview of wounding of a passenger under Article 17 of the Warsaw Convention.

It is apparent from the *cursus curiae* that a stringent standard of proof of the nature of the occurrence is insisted upon by the courts if liability of the carrier is to be established under Article 17 of the Warsaw Convention. In *Salce v. Aer Lingus Airlines*[111] the District Court for the Southern District of New York required of the plaintiff to show that the landing of the aircraft in which the plaintiff travelled was anything other than a normal landing. The plaintiff averred that he had received personal injuries due to the hard landing of the aircraft. In the absence of clear evidence of a hard landing, the court would presume that the landing performed by

---

[105] *Rosman v. Trans World Airlines Inc.* 34 NY 2d. 385 (1974).

[106] *Husserl v. Swiss Air Transport Co. Ltd.388 F. Supp 1238 (SDNY 1975).*

[107] *Krystal v. BOAC 403* F. Supp. 1332 (DC Cal. 1975). *Karfunkel v. Cie Nationale Air France 427* F. Supp. 971 (SDNY. 977). *Borham v. Pan American World Airways Inc.* Avicas. 18, 236 (SDNY 1977).

[108] See Abeyratne (1986), pp. 338–360.

[109] 398 S.W.2d. 170.

[110] *Id.* at 273–274.

[111] 19 Av. Cas (CCH) 17, 377 (SDNY 1985).

the aircraft in this instance was not an unexpected or unusual event that would satisfy the requirements of an accident under the Warsaw Convention.

However, when facts are self evident as in the case of *Salerno v. Pan American World Airways*[112] the courts would not hesitate to award damages to a plaintiff passenger. In this case, the District Court for the Southern District of New York held that knowledge of a bomb threat which subsequently caused a miscarriage to a passenger came within the meaning of scope of the word accident. The plaintiff, together with her two children, were passengers aboard a PANAM flight from Miami to Uruguay. The cockpit crew, after take off, instructed the cabin crew to look for a bomb which the former had been informed by air traffic control to be on board. The crew notified the passengers including the plaintiff. She suffered a miscarriage 24 h after having been informed of the alleged bomb on board and having watched the cabin crew looking for the object. The court held that an accident within the meaning of the Warsaw Convention caused the plaintiff's injuries because a bomb threat is external to the passenger and in an unexpected and unusual event outside the usual, normal and expected operation of the aircraft.[113]

The above discussion surfaces the salutary principle that the word accident is considered far more liberally in modern air law than is done under other areas of common law. It also underscores the fact that courts would be more inclined to treat acts of omission on the part of airlines as an accident as was shown in *Seguritan's* case.[114] The airline is presumed liable for an accident where a drunken passenger assaults another, or where a passenger suffers a heart attack and is not given the necessary medical attention in flight as is possible, just to name two instances. Of course, the claimant has to adduce clear evidence of the event and the ensuing injury.

4.1.5.1 Embarking and Disembarking

Both Conventions provide that the accident which causes the damage should take place on board the aircraft or in the course of any of the operations of embarking or disembarking. The first alternative—that of being on board is self explanatory and does not require discussion. The second, which involves the operations of embarking or disembarking has been subject to sustained judicial discussion and analysis. Although *ex facie*, the words on board the aircraft are not problematical, the phrase has been interpreted at least once to engulf time spent by passengers in a hotel consequent to a hijacking.[115] The argument in this case was that the

[112] *606 F. Supp. 656 (SDNY 1985).*

[113] *Ibid.*

[114] *Supra* note 93 (this chapter).

[115] *Husserl v. Swiss Air transport Co. Ltd.* Op cit. See also, *People of the State of Illinois v. Gilberto* 383 NE 2d 977.

passengers would have been on board if not for the hijacking. This is an extreme interpretation which seems to say that the airline is liable for all accidents within that period of time from the start of the embarkation process to the end of the disembarkation.

Current law on the subject seems to favour the test known as the *Day-Evangelinos* test which was developed as a consequence of a series of terrorist acts on passengers in airport departure lounges. This is a tripartite test which has the three elements of consideration—the location of the passenger, the nature of his activity at the time of the accident and the degree of control exercised by the airline at the relevant time. A number of United States cases have accepted this test.[116] This test clearly establishes the fact that unless the passenger is under the control or direction of the airline at the terminal there is no liability for injury or death caused to the passenger under the provisions of the Warsaw Convention. A case which brings out the significance of this test is *Adler v. Austrian Airlines* where a passenger slipped on some ice and fell between the terminal building and the aircraft by bus. The bus was operated by the airport staff and not by the airline. A Brussels court, applying a test similar to the *Day-Evangelinos* test held that the passenger was not under the control of the airline and was thereby precluded from invoking the provisions of Article 17 of the Convention.

The test itself obviates the need to painstakingly go through every possible exigency in the light of the requirement that the accident should occur during the process of embarkation or disembarkation. Prior to the adoption of this test there was no uniformity in the judicial reasoning behind the definition of embarkation and disembarkation. It was left to each individual court to determine whether a given situation would fall within the scope of chronology of these two extremities. Now, the tripartite test has made the task of the Courts much easier.

When there is incontrovertible evidence of a person contracting a disease as a result of being contaminated in an aircraft whilst on board, liability issues pertaining to the airline arising from the incident may involve principles of private air carrier liability. The new *Montreal Convention* of 1999 which emerged consequent to the Diplomatic Conference on Private Air Law of the International Civil Aviation Organization held from 10 to 28 May 1999, provides that the carrier is liable for damage sustained in the event of death or bodily injury of a passenger upon condition only that the accident which caused the damage so sustained took place on board the aircraft or in the course of any of the operations of embarking or disembarking. The *Warsaw Convention* of 1929[117] provides that the carrier is liable for damage sustained in the event of death or wounding of a passenger or any other bodily injury suffered by a passenger, if the accident which caused the damage so

---

[116] *Day v. Trans World Airlines Inc.* 528 F 2d. 31 (2nd Circ. 1975); *Evangelinos v. Trans World Airlines Inc.* 550 F2d 152 (2d. Circ. 1977); *Leppo v. Trans World Airlines Inc.*392 NYS 2d 660 (AD 1977); Rolnick v. *El Al Israel Airlines Ltd.551 Supp. 261 (EDNY 1982).*

[117] Convention for the Unification of Certain Rules Relating to International Carriage by Air, signed at Warsaw on 12 October 1929.

sustained took place on board the aircraft or in the course of any of the operations of embarking or disembarking. Both these Conventions have similar wording, admitting only of death or bodily injury or wounding. Of course, on the face of the provision, the words wounding and bodily injury do not necessarily lend themselves to be associated with infection. *A fortiori*, according to the Montreal Convention, the bodily injury must be caused as a result of an accident, and, according to the Warsaw Convention, the wounding or injury must be caused by accident which is not typically a synonym for infection in both cases. However, the recent decision in *El Al Isreal Airlines Limited* v. *Tseng*[118] introduced a new dimension to the word accident under the Warsaw Convention by giving it pervasive scope to include such acts as security body searches performed by the airlines. In this context, the word accident its fortuity and it becomes applicable to an expected or calculated act. Thus, if an airline knows or ought to have known that an infected passenger was on board its flight, causing others on board to be infected, it may well mean that the act of the airline would be construed by the courts as an accident within the purview of the Warsaw Convention. Later in this discussion, liability issues under the Montreal/Warsaw regime will be addressed with a view to determining whether, in general terms, the infection of a passenger with tuberculosis could be considered an accident if the negligence of the carrier can be shown.

Incontrovertibly, whenever an airline issues a ticket to a passenger in return for the payment of money, the airline undertakes to carry the passenger to the destination given in his ticket with reasonable care and within a reasonable time.[119] Courts have accordingly followed the principle that the airline is contractually bound to carry a passenger who has been issued with a passenger ticket[120] and that the airline is bound by the following legal parameters: to carry any person, without the imposition of any unreasonable conditions, if he is not in an unfit condition, and has paid his fare; to carry the passenger safely; to treat the passenger with due care throughout his/her flight; and, to ensure that the passenger is given the benefit of the speed inherent in air transport. Of course, the parties to a contract of carriage of passengers may generally incorporate in the contractual document (the airline passenger ticket) any terms and conditions upon which they agree, as long as such conditions are not contrary to the provisions of the Warsaw Convention.[121]

The terms and conditions in an airline ticket are not binding upon the passenger unless he consents to them either expressly or impliedly.[122] Although generally, if a passenger accepts his ticket without objection, he will be taken to have accepted the

---

[118] 1999 Westlaw 7724 (January 12 1999).

[119] Corrigan (1978), p. 25.

[120] *Clark* v. *West Ham Corp*., (1909) 2 K.B. 858, *Readhead* v. *Midland Railway Co*., (1869) L.R. 4 Q.B. 382, *Overseas National Airways* v. *C.A.B*., 307 F. 2d. 634, *U.S*. v. *Stephen Bros. Lines*, 384 F. 2d. 118, *S.M.T. Ltd*. v. *Ruch*, 50 C.R.T.C. 369, *Roussel* v. *Aumais* 18 Que. S.C. 474, *Thibault* v. *Garneau* (1959) Que. P.R. 377.

[121] *Warsaw Convention*, Article 32.

[122] *Watkins* v. *Rymill* (1883) 10 Q.B. 178.

conditions in his ticket, a court would be very reluctant to accept strictly a condition in a ticket that totally exonerates an airline where it fails to carry a passenger with reasonable dispatch. Also, as the *Lisi* case held, the passenger cannot be taken to have assented to conditions of carriage incorporated in his ticket if he has no opportunity to examine his ticket or dissent from the provisions thereof,[123] although the airline could, in its defence, prove that the passenger knew of the condition in the ticket even though he did not read the provisions incorporated in his ticket or that the passenger was given notice of the existence of the conditions in the ticket.[124]

Clearly, "bumping" of passengers by airlines is therefore not an eventuality that the Warsaw Convention has considered *stricto sensu*. The American courts have been prolific in the view that Article 19 of the Warsaw Convention relates exclusively to delay and not to bumping *per se*. It would therefore be largely left to the courts in a given jurisdiction to interpret the Convention on this point with the help of Article 19, the analogous case law, and the general principles of the law of contract.

Admittedly, it would be extremely difficult for an airline to determine latent illnesses such as tuberculosis of its passengers. Therefore, instances of negligence pertaining to an airline accepting for travel a person infected with the TB bacterium may be rare. However, it would not be uncommon to critically evaluate the conduct of an airline after the fact *i.e.* by an assessment of the quality of air in the cabin and assistance offered to those infected in flight. Airlines have to carefully follow the guidelines issued by the World Health Organization as discussed earlier, and take initiatives on their own, such as those discussed in the introduction of this discussion, so that they can convince a court that they acted like prudent, caring business enterprises in the face of a calamity.

As for considering infection an accident particularly where it is established that the airline knew or ought to have known of the risk involved, it is highly arguable that, on the strength of the broad interpretation of the Tseng case[125] courts may be inclined to treat the contracting of tuberculosis or any disease as an accident. Also relevant to this issue is the 1996 Supreme Court decision in *Zicherman* v. *Korean Air Lines*[126] which ruled that it was quite evident that the English word damage or harm, has a wide application and was, in fact, used by the Warsaw Convention drafters in its classical French law sense of legally cognizable harm.

It must be emphasized that, in selling an airline ticket for travel by air, an airline offers a composite service, not only to carry a passenger from point A to B, but also

---

[123] See also generally the common law decision of *Thornton* v. *Shoe lane Parking* (1971) 2 Q.B. 163, which upheld the contractual principle that for a contractual clause to be considered applicable the parties to the contract should have the opportunity to read the clause and to dissent from it.

[124] *Montreal Trust and Stampleman v. CP Air* (1976) 72 D.L.R. (3d) 282.

[125] *Supra*, note 118 (this chapter).

[126] 116 *S. Ct.* 629 (1996).

to ensure that transportation is accomplished in a safe and sanitary manner. Therefore, the services offered by the airline in the area of clean air in the cabin become extremely relevant and critical to the issue.

4.1.6 Mental Injury

The issue of compensation for injury caused as a result of an aircraft accident is not one for judicial naiveté, nor is it one for judicial ineptitude and feckless insouciance. For over 83 years, courts have vacillated between a gross disregard for finding the true nature of mental injury, mostly by ignoring its relevance to recovery as a separate head of liability, and claims to pretention that mental injury is something completely comprehensible to the judicial mind as a "non-bodily injury". This article argues that, both in their interpretation of the Montreal Convention's Article 17; its intent, and their gross disregard of medical science, the courts have got their act upside down, and that it is time for a paradigm shift with regard to recovery for mental injury, which should be spearheaded by the judiciary as well as legal and medical experts.

The discussion that follows offers a realistic and practical approach by which mental injury can be categorized and classified based on scientific and medical methodology that could unequivocally result in mental injury being accepted as a compensable element through an inextricable link between mental and bodily injury.

For more than 80 years, courts have largely refused to entertain any claim for mental injury caused in an aircraft accident. The author finds the following statement interesting:

> Although the 1999 Montreal Convention retained "bodily injury" language, a close study of the treaty's history and more importantly, the negotiations among the signatories' delegates suggests that the great majority of nations intended to broaden allowable recovery beyond strict bodily injury and that many had already interpreted the phrase to include mental injury. As a result, courts interpreting "bodily injury" under the new treaty should closely review the intent of the signatories before adopting the previous treaty's precedent.[127]

Indeed, if this were the case, the first measure that could ascertain this statement would be an inquiry into the discussions of the Legal Committee of the Council of ICAO and other groups involved that developed the principles of the Montreal Convention. This article will discuss in chronological and analytical detail the preparatory work of these bodies, as well as the Diplomatic Conference that followed. However, this analysis *per se* would not be sufficient to conclude that the Montreal Convention hints at allowing compensation for mental injury. There has to be an inquiry as to the nature of mental injury and its association with bodily injury. A compelling weakness of judicial determination of this issue is that judges

[127] Cunningham (2008).

have disregarded the scientific and medical aspects of mental injury and blatantly ignored the possibility that mental injury could, in some instances, be a physical injury and therefore, recovery could follow.

On the positive side, in the 1999 case of *Weaver* v. *Delta Airlines Inc.*[128] where the plaintiff experienced terror and felt physical manifestations of that terror during an emergency landing, which necessitated her seeking medical treatment for emotional and physical problems attributable to her flight experiences, and she was diagnosed with, and received treatment for post-traumatic stress disorder, the court held that no floodgates of litigation will be opened by allowing for claims such as the plaintiff's, which are based on a definite diagnosis of a disorder that arises from a physical injury that is medically verifiable. The court held that fright alone is not compensable, but brain injury from fright is.

On the negative side, and this has been the prevalent judicial trend, in the case of *In re Crash at Little Rock, Arkansas on 1 June 1999*[129] although the trial judge held that from the evidence presented, it could be accepted that post-traumatic stress disorder is a biological/physical as well as psychological injury, this contention was rejected at the appellate level on the basis that the evidence was not sufficient to establish a physical change to the plaintiff's brain.

The glaring neglect of the judiciary in ascribing to itself the status of expert on mental injury on the presumption that mental injury is simplistically an injury of the mind which is devoid of and completely separate from bodily injury, leaves the question unanswered as to whether a more enlightened approach is called for, if only to involve the medical fraternity in the determination of this issue. This article will attempt to address this issue as well.

4.1.6.1 Historical Perspectives of Mental Injury Relating to Montreal 1999 Discussions

From 10 to 21 May 1999, representatives from over 50 states attended the International Conference on Air Law, which was held in Montreal under the auspices of the International Civil Aviation Organization (Montreal Conference). The purpose of the Montreal Conference was to consider and adopt a new international legal instrument to consolidate and modernize the carrier liability regime under the "Warsaw System."[130] The specific legal instrument was the *Draft Convention for the Unification of Certain Rules for International Carriage by Air* (Draft Convention by Legal Committee LC/30 and SGMW), as approved by the 30th session of

[128] (1999) 56 F Supp 2d 1190.

[129] (2000) 27 Avi 18,428.

[130] This instrument was later to be known as the Convention for the Unification of Certain Rules for International Carriage by Air, signed at Montreal on 28 May 1999, ICAO Doc 9775-DC/2 Volume I at 1 (Montreal Convention of 1999 as referred to earlier in this book).

the ICAO Legal Committee and amended by the Special Group on the Modernization and Consolidation of the "Warsaw System" (SGMW).

Among the topics discussed in the Montreal Conference was whether to expand the scope of the carrier's liability to include compensation for mental injuries suffered by a passenger. This issue was a significant topic of discussion not only in the proceedings of the Montreal Conference, but also in the deliberations leading to the Montreal Conference, including those of the Secretariat Study Group, the 30th session of the Legal Committee, and the SGMW.

The relevant Article in the Draft Convention developed by the Legal Committee and SGMW and its preceding versions was that of article 16, paragraph 1.

> Article 16 – Death and Injury to Passengers – Damage to Baggage
> The carrier is liable for damage sustained in case of death or bodily injury of a passenger upon condition only that the accident which caused the death or injury took place on board the aircraft or in the course of any of the operations of embarking or disembarking. However, the carrier is not liable to the extent that the death or injury resulted from the state of health of the passenger.

Throughout these proceedings, there were three recurrent issues in article 16, paragraph 1. Firstly, whether to insert the term "mental liability" or another variation in the first sentence; secondly, regarding the scope of carrier liability, whether to include the term "accident" or "event"; and thirdly, regarding the prevention or exoneration of carrier liability, whether to include the term "solely" or "to the extent that" in the second sentence of that paragraph. These discussions illustrated the delegates' diverse viewpoints on not only the specific issue of mental injury, but the general scope of carrier liability during the modernization of the "Warsaw System."

As already stated, this article will highlight the consideration of compensation for mental injury leading up to the Montreal Conference, identifying the issues and themes; and explain the position of delegates during the Montreal Conference proceedings. It will also address the medico-legal considerations that are brought to bear by this issue under the Montreal Convention.

4.1.6.2 Evolution of "Mental Injury" in the Proceedings Leading to the Montreal Conference

On 15 November 1995, the ICAO Council, during its 146th session, decided that a Secretariat Study Group be established to assist the Legal Bureau in developing a mechanism within the framework of ICAO to accelerate the modernization of the "Warsaw System."[131] The Council requested that the Secretariat Study Group assist the Legal Bureau to create a draft legal instrument for review.

---

[131] Convention for the Unification of Certain Rules for International Carriage by Air, signed at Montreal on 28 May 1999, ICAO Doc 9775-DC/2 Volume III at 1. This document will hereafter referred to as Montreal Convention).

On June 10, 1996, the Secretariat Study Group held its second meeting in which the draft text of the new legal instrument was presented. According to that draft, in its initial stages, Article 16, paragraph 1 read as follows:

Article 16 – Death and Injury to Passengers – Damage to Baggage

1. The carrier is liable for damage sustained in case of death or *[personal][bodily]* injury of a passenger upon the condition only that the event *[accident]* which caused the death or injury took place on board the aircraft or in the course of any of the operations of embarking or disembarking. However, the carrier is not liable if the death or injury resulted [solely] from the state of health of the passenger, *[or from the normal operation of the aircraft, or both]*.

The issue of mental injury was not discussed in those proceedings. Subsequently, the Chairman of the Legal Committee decided to appoint a Rapporteur to review and revise the draft and report to the 30th session of the Legal Committee.[132]

The rapporteur's report discussed the three issues highlighted in the draft prepared. In reference to mental injury, the rapporteur stated that it would be fair and equitable to compensate for impairment of health (i.e. both physical and mental injury) and that it may be preferable to define personal injury as such.[133] In relation to the second issue of the term "accident", the rapporteur argued that so as to not lose the considerable precedent developed over the years on the scope of liability created by article 18 of the Warsaw Convention (Damage to Cargo), the term "accident" be retained. In reference to the third issue, "solely", in order to prevent the carrier from avoiding liability because the passenger's state was merely a contributing factor, the word "solely" should be retained.[134]

From 28 April to 9 May 1997, the 30th session of the Legal Committee LC/30 (Legal Committee LC/30) discussed the report of the rapporteur and the draft new Warsaw instrument.[135] The Legal Committee LC/30 examined the three issues raised by Secretariat Study Group and the Legal Bureau in article 16, paragraph 1.

The Chairman of the Legal Committee stated that there was a clear majority in favour of the term "personal", rather than "bodily", subject to further definition of this particular term as proposed by one delegation.[136] There was also an overwhelming majority in favour of "accident" with respect to the terms "event" and "accident." Concerning the term "solely" in the penultimate line of the paragraph, there were two divergent points of view: some delegations supported its deletion, while other declared that it should be retained.[137] As a result of the above, article 16 was amended to read as follows:

Article 16 – Death and Injury of passengers – Damage to baggage

---

[132] Id at 29.

[133] Montreal Convention at 65.

[134] Ibid.

[135] Montreal Convention at 159–171.

[136] Id at 170.

[137] Ibid.

> 1. The carrier is liable for damage sustained in case of death or bodily injury or mental injury of a passenger upon condition only that the accident which caused the death or injury took place on board the aircraft or in the course of any of the operations of embarking or disembarking. However, the carrier is not liable if the death or injury resulted solely from the state of health of the passenger.[138]

From 26–27 January 1998, the Secretariat Study Group reviewed the work of the Legal Committee LC/30.[139] The moderator of the Secretariat Study Group, in reference to article 16, paragraph 1, summarized that the majority of members of the group were of the opinion that as then drafted, the term "mental injury" was too wide; some members even preferred to delete the term completely. However, the moderator added that if a proper way could be found to circumscribe by qualification or definition, then it might be acceptable to retain the concept. Overall, the Secretariat Study Group acknowledged that no consensus could be found on the point and that it would meet with the SGMW on 14–18 April 1998 to discuss the matter further.

A joint meeting of the SGMW and the Secretariat Study Group was held from 14 to 18 April 1998. The Chairperson presented the revised draft of articles 16 in (WP/26):

> Article 16 – Death and Injury of Passengers – Damage to Baggage
> The carrier is liable for damage sustained in case of death or bodily injury of a passenger upon condition only that the accident which caused the death or injury took place on board the aircraft or in the course of any of the operations of embarking or disembarking. However, the carrier is not liable to the extent that the death or injury resulted from the state of health of the passenger.[140]

The chairman proposed the deletion of "mental injury" and the replacement in the second sentence of "However, the carrier is not liable if the death or injury resulted *solely* from the state of health of the passenger" with "However, the carrier is not liable *to the extent* the death or injury resulted from the state of health of the passenger." The Chairman justified these deletions as part of a package to accommodate states that had difficulty in agreeing to unlimited liability with the burden of proof on the carrier.[141]

With respect to the deletion of a reference to "mental injury" in paragraph 1, many delegations and SGMW members favoured the text presented by the Chairman. One delegation and SGMW member believed that retention of the concept of purely mental injury would lead to unforeseeable risks to the carrier in the United States because there was no standard measurement to assess the damages for pure mental injury which would have to be assessed by juries.[142] A few

[138] Id at 212.
[139] Id at 323.
[140] Id at 385.
[141] Id at 250.
[142] Ibid.

delegates, while expressing a preference for retaining the words "mental injury", would nevertheless accept in a spirit of compromise the changes proposed.

With respect to the last sentence of draft paragraph 1 of article 16, some delegations and SGMW members agreed with the text presented by the Chairman to delete the term "solely" and replace it with "to the extent that."

In the end, the draft text approved SGMW on Article 16 was as follows:

> Article 16 – Death and Injury of Passengers – Damage to Baggage
> 1. The carrier is liable for damage sustained in case of death or bodily injury of a passenger upon condition only that the accident which caused the death or injury took place on board the aircraft or in the course of any of the operations of embarking or disembarking. However, the carrier is not liable to the extent that the death or injury resulted from the state of health of the passenger.[143]

4.1.6.3 Proceedings of the Montreal Conference

On 12 May 1999—the third day of the Diplomatic Conference's consideration of draft text, the delegates from Sweden and Norway presented their commentaries on the Draft Convention by Legal Committee LC/30 and SGMW, in particular article 16, paragraph 1.[144]

The first paper dealt with the first sentence of article 16, paragraph 1, while the second paper dealt with the second sentence of article 16, paragraph 1. The papers articulated historical and policy based reasons to justify textual changes to article 16, paragraph 1.

In the first paper, Norway and Sweden recommended that the words "or mental" be introduced to the first sentence. According to this proposition, the article would read: "the carrier is liable for damage sustained in the case of death or bodily injury or mental injury of a passenger." The Swedish delegate elaborated that the purpose of this inclusion was to confer the right to recover for mental injury suffered, whether or not the passenger also suffered a bodily injury.[145]

Firstly, Norway and Sweden based their argument on the drafting history of the Warsaw Convention regarding mental injury. The original Warsaw Convention, which expressly covered only "bodily injury", was changed by the Guatemala Protocol to "personal injury"—a concept that subsequently was deemed too broad. The Legal Committee then found a compromise between the two opposites in the form of "bodily and mental injury"; however, it was subsequently rejected by the SGMW.[146] Therefore, the proposal of Sweden and Norway sought to return the text to what they considered the fair and reasonable solution of the Legal Committee.

---

[143] Id at 276.

[144] Convention for the Unification of Certain Rules for International Carriage by Air, signed at Montreal on 28 May 1999, ICAO Doc 9775-DC/2 Volume II at 97–99.

[145] Id at 97.

[146] *Ibid.*

Secondly, the two States argued that the effect of a mental injury can be as serious as that of a bodily injury. They considered that it would be unfair if two people, disabled to the same extent as a result of an accident, receive different compensation because one of them is physically injured while the other is not. The effects of a mental injury can be as debilitating as that of an accident that renders a victim physically disabled. In other words, Sweden and Norway rejected the stigma attached to mental injury, which is often trivialised by mainstream society.[147]

Thirdly, they argued that the present draft discriminated between victims. The likelihood of suffering mental injury is higher for children and young people because a person uses his experience to deal with trauma—meaning that adults, who have more experience, overcome trauma easier. In other words, exclusion of mental injuries results in age discrimination.[148]

Fourthly, it was contended that the exclusion of mental injuries does not promote unification of legal systems, which is one of the main objectives of this modernization process, because the term "bodily injury" is not construed the same way across legal systems. Since some jurisdictions interpret bodily injury or its equivalent to encompass both physical and mental injury, opportunities for forum shopping would arise. Therefore, one standard scope of injury should be expanded across all jurisdictions to create uniformity and an equal playing field.

In the second paper, Norway and Sweden argued for the deletion of the last sentence of article 16, which read: "However, the carrier is not liable to the extent that the death or injury resulted from the state of health of the passenger."

Norway and Sweden based their argument on the drafting history of the Warsaw Convention. The original Warsaw Convention, which made carriers liable in case of *accidents*, had no qualification concerning the state of health of the passenger. In the Guatemala City Protocol, liability was broadened to cover "events"—not just accidents; however, to ease carrier's concerns, carriers were not liable in cases where the damage was *solely* caused by the state of health of the passenger. Norway and Sweden argued that since the new Convention returns to the term "accident" of the original Warsaw Convention, every reference to the state of health of passengers should be omitted.[149] The reason is that since the carrier's liability restricted to a narrower range of events, it would be unfair to maintain that that liability should be further limited to the extent that the damage/injury suffered is caused by the state of mind of the passenger.

The above argument was reiterated by the delegate of Norway during the Proceedings of the Commission of the Whole at the Montreal Conference. The delegate argued that the basis of liability was now narrower than the Warsaw Convention as amended by the Guatemala City Protocol. If the reservation concerning the state of health of the passenger was retained, he argued that the carrier would be given a double concession, detrimental to the interest of the passenger and contrary to the

[147] *Ibid.*

[148] *Ibid.*

[149] *Id* at 99.

spirit of modernization and improved customer protection as presented in the draft preamble to the Convention.[150] The effect of Norway's proposed amendment would be to ensure equal treatment to all passengers in the determination of the liability of the carrier irrespective of any pre-existing state of health. The carrier would be prima facie liable for the full injury sustained by the passenger, despite that the damage suffered was aggravated by a pre-existing condition. Although not stated by Norway, its proposed amendment reflects the common law tort principle of the egg-shell skull rule.

The second argument of Norway and Sweden was that article 16, paragraph 1, may leave a fairly large group of passengers, including the sick and the handicapped, without protection, because insurance companies might refuse to sell them insurance.

Thirdly, it was argued that the article in its present form would give rise to a large number of proceedings on the question of what extent the state of health of contributed to the damages or injury suffered. The cost, time, and complexity of these proceedings would be significant. Fourthly, article 19, which exonerates the carrier from liability if the passenger was negligent, would exonerate the carrier from liability if it was regarded as negligent for a sick or disabled passenger to travel by aircraft.[151]

4.1.6.4 Proceedings of the Commission of the Whole

In the course of the proceedings of the Commission of the Whole, a variety of positions and counter positions were made that represented a wide divergence of views on the subject. However, at the general level, States grouped themselves under three separate camps: firstly, in support of the Norway-Sweden proposal; secondly, against the Norway-Sweden proposal; and thirdly, in support of compensation of mental injury only if it arises from physical injury.

States such as Chile, Denmark, United Kingdom, Dominican Republic, Panama, Lebanon, Namibia, Columbia, Switzerland, Finland and Spain, Australia, United States supported the Norway-Sweden proposal.[152]

Chile stated that it was not possible to divide human beings up into purely physical or mental elements. The delegate of Denmark agreed with Chile, as mental injury could be as serious as bodily injury. The United Kingdom also supported the above, reiterating that one could not possibly distinguish between passengers who had suffered solely a physical injury from those who had suffered solely a mental injury. The Dominican Republic, Panama and Namibia supported the inclusion of mental injury. Namibia further stated that as a matter of policy and law, it was inconceivable that in the interest of modernization one could justify a

[150] Montreal Convention 1999, hereafter referred to as Montreal 2.

[151] Ibid.

[152] Larsen et al. (2012), p. 410.

differentiation between aviation accidents solely on the basis of the type of injury suffered, despite the fact that both types of injury resulted from the same accident. Colombia, Switzerland, Finland also supported the inclusion of mental injury.[153]

Saudi Arabia, Austria, Madagascar, Ethiopia, Mauritius, Algeria and India expressed reservations on or outright rejected the inclusion of mental injuries in the text.[154] Austria stated that mental injury was an extremely vague concept that could lead to the unlimited compensation of passengers in case of injury.[155] Madagascar stated that carriers and the industry wished to quickly resolve any litigation ensuing from an accident and therefore, rejected the proposal. Ethiopia claimed that "mental injury" would be difficult to prove or disprove, and therefore increase litigation. Mauritius supported the deletion of the term because of potential abuse flowing from the subjective and open-ended meaning. Algeria argued that it would be difficult to determine compensation. India supported the deletion of the term "mental injury" by arguing that severe mental injury would generally manifest in a physical form and would be adequately covered by "bodily injury."

The observer from the International Union of Aviation Insurers (IUAI) also supported the deletion of "mental injury." He argued that fear of flying was a well-recognized phenomenon without significant parallel in other modes of transport and could be easily construed by a sympathetic medical opinion as an injury. The existence, or otherwise, of mental injury was very difficult to prove, giving rise to the possibility of fraud and expensive protracted litigation. The IUAI observer further argued that with respect to the inclusion of mental injury in other Conventions, the 1980 Berne Convention of International Carriage by Rail included strict liability for personal injury and specifically included mental injury in that definition. However, it was important to note that the Berne Convention contained limited liability without any provision, such as wilful misconduct, to permit that limit to be broken.[156]

The observer from IATA stated that it would not be appropriate to include mental injury as a separate compensable injury for which damages may be awarded. He cited the 1991 United States Supreme Court decision in the case *Eastern Airlines v Floyd*[157] which held that mental injury unaccompanied by bodily injury of passengers was not compensable under Article 17 of the Warsaw Convention. The observer further argued that expert studies had indicated that approximately one-half the passengers on any given flight experienced fear of flying and any untoward event during flight could potentially cause mental injury to such passengers.[158]

---

[153] Montreal 2 at 67–74.

[154] Larsen et al. (2012), p. 410.

[155] Montreal 2 at 70.

[156] Id at 69.

[157] 499 U.S.530, 111 S.Ct.1489 (1991).

[158] Id at 73.

The third group of states, which included China, Cameroon, Egypt, Italy, Senegal and Yemen, suggested that the convention should limit carrier liability to situations in which mental injury resulted directly from bodily injury.[159]

The delegate of China argued that in order to strike a balance between the interest of the carrier and the passenger, consideration should be given to the wording: "bodily injury and/or mental injury resulting directly from bodily injury". The delegates of Cameroon, Egypt, Italy, Senegal and Yemen shared these views.

4.1.6.5 Friends of the Chairman Group Proceedings

The divergent views of the delegates over the course of proceedings, prompted the President, Dr Kenneth Rattray to create a working group entitled the Friends of the Chairman, comprised of 28 delegates including the United States. The purpose of this group was to manage the "central issues" of the Draft Convention by Legal Committee and SGMW, including article 16. However, the chairman reiterated that this issue could not be considered in isolation from other issues under the Draft Convention by Legal Committee LC/30 and SGMW.[160]

In the beginning of the Group's proceedings, the Chairman stated that the deliberations of the Commission of the Whole seemed to indicate that there was still no general consensus, even in jurisprudence, as to the extent to which the "Warsaw System" recognized liability for damage in respect of mental injury. While there was liability arising in respect of mental injury in some jurisdictions, the drafting history of the Warsaw Convention had not contemplated pure psychological injury. Therefore, the Chairman stressed the need for the Conference to clearly express in the "travauxpré preparatoires" its intentions regarding the scope of liability for the future interpretations of the Montreal Convention.[161]

The delegates of Chile, United Kingdom, New Zealand, Sweden and Switzerland proposed that indemnification for mental injury as a stand-alone injury should be seriously considered as a viable and compensable injury. The delegate of Chile suggested that the best solution was to refer to "mental injuries which had an adverse effect on health." He underscored that a mere reference to bodily injury and to mental injury associated with bodily injury excluded some situations where a passenger suffered adverse mental trauma without bodily injury. The terms "adverse effect" could be qualified as being "significant", which would restrict the scope of liability by preventing frivolous claims.[162] This was reiterated by the delegate of the United Kingdom who also supported an inclusion of indemnification for mental injury alone. The delegates of New Zealand, Switzerland, Slovenia and Lebanon, Syria, and Sweden further supported Chile's proposal.

---

[159] Larsen et al. (2012), p. 410.

[160] Montreal 2 at 110.

[161] *Id* at 111.

[162] *Id* at 110.

Canada opposed the language proposed by Chile by arguing that the excessively general language of its proposal had a real potential for abuse as the qualifier "serious" was insufficient to tighten the wording of the proposal. Instead, Canada proposed that the qualifier "long term" be incorporated as it would then be clear that to be recoverable the damage sustained would have to have long term effects on health.[163] As a compromise solution that would avoid abuse of the kind referred to by the delegate of Canada, the delegate of Singapore suggested that reference be made in article 16, paragraph 1, to "mental injury that substantially impairs a person's health."[164] This suggestion was supported by the delegate of Japan.

The Chairman indicated that a consensus in principle had emerged: that bodily injury resulting in mental injury would be covered; however, mental injury standing alone as a separate type of recoverable damage would only be covered if it has a significant effect on the health of the passenger.[165] In other words, the terms "adverse effect", "serious", "significant" and "long term" meant the same thing: that the damage was significant. The Chairman also underscored that the burden of proof would rest with the passenger and not with the carrier. The passenger making the assertion for the purpose of claiming damage would have to produce the evidence to show that the impact on his health was significant.

The Chairman then referred to the last sentence of Article 16, paragraph 1, dealing with the extent of the liability of the carrier where death or injury resulted from the state of health of the passenger. He recalled that different views had been expressed, with some delegates supporting the text in its present form and others suggesting that the word "solely" be inserted in the sentence, so that the carrier would not be liable for death or injury that resulted "solely" from the state of health of the passenger. The Chairman indicated that, while support was expressed for the introduction of the word "solely", there had been equal support for the retention of the text in its present form. The chairman stressed the need to examine the above considerations in the context of the new regime.[166]

The delegate of Sweden wished to exclude the last sentence to preserve the liability regime currently in force under the Warsaw Convention. He noted that article 19 (Exoneration) proposed to introduce a defence in which carriers could defend themselves when damage was solely caused by the passenger's negligence. He suggested that this would strike a fair balance for the carrier because the liability carriers would face for a passenger's pre-existing injury would be offset by that passenger's negligence.[167] The delegate of Singapore argued that the principle of apportionment would serve as a fair criterion because in jurisprudence it was accepted that each side should be liable to the extent that it was responsible for the damage sustained.[168] The delegate of the United Kingdom argued that once the

---

[163] *Id* at 113.

[164] *Id* at 114.

[165] *Id* at 117.

[166] *Ibid.*

[167] *Id* at 118.

[168] *Ibid.*

new convention was in force, the last sentence would result in cases where people who previously could have recovered damages would not be able to do so, because the carrier would invoke article 19 to limit liability. However, he stated that a compromise could be struck by removing the words "to the extent that" and reinserting the word "solely".[169] The delegate of the United States argued that in the interest of a fair balance between the interests of passengers and carriers, the word "solely" be inserted in the place of "to the extent that", to offset the advantage that carriers would enjoy as a result of the insertion of the word "accident" instead of "event", which results in limiting the scope of carrier liability.[170] The delegate of France disagreed with the proposal of the United Kingdom and the United States because the insertion of the word "solely" would make it impossible for carriers to prove the injury resulted solely from the state of the passenger.

The Chairman noted the above observations and stated that the last sentence of the paragraph, in its present form, did not preclude the recovery by a passenger of a certain amount of damages where that passenger had died or suffered injury. The phrase "to the extent that" recognized that where the state of health of the passenger was a contributory factor, it would be taken into account in determining the apportionment of liability and damages. Paragraph 1 of article 16 was not intended to deprive the passenger of damages as such; rather, it did no more than recognize that there would be circumstances in which the extent of injury would not have been so great had it not been for the state of health of the passenger.[171]

In the end, the Group agreed to retain the present wording of the last sentence of article 16, paragraph 1, on the understanding that the state of health of a passenger would merely be taken into account insofar as the intensity of the injury was concerned.

However, that deal would prove short-lived. As the discussions continued and the interdependence of issues became ever more apparent, including those of article 20 (Compensation in case of death or injury) and article 27 (Jurisdiction), it became clear that none of these issues could be resolved in isolation.[172]

4.1.6.6 Consensus Package

On 19 May 1999, in reference to the protracted negotiations on article 20 and 27 and reservations expressed by delegates on the previous consensus on article 16, the Chairman stated that it became necessary to begin the process of generating a draft that would be a basis for consideration of the following provisions: article 16, in particular, in resolving the issue of mental injury and other related issues; the provisions of Article 20, taking into account the proposals that had been made in

[169] *Id* at 119.

[170] *Ibid.*

[171] *Id* at 122.

[172] *Id* at 140.

terms of liability issues and the possibility of different tiers being applicable; and the provisions of article 27, in particular, that relating to the fifth jurisdiction, in light of the group's discussions and the varying proposals that had been made.[173]

On 20 May 1999, the chairman presented "Draft Consensus Package", which contained a preliminary proposal for eventual submission to the Commission of the Whole regarding articles 16, 20, 21A, 22A and 27. Referring to article 16, the chairman recalled that in dealing with that article at previous meetings, it had been agreed that mental injury would include not only mental injury associated with or resulting from bodily injury, but also mental injury alone that significantly impaired the health of the passenger.

The Draft Consensus Package, in article 16, paragraph 1, referred to "death or injury" and deleted the word "bodily", and a new paragraph 2, clarified the meaning of the word "injury" as "bodily injury, or mental injury which significantly impairs the health of the passenger."[174]

> Article 16 – Death and Injury of Passengers – Damage to Baggage
> 1. The carrier is liable for damage sustained in case of death or injury of a passenger upon condition only that the accident which caused the death or injury took place on board the aircraft or in the course of any of the operations of embarking or disembarking. However, the carrier is not liable to the extent that the death or injury resulted from the state of health of the passenger.
> 2. In this article the term "injury", means bodily injury, or mental injury which significantly impairs the health of the passenger.

On 25 may 1999, the Chairman presented the second draft consensus package.[175] According to this package, the word "bodily" had been inserted before the word "injury" in the text of Article 16 appearing in DC-FCG No. 1 (Revision 2) so that it now read "death or bodily injury", as was the case in the Draft Convention; secondly, the last sentence of Article 16, paragraph 1 in the Draft Convention and DCW-FCG No. 1 (Revision 2) (However, the carrier is not liable to the extent that the death or injury resulted from the state of health of the passenger.") had been deleted.[176]

The Chairman justified these changes by arguing that the Group had concluded that the concept of death or bodily injury as contained in the Warsaw Convention and as reflected in the Draft Convention would indeed be an adequate reflection against the background of the jurisprudence that existed in relation to the question as to the circumstances in which mental injury might be recovered.[177]

To paraphrase Larsen, Sweeney and Gillick, the delegates accommodated the different positions on the subject of mental injury by adopting a so-called "consensus package" as part of which they did little more than adhere to "the concept of

---

[173] Id at 164.

[174] *Id* at 167.

[175] *Id* at 199*; DCW – FCG No. 1 (revision 2).*

[176] *Id* at 201.

[177] *Ibid.*

death or bodily injury…contained in the Warsaw Convention." They did so by ultimately approving a liability provision in the new Montreal Convention, that a "carrier is liable for damage sustained in case of death or bodily injury."[178]

4.1.6.7 Legal Issues

Article 17 (1) of the Montreal Convention[179] reads as follows:

> Article 17 – Death and injury of passengers – damage to baggage
> 1. The carrier is liable for damage sustained in case of death or bodily injury of a passenger upon condition only that the accident which caused the death or injury took place on board the aircraft or in the course of any of the operations of embarking or disembarking.

It is clear, therefore that the Montreal Convention has excluded compensation for mental injury from its purview. The only discussion remaining would relate to the inquiry as to whether mental injury could be identified as a physical or pathological injury or whether mental injury would be compensable if it were to be the result of bodily injury.

Mental distress, or nervous shock as it is sometimes called, has since the nineteenth century consistently been judicially explained as an unfathomable abstraction.[180] Today, it is no different. When it comes to examining compensation for mental injury caused to a person during air travel, it becomes even more complex as it is not even mentioned in prevailing texts of multilateral treaties covering injury caused during carriage by air. As earlier stated, the most recent treaty on this subject—the Montreal Convention of 1999—states that the carrier is liable for damage sustained in the death or *bodily* (my emphasis) injury of a passenger upon condition only that the accident which caused the death or injury took place on board the aircraft or in the course of any of the operations of embarking or disembarking.[181] The Montreal Convention's predecessor—the Warsaw Convention of 1929—in Article 17 states that the carrier is liable for damage sustained in the event of the death or wounding of a passenger or any other bodily injury suffered by a passenger, if the accident which caused the damage so sustained took place on board the aircraft or in the course of any of the operations of embarking or disembarking.

---

[178] Larsen et al. (2012), p. 412

[179] Convention for the Unification of Certain Rules Relating to the International Carriage by Air, signed at Warsaw on 12 October 1929. The Convention entered into force on 13 February 1933 and has 152 signatories. The Montreal Convention of 1999 superseded the Warsaw Convention and entered into force on 4 November 2003.

[180] In 1861, Lord Wensleydale said: "mental pain or anxiety the law cannot value, and does not pretend to redress, when the unlawful act complained of causes that alone". See *Lynch* v. *Knight* (1861) 9 H.L.C.577 at 598.

[181] *Id.* Article 17.1.

From its inception, this provision has proved contentious in its application as courts adjudicating a claim under Article 17 have been consistently constrained to interpret the words "bodily injury" either as pure physical injury or as mental suffering accompanied by physical injury where the latter was a causative factor in bringing about the former. These rulings held that there could not be compensation under Article 17 for pure mental shock, psychic trauma, anxiety or mental discomfort. In the 1991 case of *Eastern Airlines Inc.* v. *Floyd et al.,*[182] the United States Supreme Court concluded that there must at least be physical manifestation of injury, if not death or physical injury, in order for a claimant to successfully sue an air carrier under Article 17. The court, however, did not address the issue as to whether mental injury accompanied by physical injury was a compensable element. The *Floyd* decision is consistent with its precursor—the 1974 case of *Rosman (and Herman)* v. *Trans World Airlines*[183] which related to a hijacking incident—holding that there have to be palpable, objective bodily injuries, including those caused by psychic trauma related to the incident, and for the damage flowing from the bodily injuries and not from the trauma *per se*. The *Rosman* decision followed in the wake of a 1973 decision[184] which held the same.

The inclination of the courts to insist on pure physical injury as an essential element of compensability is arguably due to the reason that courts took refuge in the original French terminology of the Convention which was "*lésion corporelle*" which means in the French language "physical wound" as against "*lésionmentale*" which means "mental wound".[185]

A diametrically opposed view emerged in a *cursus curiae* which ignored the connotations of the French language and visited the original intention of the drafters of the Convention. In *Husserl* v. *Swiss Air Co. Ltd.,*[186] also a case concerning the hijacking of an aircraft, the court observed the lacuna relating to the absence of reference to emotional injury in Article 17 and deemed fit to construe the provision broadly, to include injuries other than pure physical injury But the question is, is it arguable that mental injury could indeed be construed as bodily injury? How would the Courts interpret Article 17? Would there be, on the other hand separate compensation for mental injury that follows a bodily or physical injury?

Let us first take the issue of the interpretation by the Courts of Article 17. Starting with the exclusivity of bodily injury, the courts would have to bring in mental injury as a bodily injury or as the result of physical injury resulting from an accident. The first issue therefore is "is mental injury a bodily injury?"

---

[182] 17 April 1991, 23 *Avi.* 17,367.

[183] 358 *N.Y.S.* 2d 97 (1974), 13 *Avi.* 17,231.

[184] 368 *F Supp.* 1152 (1973), 12 *Avi* 18,405.

[185] For a detailed discussion on this subject see Desbiens (1992), p. 153 at pp. 159–166.

[186] (D.C.N.Y. 1972) 351 F. Supp. 702.

4.1.6.8 Bodily Injury and Mental Injury

As the historical discussion in Part A showed the only trend that one could glean from the approach of the draughts persons of the Montreal Convention to mental distress is a thinly veiled reluctance to convey an unequivocal statement on whether mental injury is in or out of Article 17. In other words, the record of the work of the draughts persons on the Montreal Convention does not convince one whether they clearly defined mental distress and ruled it out. The inclusion of the words "bodily injury"—a mere repetition of the Warsaw provision—therefore leaves some room for the Courts to rule out the fact that the forefathers of the convention clearly meant that mental injury is distinct from bodily injury. The question now remains: Do the courts have the flexibility to inquire into this further and determine whether mental injury could constitute bodily injury.

The answer, giving the benefit of the doubt to this drafting anomaly, is yes.

The nature of mental distress, as judicially determined, and its status as a compensable head of damage in aviation claims has already been discussed extensively[187] and as such will not be addressed in this discussion. However, what will be discussed is the extent to which, from a medical perspective, the courts should inquire into its compensability as a separate head of liability. The best the judiciary has done so far is to dismiss the mind as an unfathomable extraction.

It is incontrovertible that, in an aircraft accident, acute mental distress precedes physical injury or death. This makes the *cursus curiae* on Article 17 of the Montreal Convention—that mental injury can be compensated only if it is the result of physical injury caused by an aircraft accident—an absurdity in itself. One could argue that the judiciary in this instance has got it upside down, or in other words, put the cart before the horse. Mental distress experienced by a passenger prior to an accident could lead to palpable personal injury or trauma. One commentator provides the analogy of the soldiers in Iraq and Afghanistan who suffer traumatic stress without physical injury and who are not awarded the Purple Heart—an honour bestowed for sacrifice rather than valour:

> Since the beginning of the current wars in Iraq and Afghanistan, growing numbers of veterans have been diagnosed with post-traumatic stress disorder and treated for the condition. Symptoms include recurring nightmares, severe depression and suicide. At least three hundred thousand veterans reportedly suffer from traumatic stress or major depression... Advocates for these veterans have proposed that they too, should qualify for the purple heart since psychological injuries can be at least as debilitating as physical ones, they argue, soldiers who suffer these wounds should receive the medal.[188]

The commentator—a professor at Harvard University—offers the interesting philosophical suggestion based on Aristotelian moral logic that compensation might be awarded on the basis of the virtues the medal honours.[189] This approach,

---

[187] Abeyratne (2000), pp. 225–261. See also, Cunningham (2008).

[188] Sandel (2009), p. 10.

[189] *Id.* 11.

when applied analogically to mental injury under the Montreal Convention would inevitably ask the question: What is the purpose of awarding compensation to victims under Article 17? Is it to compensate for a particular category of injury excluding others; or is it to make pecuniary amends for harm caused by the accident? This is a question hitherto not asked by the learned judges who have adjudicated this issue both under the Warsaw regime as well as the Montreal Convention since 1929. I would argue that, if it is the latter, then clearly mental injury has to be compensated.

Early judicial thinking shows much more sense. The 1925 case of *Hambrook* v. *Stokes Brothers*[190] involved a runaway truck which was apparently out of control negotiating a bend at high speed and which crashed in front of a pregnant woman who became hysterical after suffering nervous shock. She later suffered a miscarriage which caused her death. This case illustrated the fact that in an accident situation, nervous shock could well precede the physical damage. In *Bowman* v. *Williams*[191] a man was permitted to recover for physical illness suffered consequent to nervous shock at witnessing through his drawing room window a large truck crashing through the basement of his home in which his two little children were playing. In *Price* v. *Yellow Paper Mill Co.*,[192] a plaintiff, who suffered a miscarriage from shock caused when the defendant's general manager brought home her husband's bloody and bruised body, was awarded compensation for emotional injury. As far back as 1936, it was clear that:

> All in all, it is fair to say that the courts have already given extensive protection to feelings and emotions. They have shown a notable adaptability of technique in redressing the more serious invasions of this more important interest of personality. No longer is it even approximately true that the law does not pretend to redress mental pain and anguish "when the unlawful act complained of causes that alone". If a consistent pattern cannot yet be clearly discerned in the cases, this but indicates that the law on this subject is in the process of growth.[193]

The above cited view was rendered in 1936. Three years later, Prosser wrote of the infliction of mental injury:

> It is time to recognize that the courts have created a new tort. It appears, in one disguise or another, in more than a hundred decisions, the greater number of them within the last two decades.[194]

4.1.6.9 Psychosomatic Injury

Psychosomatic injury is an accepted category of injury by the medical profession, and has been for long. It is defined as: "pertaining to mind-body relationship'

[190] [1925] 1 K.B. 141 (C.A).

[191] 164 Md.397, 165 Atl. 181 (1933).

[192] 240 S.W. 588 (Tex. Civ. App. 1922).

[193] Magruder (1936), p. 1067.

[194] Prosser (1939), p. 874.

having bodily symptoms of psychic, emotional or mental origin".[195] In an aircraft accident, although a passenger in the aircraft may not suffer physical injury at all, he could suffer what is medically identified as *Somatoform Autonomic Dysfunction* (SAD) which the World Health Organization describes as follows:

> Symptoms are presented by the patient as if they were due to a physical disorder of a system or organ that is largely or completely under autonomic innervations and control, i.e. the cardiovascular, gastrointestinal, respiratory and urogenital systems. The symptoms are usually two types, neither of which indicates a physical disorder of the organ or system concerned. First, there are complaints based upon objective signs of autonomic arousal such as palpitations, sweating, flushing, tremor, and expression of fear and distress about the possibility of a physical disorder. Second, there are subjective complaints of a non-specific or changing nature such as fleeting aches and pains, sensations of burning, heaviness, tightness, and feelings of being bloated or distended, which are referred by the patient to a specific organ or system.[196]

SAD is a system where a mental condition brought about by stress may result in palpable physical discomfort and disease, and which could well result from an aircraft accident. WHO lists some of the physical diseases that could result from SAD as cardiac neurosis, gastric neurosis, neurocirculatory asthenia, and psychogenic forms of aerophagy, cough, diarrhoea, dyspepsia and irritable bowel syndrome.[197]

The WHO also lists *Persistent Somatoform Pain Disorder* which is a predominant complaint of severe, persistent and distressing pain which cannot be explained fully by a physiological process or a physical disorder, and which occurs in association with emotional conflict or psychosocial problems that are sufficient to allow the conclusion that that they are the main causative influences. The result is a marked increase in personal or medical attention given to the patient.[198]

Another glaring fact of judicial inadequacy in the determination of the nature of "injury" under firstly the Warsaw Convention of 1929—which had the exact provision as the Montreal Convention's Article 17—and then the Montreal Convention, is blatant in that the judiciary has gone on the theoretical definition of "bodily injury" which is "damage to the body" which seemingly has been relegated by the judiciary to tissue damage. This theoretical definition cannot be applied to mental injury in judicial parlance. However, this is where the legal and medical synergy should occur. As some authors correctly point out:

> . . .a sexual assault which results in no tissue damage but from which the victim experiences severe depression, will only be covered by the theoretical definition if the scope of bodily damage is broadened to include psychological damage. There would seem to be a case for such harm to be included in a theoretical definition given that significant numbers of those in injury research and practice consider this a legitimate area of concern for the field. Moreover in New Zealand (population of 4 million) at least the agency, Accident

[195] *Dorland's Illustrated Medical Dictionary* (31st Edition), Saunders: Elsevier, 2007 at 1573.

[196] ICD-10 (1992), p. 349.

[197] *Ibid.*

[198] ICD-10 (1992), p. 350.

> Compensation Corporation, which has the primary mandate for injury prevention, rehabilitation and compensation, compensates victims who suffer such harm.[199]

One way of reaching a reasonable outcome to this inept judicial approach is for both the judicial and medical fraternities to put their heads together when a case comes up for adjudication. In doing so, they could consider the relevance of "accident neurosis"—also called compensation or litigation neurosis-which is essentially a disabling subjective functional nervous disorder that sometimes occurs without physical injury of any kind.[200] Miller states of accident neurosis:

> Symptomatology is described with a wealth of superlatives and often with a palpable determination to emphasize the seriousness of disablement. Headache, dizziness, irritability, loss of memory, anxiety, easy startling, bouts of depression,[201] blackouts, and nightmares about the accident. . .loss of interest and an unshakable conviction of unfitness for work usually comprise the picture.[202]

Miller also inquires into the possibility of schizophrenia being caused by accident neurosis without any attendant physical injury. He states:

> Schizophrenia is rarely provoked by head injury and here the paramount role of constituent factors cannot be doubted. In a personal case the traumatic delirium of a severe injury shaded imperceptibly into a frankly schizophrenic illness which persists several years later – a development that was strikingly clarified when I discovered that the patient's elder brother had already spent several years in hospital with a schizophrenic illness unassociated with any injury.[203]

4.1.6.10 The Car Accident Analogy

It is well recognized that car accidents can not only cause physical injuries but mental injuries as well without physical injury. Mental injury claims are called personal injury claims for pain and suffering, which could include depression; temporary and permanent limitations on the victim's daily activities and post traumatic stress disorder linked to memories of the accident:

> At the less severe end of the spectrum, mental and emotional injuries can include such problems as mental anguish, emotional distress, fear, anger, humiliation, anxiety, shock, or embarrassment. People with even mild cases of mental or emotional distress can experience bouts of crying, severe anger, loss of appetite, weight fluctuations, lack of energy, sexual dysfunction or loss of interest in sex, mood swings, and/or sleep disturbances. A mild case of mental or emotional distress might go away relatively quickly, but a more severe case might require professional medical or psychological assistance.

---

[199] Langley and Brenner (2001).

[200] Miller (1966), p. 258.

[201] Accident neurosis must be distinguished from examples of true depressive illness of endogenous pattern beginning shortly after cerebral trauma.

[202] *Ibid.*

[203] *Ibid.*

> More severe mental and emotional injuries can have specific diagnoses such as acute stress disorder or even post-traumatic stress disorder (PTSD). You may have thought that PTSD only affects soldiers or crime victims, but it can affect car accident victims as well. Some people keep replaying the accident over and over in their head or begin to fear certain types of driving situations. For example, let's say that you were rear ended pretty badly. After you got back on the road again, you might start worrying that you would get rear ended any time that you stopped in an unusual situation, like to let a child cross the road in the middle of a block. PTSD can become paralyzing for those who don't get help sooner rather than later[204]

It goes without saying that what applies to car accidents can inevitably apply to aircraft accidents as well. Fear of driving or being in a car after an accident is much the same as a result as the fear of flying after an aircraft accident which could deprive a victim who is a businessman or professional whose work requires frequent travel his means of livelihood, or a pilot his job.[205]

Research carried out in the United States in 2003 on 161 car accident victims revealed that 110 of the victims were diagnosed with post traumatic stress disorder (PTSD), and 33 were diagnosed with subclinical PTSD. Approximately 60 % of the participants in this group who were diagnosed with PTSD were also diagnosed with major depression. A questionnaire calculated to assess anxiety created by car travel, revealed that nearly 95 % of the participants diagnosed with PTSD suffered from noticeable to high levels of anxiety when driving in some situations and avoided certain driving situations (i.e. night time driving, driving on the highway or in bad weather).

PTSD caused by motor vehicular accidents and other psychological traumas is a serious mental health problem and must be appropriately diagnosed. The authors of the report that followed the study have stated that if such damage is not diagnosed in time or is misdiagnosed, the trauma victim could suffer from lifelong PTSD. The authors address this issue by offering some new perspectives on treatments for chronic PTSD, co morbid depression and anxiety based on a randomized controlled trial of a cognitive-behavioral therapy (CBT) protocol.[206]

4.1.6.11 The Need for Judicial Clarity

Judicial reasoning in insisting that mental injury must follow physical injury in aircraft accidents has got one thing right—that the danger of numerous arbitrary claims without merit is real and must be addressed. However, to reject the notion of mental injury as a stand-alone injury merely because of this reason is irresponsible and fecklessly naive. They must distinguish true claims from false ones; not pretend to be psychiatrists or medical experts, but solicit the assistance of the medical

[204] http://www.all-about-car-accidents.com/resources/auto-accident/car-accident-injuries/mental-emotional-injuries-from-car-wreck.

[205] See Abeyratne (2008), pp. 45–66.

[206] http://www.google.com/firefox?client=firefox-a&rls=org.mozilla:en-GB:official.

fraternity. The driving criteria should be both the intention of the parties who drafted the Montreal Convention, which, as discussed, ended in a reluctant compromise and a somewhat timid and insouciant reliance on the Warsaw terminology of "bodily injury", and a well reasoned determination whether, in medical parlance, mental injury causes palpable harm to a victim even when unaccompanied by physical injury in certain circumstances.

The problem lies in the fact that competing paradigms—physical injury and mental injury—are incommensurable and the courts have generally gone on the basis that, as Justice Oliver Wendell Holmes said:

> The life of the law has not been logic: it has been experience. The felt necessities of the time, the prevalent moral and political theories, intuitions of public policy, avowed or unconscious, even the prejudices that judges share with their fellow men, have had a good deal more to do than the syllogism in determining the rules by which men should be governed.[207]

This view, offered by one of the most eminent jurists of all time, was reiterated in 1994 in *Browns* v. *LumbermansMut. Cas. Co.*[208]—that public policy considerations may preclude tort liability that would otherwise be imposed.[209] This is indeed unfortunate. Pandering to public policy or prejudices should not interfere with independent judicial thinking in the dispensation of justice. It is, and remains a matter for fair, equitable, courageous and unbiased judicial interpretation. Another misleading statement is found in the *Vienna Convention on the Law of Treaties,* which provides that a provision of a treaty should be interpreted according to the ordinary meaning to be given to the terms of the treaty in their context and in the light of its object and purpose.[210] In this context, the words in question are "bodily injury" which would prompt the courts to simplistically accept the common man's distinction between bodily injury and mental injury. As discussed above, bodily injury has necessarily to be associated *inter alia* with diseases such as accident neurosis which cause such bodily harm as headache, dizziness, irritability, loss of memory, anxiety, easy startling, bouts of depression. The simplest explanation is that in almost all cases of physical injury the patient is administered with medication. If medication is prescribed for a headache, would that not be a bodily injury that follows mental injury as a result of an accident?

The answer to this fundamental difficulty in Article 31 of the Vienna Convention lies in the Convention itself. Article 32 states clearly if:

> Recourse may be had to supplementary means of interpretation, including the preparatory work of the treaty and the circumstances of its conclusion, in order to confirm the meaning

---

[207] Oliver Wendell Holmes, The Common Law (Mark DeWolfe ed.) 1963 at 5.

[208] 517 N.W. 2d. 432.

[209] *Id.* 439–444.

[210] *Vienna Convention on the Law of Treaties,* done at Vienna on 23 May 1969, Article 31.1. The Convention entered into force on 27 January 1980. See United Nations Treaty Series, vol. 1155 at 331.

resulting from the application of article 31, or to determine the meaning when the interpretation according to article 31:

(*a*) leaves the meaning ambiguous or obscure; or

(*b*) leads to a result which is manifestly absurd or unreasonable.

The International Law Commission has stated that "the interpretation of documents is to some extent an art, not an exact science".[211] The issue is, when considering the *travaux préparatoire* discussed above, which creates some doubt as to whether the "bodily injury" compromise was unequivocal, judges could not invoke both (a) and (b) of Article 31. The discussions already indulged in, and those to follow could well have the answer.

This is one instance where judges have to dispense justice and award compensation for a palpable injury rather than resorting to hiding behind their intuition or preferences. The transition change between competing paradigms in judges' minds must not be a slow process but must be arrived at in accordance with the dictates of reason and scientific analysis:

> Judges recognize that the choices they make among competing principles are inevitably influenced by that which is unconscious and semi-conscious. Nearly one hundred years ago, Homes cautioned that judicial decisions "depend on a judgment or intuition more subtle than any articulate major premise" are more generally "the unconscious result of instinctive preferences and inarticulate convictions", and often "express an intuition of experience which outruns analysis and sums up many unnamed and tangled impressions; impressions which may lie beneath consciousness without losing their worth. . ."[212]

The overarching principle in the recovery of damages should be full compensation for foreseeable injury, without prejudice to the rights of the victim to be compensated for damage suffered irrespective of labels, judicial convenience and lip service to an ongoing trend.

Sarah B. Rasool, in her doctoral thesis in 2007 stated:

> Research in the field of trauma has shown that accidents can lead to the onset of an array of psychiatric disorders, not only PTSD and acute stress disorder (ASD). Depression, anxiety and specific phobias, substance misuse, and insomnia have all been implicated in the psychological aftermath of Road Traffic Accidents (RTAs).[213]
>
> The concept of psychological trauma has evolved over the 20th century. From a psychodynamic perspective, a traumatic event is one that breaks through, or overrides the mind's filtering process and floods the mind with a degree of stimulation that it cannot manage. There is a colossal disruption in functioning, amounting to a kind of breakdown of the defence organisation, leaving individuals vulnerable to intense overwhelming anxieties from internal sources which have been provoked by an external event. The associated treatment guided by this conceptual model is long term psychotherapy. The aim of which is to make the unconscious conscious and resolve conflicts that the current trauma may have unearthed.[214]

---

[211] ILC Commentary, p. 218 para (4) cited in Aust (2000), p. 184.

[212] Wise (1998), p. 40.

[213] Rassool (2007), p. 113.

[214] *Id.* 117–118.

One way of evaluating the extent of mental injury is to follow the New Zealand model of assigning external cause codes to cover mental health outcomes where psychological injury could be ascertained through injury events which can in turn be paired with particular mental disorders. This system could be implemented through the maintenance of hospital records that assign codes at discharge from the hospital even if there is no linking physical damage. An operational definition of injury should be developed, taking into consideration the strong link between physical and mental injury through established medical research and findings. A thorough revision should be undertaken pertaining to hitherto acceptable theoretical and operational definitions that are arrived at without basis or appropriate scientific and legal considerations. Injury statistics should be carefully considered at an international platform and ICAO medico-legal expertise should be utilized to make appropriate recommendations to the international civil aviation community.

The ICAO Council also has the opportunity, under Article 55 d) of the Chicago Convention to study any matters affecting ICAO and the operation of international air transport. Taking into consideration the above discussion, and given ICAO's new role in air transport—that of fostering the development of the civil aviation system—one is encouraged that the Organization's impetus will be increased and more guidance could be forthcoming in air transport law and economics. For instance, in accordance with the permissive functions of the Council, which provides that the Council may conduct research into all aspects of air transport which are of international importance,[215] such studies, taking into account trends such as those discussed at the commencement of this chapter, could analyze global legal trends affecting air transport and their effects on ICAO's aim to ensure the safe, regular, efficient and economical air transport. Although this approach is highly desirable, it might not be deemed by ICAO to be practical or even worthwhile. However, such an approach could largely embellish the current role of ICAO as a leader which provides expertise in matters of law and economics, rather than just being the global forum for air transport, technically, economically and legally.

The most logical step in this process would be to make the paradigm shift from bodily injury to mental injury that could result in mental injury being accepted as a compensable element and determine the inextricable link between the two, in interpreting the term "bodily injury" under the Montreal Convention.

4.2 Flight 8501, International Cooperation and State Responsibility

There are two key words that drive aviation safety—standardization and harmonization. Standardization means compliance with regulations, processes and practices. Harmonization means global consistency in regulations, processes and

[215] Chicago Convention, *op.cit*, Article 55 c).

practices. Without these words being put into practice safety in aviation would inevitably fail. At the base of these two words lie the International Civil Aviation Organization (ICAO),[216] and its aims and objectives implemented and realized through the 19 Annexes to the Convention on International Civil Aviation (Chicago Convention)[217], the ICAO manuals and guidance material, ICAO Assembly Resolutions and Council Resolutions and other statements that indicate to States their responsibilities in ensuring safety in aviation.

On Sunday 28 December of 2014 Air Asia flight QZ 8501 crashed into the Java sea on its way from Surabaya to Singapore, killing all 162 passengers and crew on board. It was later reported that the flight did not have authorization to be operated on the route on Sundays since authorization had been withdrawn by the Indonesian authorities. Later, on 2 January 2015, the Indonesian authorities suspended Air Asia flights departing from Indonesian airport and brought in strict regulations and even stricter monitoring and supervision controls in its air transport system.[218]

In ICAO terminology the two essentials of standardization and harmonization can be subsumed into one practice—safety management. At the 2nd ICAO High Level Safety Conference, convened in Montreal from 2–5 February 2015, Latvia (on behalf of member States of the European Union, the European Civil Aviation Conference and EUROCONTROL) submitted *inter alia* that a study, carried out by the ICAO Filing of Differences Task Force (FDTF), identified a widening gap between the adoption of Standards and Recommended Practices (SARPs) by ICAO and their implementation by States. It was claimed that this widening gap was cause for serious concern when the aviation community at large should be improving the overall implementation of SARPs and thus improving overall safety management. The EU working paper went on to say that the issue of interpretation and implementation has been identified as a continuing problem in this context, concluding that some form of quality control was required to ensure that the apparent lack of clarity does not have a negative impact on the proper implementation of SARPs, especially when consideration in the overall safety management context is being given to the introduction of performance-based regulations. It was observed that this was in part because there is no effective feedback process.[219]

It is a known fact that Flight QZ 8501 went down in severe adverse weather. However, this article addresses only the issue of compliance with regulations where, if they were adhered to in the instance of this particular flight, the flight

[216] The International Civil Aviation Organization, a specialized agency of the United Nations, was established by Article 44 of the *Convention on International Civil Aviation* (Chicago Convention), signed at Chicago on 7 December 1944 (ICAO Doc 7300/9, Ninth Edition, 2006). The main objectives of ICAO are to develop the principles and techniques of international air navigation and to foster the planning and development of air transport. ICAO has 191 Contracting States.

[217] Chicago Convention, *Ibid.*

[218] http://translate.google.ca/translate?hl=en&sl=id&u=http://hubud.dephub.go.id/%3Fen&prev=search.

[219] *Towards a Global Implementation of Safety Management*, HLSC/15-WP/35, 15/12/14 at 4.

would have been effectively precluded from taking off from Surabaya Airport in the first place.

4.3 Safety Oversight

4.3.1 Critical Elements of Safety Management

There are eight recognized critical elements of aviation safety management: *Primary aviation legislation*—the provision of a comprehensive and effective aviation law consistent with the environment and complexity of the State's aviation activity and compliant with the requirements contained in the Chicago Convention; *Specific operating regulations*—the provision of adequate regulations to address, at a minimum, national requirements emanating from the primary aviation legislation and providing for standardized operational procedures, equipment and infrastructures (including safety management and training systems), in conformance with the SARPs contained in the Annexes to the Chicago Convention[220]; *State civil aviation system and safety oversight functions*—the establishment of a Civil Aviation Authority (CAA) and/or other relevant authorities or government agencies, headed by a Chief Executive Officer, supported by the appropriate and adequate technical and non-technical staff and provided with adequate financial resources. The State authority must have stated safety regulatory functions, objectives and safety policies[221]; *Technical personnel qualification and training*—the establishment of minimum knowledge and experience requirements for the technical personnel performing safety oversight functions and the provision of appropriate training to maintain and enhance their competence at the desired level. The training should include initial and recurrent (periodic) training; *Technical guidance, tools and the provision of safety-critical information*—the provision of technical guidance (including processes and procedures), tools (including facilities and equipment) and safety-critical information, as applicable, to the technical personnel to enable them to perform their safety oversight functions in accordance with established requirements and in a standardized manner. In addition, this includes the provision of technical guidance by the oversight authority to the aviation industry on the implementation of applicable regulations and instructions; *Licensing, certification, authorization and approval obligations*—the implementation of processes and procedures to ensure that personnel and organizations performing an aviation

[220] The term "regulations" is used in a generic sense to include but is not limited to instructions, rules, edicts, directives, sets of laws, requirements, policies, and orders.

[221] The term "State civil aviation system" is used in a generic sense to include all authorities with aviation safety oversight responsibility which may be established by the State as separate entities, such as: CAA, Airport Authorities, Air Traffic Service Authorities, Accident Investigation Authority, and Meteorological Authority.

activity meet the established requirements before they are allowed to exercise the privileges of a licence, certificate, authorization and/or approval to conduct the relevant aviation activity; *Surveillance obligations*—the implementation of processes, such as inspections and audits, to proactively ensure that aviation licence, certificate, authorization and/or approval holders continue to meet the established requirements and function at the level of competency and safety required by the State to undertake an aviation related activity for which they have been licensed, certified, authorized and/or approved to perform. This includes the surveillance of designated personnel who perform safety oversight functions on behalf of the CAA; *Resolution of safety concerns*—the implementation of processes and procedures to resolve identified deficiencies impacting aviation safety, which may have been residing in the aviation system and have been detected by the regulatory authority or other appropriate bodies.[222]

In the case of Flight QZ 8501 the second critical safety management system pertaining to State civil aviation system and oversight, where a civil aviation authority regulates, authorizes and monitors flights, seemingly fell through the cracks. If the flight on that Sunday was not authorized, there should have been adequate supervision that due process was followed and the flight did not take off. Another critical element that was not implemented was one which pertains to licensing, certification, authorization and approval obligations, which were clearly ignored and unmonitored.

At the 2nd ICAO High Level Safety Conference alluded to above, The United States requested the conference to consider that the eight critical elements of safety oversight are the corner stone of and fundamental postulate for a State to effectively implement and conduct safety oversight of their aviation system. The United States emphasized that it was only if a State became compliant under these critical elements where that State adheres to and reaches minimum requirements of the eight critical elements will that State have the experience and ability to implement and recognize the benefits of a State Safety Programme (SSP). In this context a State's primary focus should be to successfully implement and sustain adherence to the eight critical elements of safety oversight, which was a prerequisite to the implementation of an effective SSP. The United States went on to say:

> The benefits of the SSP are expansive; an effective SSP is the mechanism that will allow the States with mature oversight abilities and procedures to further enhance decision-making and allocation of resources to mitigate risks to their aviation system. It must be stressed that the success of the SSP is dependent on the ability of the State to satisfactorily perform and maintain a strong foundation of safety oversight. Although the SSP is a mechanism by which States can manage their aviation system, the existence of an SSP is not indicative of functional and capable national safety oversight.[223]

---

[222] This would include the ability to analyse safety deficiencies, forward recommendations, support the resolution of identified deficiencies, as well as take enforcement action when appropriate

[223] *The Future of Annex 19: Determining The Role of the Eight Critical Elements of Safety Oversight,* Presented by the United States, HLSC/15-WP/22, 12/11/14 at 2.

In Conclusion the United States said that the current critical elements of safety oversight are a necessity for all States, especially those States that are not compliant with minimum ICAO Standards and for use by auditing programmes as the common foundation by which to measure and evaluate the effectiveness of their implementation. As a result, ICAO should consider a two-phased approach whereby States with a low USOAP effective implementation score first implement a system of oversight compliant with ICAO minimum Standards using the eight critical elements identified under Appendix 1 of Annex 19. Then as these States further mature then implement the additional SSP elements. However, the United States mentioned that however, those States with adequately mature safety oversight systems should strive towards implementing an SSP promptly.[224]

4.3.2 Safety Management

Many of ICAO's 191 member States are already facing problems with respect to safety oversight. A glaring fact emerging from safety audits conducted by ICAO on States is that the findings of the initial safety oversight audit conducted by ICAO relating to the three Annexes to the Chicago Convention—Annex 1– *Personnel Licensing*, Annex 6—*Operation of Aircraft* and Annex 8—*Airworthiness of Aircraft*, indicated that of the ICAO Contracting States that were audited between March 1999 and July 2004, considerable numbers of States had deficiencies in respect of a number of requirements under these Annexes. Furthermore, audit follow-up missions revealed that in many cases, significant deficiencies identified during the initial audits remain.

Several fatal accidents in August, October and December 2005 brought to bear the need for regulators and operators to re-focus on the safety of civil aviation. One of the reasons for ensuring the safety of air transport is the need to keep up with the pace at which liberalization of air transport is spreading. With liberalization of commercial air transport comes the opportunity for new carriers to compete for market access. It also gives existing carriers the opportunity to compete with each other as well as with the new entrants with a view to consolidating their positions in a liberalized market. There is an onerous burden on States to ensure that, hand in hand with unfettered competition among carriers, rigid enforcement of technical regulation ensuring the safety of flight is necessary. This calls for the need to put into operation coordinated global safety management systems. This challenge calls for renewed thinking by all concerned in the best interests of the entire aviation industry

Safety Management Systems (SMS) are processes which proactively manage the projected increase in aircraft incidents and accidents brought about by the increase in air traffic movements. SMS require vigilance in the liberalization of air transport

---

[224] *Id.* at 3.

and the correspondent increase in capacity. At the Directors General of Civil Aviation Conference on a Global Strategy for Aviation Safety, convened by ICAO in Montreal from 20 to 22 March 2006, Canada defined a Safety Management System as a businesslike approach to safety. An SMS is a systematic, explicit and comprehensive process for the management of safety risks that integrates operations and technical systems with financial and human resource management, for all activities related to an air operator as an approved maintenance organization's certificate holder[225]

Any management system, including that which involves aviation safety, would necessarily entail planning, goal setting, performance measurement and accountability. This process is systemic, in that there cannot be one factor without the other to complete the entire system of the safety management process. It also requires a symbiotic relationship between operator and regulator, who are jointly responsible for the determination of the parameters of regulations and their intent. While the regulator's role is to give a clear set of instructions to the operator reflecting the expectations of the regulator, the operator has to ensure their compliance as well as maintain close coordination with the regulator.

At the DGCA Conference, ICAO emphasized that, although the prescriptive aspect of safety management, complied with by adherence to Standards and Recommended Practices (SARPs) of the Annexes to the Chicago Convention were essential, safety management has to transcend the mere compliance with SARPs toward a performance based process.[226] In this context, ICAO suggests a three step management process: firstly, between the operator and the oversight authority where there must be compete agreement between operators and service providers on the one hand and the oversight authority on the other hand on the safety performance to be expected. Performance will be measured by both parties through established safety indicators and safety targets. Secondly, all parties concerned would agree on the safety requirements necessary to achieve the safety performance agreed upon in the first step. Finally, oversight authorities would verify achievement of the agreed safety performance or lack thereof and the operator/service provider would rectify any observed deviations.

At the DGCA Conference, ICAO also reiterated the fact that the term safety management conveys the notion that the management of safety is a business process that should be considered at the same level and along the same lines as any other business process. The safety management systems approach, as proposed, would not only be reactive to some triggering event such as an incident or reportable event, but it would also involve a proactive process of ongoing and routine collection and analysis of safety data during the course of a safety oriented work programme that an organization must pursue on a daily basis while in the business of its core functions. Such a system must essentially be performance driven and proactive.

[225] Management of Aviation Safety, Presented by Canada, DGCA/06-WP/15, 4/2/06 at 2

[226] Implementation of Safety Management Systems (SMS in States), Presented by the ICAO Secretariat, DGCA/06-WP/6, 9/01/06 at 1 and 2.

Furthermore, the term 'system' involves an integrated and systemic set of processes calculated to actively manage safety transcending departmental boundaries, thus ensuring safety from a broad perspective.

A systemic safety management system would also require a well thought out performance measurement process and indicators of performance which would be determined through organizational structure and accountability established by policies and procedures. The accountability in safety management should be enforced throughout an organization including its senior management.

It is incontrovertible that an effective safety management system has to necessarily co-exist with an efficient air transport product. If the former were to affect the latter's development and its capacity to offer services to the consumer, the fundamental purpose of safety management and its supportive role would be lost, as there would be no longer an air transport system that could adequately cater to the exponential demands made on it. The key purpose of the SMS is to adapt the growth in air transport, however evolving the operations of air transport are. A typical approach was exemplified by the United States at the ICAO DGCA Conference when they advised the delegates of four key areas of the SMS which work synergistically to maintain harmony between safety and efficiency, two of which—safety assurance and safety promotion—are critical to the efficiency and continuity of air transport. According to the United States submission the four key elements to effective SMS are: *Safety Policy*, which covers the SMS requirements, responsibilities and accountabilities for system functions; *Safety Risk Management*, which involve procedures used to identify hazards, assess current and future safety risk of the system or operation, manage the safety risk, document the mitigations selected to manage the safety risk and verify and monitor the mitigations throughout its lifecycle; *Safety Assurance*, which calls for the assurance of the safety product or service provider's system, including audits, evaluations and inspections, as well as data tracking and analysis; and *Safety Promotion*, which entails training, communication and dissemination of safety information to strengthen the safety culture and support integration of the SMS into operations[227]

At the 35th Session of the ICAO Assembly, contracting States adopted Resolution A35-7,[228] which urges all contracting States to share with other contracting States critical safety information which may have an impact on the safety of international air navigation and to facilitate access to all relevant safety information. This resolving clause is based on the premise, explicitly recognized in the Resolution, that the improvement of the safety of international civil aviation on a worldwide basis requires the active collaboration of all stakeholders. By adopting Resolution A35-7, the then 189 ICAO contracting States arrogated to themselves increased responsibility for the safety of international civil aviation, irrespective of the nature of the provider of services that ensure such safety. This strategic approach of global safety oversight was further endorsed at the ICAO DGCA

[227] Safety Management System Concept, DGCA/06-WP/12, 4/2/06 at 2

[228] *Assembly Resolutions in Force* (as of 8 October 2004), ICAO Doc 9848 at 1–60.

Conference of March 2006, where all States consensually agreed that there should be an effective implementation of Safety Management Systems (SMS) in States, established according to a performance based approach in three steps: Firstly, where oversight authorities and operators and servicc providers agree on the safety performance to be expected from them while conducting their core business functions; secondly, that oversight authorities and service providers agree on the safety requirements necessary to achieve the safety performance agreed upon in the first step; and thirdly and finally, that oversight authorities verify achievement of the agreed safety performance or its lack thereof, and operators and service providers correct observed deviations.[229]

All the above goes to demonstrate the heavy responsibility placed by States on themselves in undertaking that they and they alone will be accountable for the safety of aviation within their territories as well as cooperating on a global basis to implement a seamless air navigation system. Of course, any change in the provision of air traffic services, particularly with regard to the commercial nature of the service provider, will remain subservient to the fundamental concept of State sovereignty, which was endorsed by the Eleventh ICAO Air Navigation Conference held in Montreal from 22 September to 3 October 2003.[230]

4.3.3 Annex 19

Annex 19 to the Chicago Convention, which is entitled *Safety Management* became applicable on 14 November 2013 and is the first new ICAO Annex in over 30 years. The Annex complements existing ICAO safety-related programmes and activities, including the recently-revised Global Aviation Safety Plan and ICAO's safety management training offerings. Particular emphasis will be placed on promoting the safety management approach through the network of Regional Aviation Safety Groups around the world.

The adoption of the Annex highlights the importance of safety management at the State level; enhances safety by consolidating safety management provisions applicable to multiple aviation domains; promotes aligned implementation of SMS and State Safety Programmes (SSP) provisions; creates a process to collect and analyse feedback regarding the implementation of SMS and SSP; and facilitates the future evolution of safety management provisions.

Annex 19 requires that each State establishes an SSP for the management of safety in the State, in order to achieve an acceptable level of safety performance in

[229] *Implementation of Safety Management Systems (SMS) in States*, Working paper submitted by the ICAO Secretariat, DGCA/06-WP/6, 9/01/06 at 1–2.

[230] See Eleventh Air Navigation Conference report, *ICAO Doc 9828, AN-Conf/*11, Montreal: 20 at para. 1.2.1.2 at p. 1–1, where it is recorded that the Conference agreed that the issue of sovereignty was paramount in the operation of global air traffic management, as interpreted through the Global ATM operational concept of sovereignty.

civil aviation. The SSP is required to have four components: State safety policy and objectives; State safety risk management; State safety assurance; and State safety promotion.[231] The Annex prescribes that under an SSP, each State should require *inter alia* operators of certified aerodromes in accordance with Annex 14 to implement SMS.[232]

All these components are relevant to the withdrawal by the Indonesian authorities of authorization for Flight QZ 8501 to operate on Sundays. Such withdrawal was an integral part of State safety policy and objectives as well as safety assurance and safety promotion.

The key to Annex 19 is the implementation of SMS. The key words to management of safety are coordination, performance and accountability. Of these, performance is the most critical. While coordination ensures global interoperability of safety measures and performance measurement, accountability ensures the implementation of the safety management system worldwide. Performance measurement is inextricably linked to the most critical factors—the performance indicators. In SMS, performance should be measured within the basic premise that only what is important for safety management should be measured. One must not measure too much; and factors that impact customer satisfaction should also be measured.

A critical first step towards implementing performance management in any organization is to define a set of conceptual, high-level, key performance indicators (KPIs). In view of the level at which they are identified, KPIs do not identify targets to be achieved, deadlines to be met or any other details regarding execution of tasks. They would flow from an organization's strategic objective of ensuring safety and are described at three levels:

- *Outcomes*: through which the decision making bodies of the organization can objectively validate the strategies of the Organization, in other words: "Is the Organization doing the right things and do the overall effects of performance bring on expected results?";
- *Outputs*: through which the senior management of the organization can objectively assess the effectiveness of the Organization, in other words: "How well is the Organization doing what it set out to do?"; and
- *Measurement:* through which those accountable for key activities can independently assess the proper alignment of their resources and timely achievement of outputs.

Once the KPIs have been approved, the organization needs to develop and implement the other key components of its performance management system to enable the indicators to drive better utilization of the available resources. These components include, but are not limited to:

---

[231] Annex 19 to the Chicago Convention – *Safety Management*, First Edition: 2013, Standard 3.1.1.

[232] *Id.* Standard 3.1.3.

- The infusion of the performance management philosophy into the management culture and practices of the organization;
- The determination of realistic and measurable targets for each of the indicators;
- The alignment of data collection and interpretation mechanisms with the defined indicators and targets; and
- The traceability of operational plans and milestones to the targets.

Once implemented, the performance measurement system will be subject in its entirety to regular monitoring and use by linking it to planned budgetary allocations. The outcomes will also serve as a barometer to redefine measures, targets, outputs and resources if necessary. In order to derive full benefits from the performance measurement system, SMS may have to be realigned periodically with the changing external environment and newly identified critical objectives. This in turn will involve some realignment on the designed performance measurement system and the operational plans that are in place.

The above-mentioned approach will result in a management system where the allocation of resources and measurement of performance will bring about the timely achievement of outcomes that are most valuable to stakeholders.

The key performance indicators in a SMS should be also developed on the basis that the most critically important factor is employee involvement. The first step toward employee involvement is identification of the process flow. The fundamental issue is that if employees cannot agree on their process(es), they could not effectively measure them or utilize the output of what they have measured.

Another important consideration is to identify the critical activity to be measured. The critical activity is that culminating conduct where it makes the most sense to locate a sensor and define an individual performance measure within a process. This should immediately be followed by the recognition that the next step would be to establish performance goal(s) or standards. All performance measures should be tied to a predefined goal or standard, even if the goal is at first somewhat subjective. Having goals and standards is the only way to meaningfully interpret the results of measurements and gauge the success of management systems.

Consequent to establishing goals and standards, it is necessary to establish performance measurement(s). This helps build the performance measurement system by identifying individual measures.

Once measures are identified, the next step would be to identify responsible parties. A specific entity (as in a team or an individual) needs to be assigned the responsibilities for each of the steps in the performance measurement process. Thereafter, collection of data becomes the next consideration. In addition to writing down the numbers, the data need to be pre-analyzed in a timely fashion to observe any early trends and confirm the adequacy of your data collection system.

A corollary to the identification of goals, measurements and the collection of data is analysis of reports of actual performance. The raw data are formally converted into performance measures, displayed in an understandable form, and disseminated in the form of a report. Comparison actual performance to goal(s) is a key step, where the Organization will compare performance, as presented in the report, to predetermined goals or standards and determine the variation (if any).

Finally, to complete the development of key performance indicators, consideration of new goals is important. Even in highly successful safety management systems, changes may need to be revised in order to establish ones that challenge an organization's resources but do not overtax them. Goals and standards need periodic evaluation to keep up with the latest organizational processes. Management of the SMS and in particular the performance measurement side of things is a key factor, bringing to bear the importance of the question as to who will manage the PM system. The person responsible for measurement of performance would have to prepare performance reports and decide on the periodicity of measurement.

Good performance in SMS requires the maintenance of public confidence in the safety of international civil aviation. This is primarily achieved by the decrease in incidents and accidents involving passenger fatalities and fatalities on the ground. The measurements would evaluate the capability of the organization concerned in collection and analysis of safety information, leading to the output of implementation of a global safety data collection and analysis capability. This would in turn lead to the outcome that ensures global resources are aligned to address the most critical safety risks.

Economic liberalization as well as the evolution of business and operating practices may have serious implications for safety regulation, which need to be addressed properly. The most fundamental consideration is that safety of air transport must remain the ultimate responsibility of the State concerned and should not be compromised by economic considerations. While ICAO should continue its monitoring of liberalization within the parameter of safety, all players, including States, service providers and airlines should coordinate well in understanding their responsibilities and providing safety oversight. In this constantly evolving environment, due regard should be paid to the impact of a growing air transport industry's impact on safety so that coherent SMS are developed.

The operative criterion should be that safety of flight is a valid concern of the aviation industry whether or not air transport is liberalized and, as already discussed, liberalization and consequent competition between air carriers should alert States to the potential safety hazards that a highly competitive air transport industry brings to bear. Considering the elaborate work already accomplished by ICAO, in establishing guidance to States on the implementation of Article 83 *bis* to the Chicago Convention. in addition to Articles 12, 30, 31 and 32 a)[233] of the

---

[233] Article 12 stipulates that each contracting State is required to insure that every aircraft flying over or manoeuvring within its territory and that every aircraft carrying its nationality mark, wherever that aircraft might be, shall comply with the rules and regulations relating to flight and manoeuvre applicable there. It also provides that each Contracting States should keep its regulations consistent with the principles of the Chicago Convention. Over the high seas the rules will be as established under the Convention and States undertake, under Article 12, to prosecute any offender who does not adhere to these principles. Articles 30 and 31 refer to the need for aircraft to carry radio equipment as required and be issued with certificates of airworthiness by the State of Registry, respectively. Article 32 a) requires the pilot and other members of the crew to be provided with certificates of competency and licenses issued by the State of Registry.

Convention and the up-to-date Standards and Recommended Practices adopted from time to time as necessary, particularly in Annexes 1, 6 and 8, there is no room for doubt that the regulatory provisions are in place.

Therefore what remains is for all players concerned to move toward a performance based SMS process where accountability will drive aviation safety to its desired goals.

It is worthwhile noting that at the 38th Session of the ICAO Assembly adopted Resolution A38-2 (ICAO Global planning for safety of air navigation) which in Appendix A (Global Aviation Safety Plan) states that it is prudent to apply the safety management principles in the GASP to enhance safety by focusing action where it is most needed.

The liability of Air Asia for the death of the passengers on board Flight QZ 8501 will be dealt by the private air carrier liability regime that applies. Indonesia ratified the Warsaw Convention of 1929[234] while Singapore ratified the Montreal Convention of 1999[235] (Singapore also ratified the Warsaw Convention) which replaced the Warsaw Convention. Courts have conventionally applied the lower common denominator and it is arguable that the court hearing this case may opt to apply the Warsaw Convention principles of liability of the carrier, where if the carrier is guilty of wilful misconduct the limitation of liability of the carrier is lifted.

Of greater importance, however, is the issue as to who is accountable for letting the flight take off despite the withdrawal of authority for that particular flight on Sundays. In the context of safety management, the principles of which have been discussed above, it boils down to the question as to who had the accountability and responsibility for supervision and oversight of safety of that flight. Safety oversight is largely a matter for the State concerned. Therefore it is a matter of State responsibility. Article 28 of the Convention obligates Contracting States to provide, as far as practicable in their territories, airports, radio services, meteorological services and other air navigation facilities to facilitate international air navigation according to Standards established pursuant to the Convention.

The ICAO Safety Oversight Manual[236] states that accountability and responsibility is the state of being responsible for an undertaking, person, thing or action and for which an organization or individual or both are liable to be called to account. An ICAO Contracting State and its respective civil aviation authority are ultimately responsible for the implementation of ICAO SARPs within their State. A State may either perform these obligations or, through mutual agreement, have another organization perform and be accountable for these functions; however, the State retains the responsibility under its duties of sovereignty. The ICAO Assembly at its 29th Session in 1992, adopted Resolution A29-13 which reaffirmed that each individual

[234] Convention for the Unification Convention for the Unification of Certain Rules Relating to International Carriage by Air, signed at Warsaw on 12 October 1929.

[235] Convention for the Unification of Certain rules for International Carriage by Air, done at Montreal on 28 May 1999.

[236] Doc 9734, AN/959.

State's responsibility for safety oversight is one of the tenets of the Chicago Convention.

The ICAO Assembly, at its 37th session in 2010 adopted Resolution A37-5 (The Universal Safety Oversight Audit Programme USOAP continuous monitoring approach) which recognized that safety oversight and the safety of international civil aviation in general, is the responsibility of the Contracting States, both collectively and individually.

According to the Chicago Convention, safety oversight and a State's responsibility pertaining thereto can be subsumed into two broad areas: nationality and registration of the aircraft (The airbus a 320 operating flight QZ 8501 was registered in Indonesia); responsibilities that flow therefrom; and territorial jurisdiction of that State. Accordingly, there are certain mandatory responsibilities of a State in ensuring safety oversight. Article 17 of the Convention provides that aircraft have the nationality of the State in which they are registered. Article 18 prohibits the registration of aircraft in more than one State, which effectively precludes dual nationality of aircraft and responsibility of more than one State.

The registration of an aircraft in a particular State devolves upon that State certain safety related obligations. For example Article 12 of the Chicago Convention states:

> Each contracting State undertakes to adopt measures to insure that every aircraft flying over or manoeuvring within its territory and that every aircraft carrying its nationality mark, wherever such aircraft may be, shall comply with the rules and regulations relating to the flight and manoeuvre of aircraft there in force. Each contracting State undertakes to keep its own regulations in these respects uniform, to the greatest possible extent, with those established from time to time under this Convention. Over the high seas, the rules in force shall be those established under this Convention. Each contracting State undertakes to insure the prosecution of all persons violating the regulations applicable".

Article 37 of the Chicago Convention imposes an obligation on each contracting State to undertake to collaborate in securing the highest practicable degree of uniformity in regulations, standards, procedures, and organization in relation to aircraft, personnel, airways and auxiliary services in all matters in which such uniformity will facilitate and improve air navigation. These Standards and Recommended Practices are contained in the 19 Annexes to the Chicago Convention, all of which except Annexes 9 (Facilitation) and 17 (Security) are applicable to safety oversight either directly or indirectly.

ICAO Assembly Resolution A37-1 (Principles for a code of conduct on the sharing and use of safety information) provides that ensuring the safety of international civil aviation is the responsibility of Member States both collectively and individually. An earlier Resolution (still in force) A29-13 (Improvement of safety oversight) reaffirmed that one of the tenets of the Chicago Convention is that it is an individual State's responsibility for safety oversight. Nine years later, when ICAO established its Aviation Safety Oversight Audit Programme through Resolution A32-11 (Establishment of an ICAO Safety Oversight audit Programme), the Assembly recognized that the objectives of the ICAO safety oversight programme were to seek to ensure that Contracting States are adequately discharging their responsibility for safety oversight over aircraft operations, the licensing and training of personnel, and aircraft airworthiness. The same resolution recalled that ultimate responsibility for safety oversight rests with Contracting States, which are required to continuously review their respective safety oversight capabilities.

At the latest Assembly held in 2013 (38th session) this overarching principle of State responsibility was reiterated in Resolution A38-5 (Regional cooperation and assistance to resolve safety deficiencies, establishing priorities and setting measurable targets)—that ensuring the safety of international civil aviation is the responsibility of Member States both collectively and individually. However, although States have overall responsibility for safety in aviation, they are not alone. ICAO and other key players are required to collaborate when States implement ICAO's Global Aviation Safety Plan. The 37th Session of the assembly, held in 2010, adopted Resolution A37-5 (The Universal Safety Oversight Audit Programme (USOAP) continuous monitoring approach) which, in a preambular clause recognizes that safety oversight, and the safety of international civil aviation in general, is the responsibility of Contracting States, both collectively and individually, but it also depends on the active collaboration of ICAO, Contracting States, industry and all other stakeholders in the implementation of the Global Aviation Safety Plan. The bottom line in responsibility is explicitly stated in Resolution A32-19 (Charter on the Rights and Obligations of States Relating to GNSS Services) which recognizes that every State preserves its authority and responsibility to control operations of aircraft and to enforce safety and other regulations within its sovereign airspace.

Although ICAO Assembly Resolutions are the outcome of political compromises,[237] the fact that Indonesia has not filed a reservation to any of these resolutions would make for persuasive authority that devolve ultimate responsibility on Indonesia for letting Flight QZ 8501 take off in the face of its withdrawal of authority to operate the flight.

References

Abeyratne RIR (1986) The human stress factor and mental injury in American tort law – a patchwork quilt? Anglo Am Law Rev 15(4):338–360

Abeyratne RIR (2000) Mental distress in aviation claims – emergent trends. J Air Law Commerce 65(2):225–261

Abeyratne R (2008) The fear of flying and air rage – some legal issues. J Transport Secur 1:45–66

Aust A (2000) Modern treaty law and practice. Cambridge University Press, Cambridge

[237] *Brownlie* has expressed the view that decisions by international conferences and organizations can in principle only bind those States accepting them. *Shaw*, referring to the binding force of United Nations General; Assembly Resolutions states "*...one must be alive to the dangers in ascribing legal value to everything that emanates from the Assembly. Resolutions are often the results of political compromises and arrangements and, comprehended in that sense, never intended to constitute binding norms. Great care must be taken in moving from a plethora of practice to the identification of legal norms*". See Brownlie (1990), p. 691; Shaw (2003), p. 110.

Brownlie I (1990) Principles of public international law, 4th edn. Clarendon, Oxford

Cheng B (1977a) Wilful misconduct, from Warsaw to the Hague and from Brussels to Paris. Ann Air Space Law 11

Cheng B (1977b) ...From Warsaw to the Hague... Ann Air Space Law 11:55

Cheng B (1989) What is wrong with the Montreal additional protocol no. 3? Air Law XIV(6):220

Corrigan JM (1978) The right of the air carrier to refuse carriage. Ann Air Space Law III (McGill)

Cunningham M (2008) The Montreal Convention: can passengers finally recover for mental injuries? Phoenix School of Law, From Selected Works of McKay Cunningham, January 2008. http://works.bepress.com/mckay_cunningham/1

Desbiens C (1992) Air carrier's liability for emotional distress under Article 17 of the Warsaw convention: can it still be invoked? Ann Air Space Law XVII(Part II):153

Drion H (1954) Limitation of liabilities in international air law. Martinus Nijhoff, The Hague

Guldimann W (1991) A future system of liability in air carriage. Ann Air Space Law XVI:93

Horner RC, Legrez D (1975) Minutes, Second International Conference on Private International Law, 4–12 October 1929, Warsaw (New Jersey: Fred B. Rothman, 260, 1975)

ICD-10 (1992) International statistical classification of diseases and related health problems, 10th revision, vol 1. World Health Organization, Geneva

Kilbride DA (1989) Six decades of insuring liability under Warsaw. Air Law XIV(4/5)

Langley J, Brenner R (2001) What is an injury? – a clear definition is needed, injury prevention. http://www.ncbi.nlm.nih.gov/pmc/articles/PMC1730061/

Larsen PB, Sweeney JC, Gillick JE (2012) Aviation law: cases laws and related sources. Martinus Nijhoff, The Hague

Magruder C (1936) Mental and emotional disturbance in the law of torts. Harv Law Rev XLIX(7)

Mankiewicz RH (1989) From Warsaw to Montreal with certain intermediate stops... Air Law XIV

Martin P (1979) 50 years of the Warsaw Convention, a practical man's guide. Ann Air Space Law IV:233

Martin P (1983) Intentional or reckless misconduct, from London to Bangkok and back again. Ann Air Space Law VIII:145–149

Matte NM (1983) The Warsaw system and the hesitation of the United States Senate. Annals of Air and Space Law VIII:151

Milde M (1989) ICAO work on the modernization of the Warsaw system. Air Law XIV(4/5):193

Miller H (1966) Mental after effects of head injury, proceedings of the royal society of medicine. Sect Psychiatry 59

Miller G (1977) Liability in international air transport. Kluwer, Deventer

Prosser WL (1939) Intentional infliction of mental suffering: a new tort. Mich Law Rev 37

Rassool SB (2007) Psychological trauma and road traffic accidents. Doctoral thesis on clinical psychology University of Hertfordshire, Chapter Four, Literary Review: 9 November 2007, 113. www.kellyconnor.com/research/f/phd_thesis.pdf

Sandel MJ (2009) Justice – what's the right thing to do? Farrar, Straus and Giroux, New York

Shaw MN (2003) International law, 5th edn. Cambridge University Press, Cambridge

Tobolewski A (1978) Against limitation of liability, a radical proposal. Ann Air Space Law III:261

Wise SM (1998) Hardly a revolution – the eligibility of non human animals for dignity-rights in a liberal democracy. Vt Law Rev 21

Chapter 5
Conclusion

The fundamental human rights of a passenger who is carried by air are primarily embodied in Article 44 d) of the Chicago Convention which devolves responsibility on ICAO to ensure that the needs of the people of the world are met that obtain for them safe, regular, efficient and economical air transport. Thus the needs of the passenger are entrenched both in safe and economical air transport that is provided regularly and efficiently. These four elements are a good spread of the composite needs of the passenger with attendant subsets that allow him/her to travel without discrimination and in basic comfort, whatever the disability or illness the passenger might suffer from.

In order to ensure the above needs, the primary tool that the airline would use is prudent management. The States which are responsible for the safe and economic delivery of the air transport product through the airlines have to ensure proper oversight and risk management that assure the passenger that when he/she steps into the aircraft, adequate safety and security measures have been taken.

This book was written at a juncture where three outrageous and horrendous aircraft accidents took place the previous year, in addition to other smaller but equally disastrous and significant accidents with grave loss of life. In 2014 Flights MH 370, MH 17 and QZ 8501 all showed that there was some degree of dereliction of duty on the part of the key aviation players involved in that flight. The result was loss of life, guaranteed to every human being by the Universal Declaration of Human rights, and untold grief to the relatives of those killed often resulting in outrage.

Fourteen years have passed since the catastrophic events of 11 September 2001. Before then, we were apathetic to risks. After 9/11 we had no alternative but to change our thinking. None of us can get over the outrage we felt at the dastardly terrorist acts which killed nearly 3,000 people that day. We felt an amalgam of concern, fear, and anger. We became motivated to take precautions. Overnight, we converted ourselves from being laid back and reactive to being pre-emptive and preventive so that we could make sure such things never happened to us again. Our pre-emption was to guard ourselves against any ac, which is imminent and

R. Abeyratne, *Aviation and International Cooperation*,
DOI 10.1007/978-3-319-17022-0_5

potentially harmful to us. Our prevention was to shield us from an act, which is potentially harmful to us and is inevitable. We resorted to military action. We had a new limb of law we now call the law of outrage.

Added to this is what the future portends for aviation and air transport. There are going to be crowded skies and the annual carriage of passengers globally, which, at the time of writing was 3.3 billion, would double in just a few years. There are going to be thousands more aircraft in the skies. *Airbus Industrie*, in its forecast from 2011 to 2030 envisions the need for more than 26,900 passenger airliners with seating capacities of 100 seats and above, along with over 900 new factory-built freighter aircraft. In the same timeframe, the world's overall passenger aircraft inventory will more than double from today's 15,000 to more than 31,500 by 2030.[1] Boeing in its forecast for 2011–2031 states that the long-range forecast for 2011 anticipates delivery of 33,500 new airplanes over the next 20 years, valued at more than $4.0 trillion. Retrospectively, Boeing's long term forecasts have tended to be conservative compared to actual industry performance, but well thought through.

The figures of both these "predictions" would cater mostly to the megacities of the world (26 in number, according to the IATA Report), which have a population of more than 10 million and which account for more than 20 % of global air travel. 62 cities, with more than 5 million population will account for another 40 %.

With regard to European traffic forecasts, EUROCONTROL[2] forecasts that there will be 16.9 million IFR movements in Europe in 2030, which is approximately 2 % more than the movements in 2009. The number of flights concerned is estimated between 13.1 and 20.9 million flights during 2030.

Although at first glance this rosy picture seemingly suggests that air transport is progressing well, but there is something wrong in this scenario. And that is that we are perpetuating the "predict and provide" principle that we have followed over the past 70 years of the formal existence of commercial air transport if we do not decide what to do about this growing and increasing trend. What is wrong is that, while the "predict" part of the equation is fine, the "provide" part of it, if blindly catered to without looking at the systemic consequences, would lead us to all sorts of problems, particularly in the safety, security and environmental fields.

The 1983 principle of predict and provide for transport planning, advocated by the Economist Adams, was a disaster for the simple reason that it took as inevitable perpetual transport growth at face value. Europe is one area which is planning for predicted increases in air transport up to 2050. The European Commission's *Flightpath* (Vision for 2050) states:

[1] http://www.airbus.com/company/market/forecast/.

[2] EUROCONTROL is the European Organisation for the Safety of Air Navigation, is an intergovernmental organisation made up of 39 Member States and the European Community. Together with its partners, EUROCONTROL is committed to building a Single European Sky that will deliver the ATM performance required for the twenty-first century and beyond. Founded in 1960, it is a civil-military organisation that has developed into a vital European repository of air traffic management (ATM) excellence, both leading and supporting ATM improvements across Europe.

> The European aviation market will be larger than in 2011. Within Europe the number of commercial flights 8 is up to 25 million in 2050 compared to 9.4 million in 2011. An air traffic management system is in place that provides a range of services to handle at least 25 million flights a year of all types of vehicles (fixed-wing, rotorcraft), and systems (manned, unmanned, autonomous) that are integrated into and interoperable with the overall air transport system with 24-hour efficient operation of airports...A coherent ground infrastructure is developed including: airports, vertiports and heliports with the relevant servicing and connecting facilities, also to other modes.[3]

The preferred policy is "predict and decide" and this should be approached on two fronts: economic and environmental.

At the centre of all this is the consumer of the air transport product.

In view of the above, a humanitarian approach to this gigantic commercial juggernaut is advocated that would focus on the interests of the consumer as the centrifugal point of the equation.

ICAO has not helped at all on the subject of consumer protection except for an ambivalent and meaningless nod to States at its 5th Air Transport Conference 12 years ago:

> As a premise in addressing consumer interests issues, States need to carefully examine what elements of consumer interests in service quality have adequately been dealt with by the current commercial practices of airlines (and service providers, if applicable) and what elements need to be handled by the regulatory and/or voluntary commitment approaches. States need to strike the right balance between voluntary commitments and regulatory measures, whenever the government intervention is considered necessary to improve service quality... [4]

The United Nations on the other hand has adopted guidelines on consumer protection which apply in general to the consumer. By Resolution A/RES/39/248 of 16 April 1985 (Consumer Protection) to which is Annexed the guidelines the United Nations recognized that: in taking into account the interests and needs of consumers in all countries, particularly those in developing countries; recognizing that consumers often face imbalances in economic terms, educational levels, and bargaining power; and bearing in mind that consumers should have the right of access to non-hazardous products, as well as the right to promote just, equitable and sustainable economic and social development, these guidelines for consumer protection have the following objectives: to assist countries in achieving or maintaining adequate protection for their population as consumers; to facilitate production and distribution patterns responsive to the needs and desires of consumers; to encourage high levels of ethical conduct for those engaged in the production and distribution of goods and services to consumers; to assist countries in curbing abusive business practices by all enterprises at the national and

---

[3] *Flightpath 2050 – Europe's Vision for Aviation* – Maintaining Global Leadership & Serving Society's Needs, Report on the High Level Group on Aviation Research, 2011 http://ec.europa.eu/transport/air/doc/flightpath2050.pdf.

[4] Report of the Worldwide Air Transport Conference, Challenges and Opportunities of Liberalization (Montreal, 24–28 March 2003), *ICAO Doc 9819; ATConf/5 2003* at 42.

international levels which adversely affect consumers; to facilitate the development of independent consumer groups; to further international co-operation in the field of consumer protection; to encourage the development of market conditions which provide consumers with greater choice at lower prices.

The Resolution also states that Governments should develop, strengthen or maintain a strong consumer protection policy, taking into account the guidelines set out below. In so doing, each Government must set its own priorities for the protection of consumers in accordance with the economic and social circumstances of the country, and the needs of its population, and bearing in mind the costs and benefits of proposed measures.

The Resolution furthermore provides that the legitimate needs which the guidelines are intended to meet are the following: the protection of consumers from hazards to their health and safety; the promotion and protection of the economic interests of consumers; Access of consumers to adequate information to enable them to make informed choices according to individual wishes and needs; consumer education; availability of effective consumer redress; freedom to form consumer and other relevant groups or organizations and the opportunity of such organizations to present their views in decision-making processes affecting them.

In the Guidelines is the fundamental principle that Governments should adopt or encourage the adoption of appropriate measures, including legal systems, safety regulations, national or international standards, voluntary standards and the maintenance of safety records to ensure that products are safe for either intended or normally foreseeable use. Also in the Guidelines is a statement to the effect that Government policies should seek to enable consumers to obtain optimum benefit from their economic resources. They should also seek to achieve the goals of satisfactory production and performance standards, adequate distribution methods, fair business practices, informative marketing and effective protection against practices which could adversely affect the economic interests of consumers and the exercise of choice in the market-place.

The Guidelines go on to say that Governments should intensify their efforts to prevent practices which are damaging to the economic interests of consumers through ensuring that manufacturers, distributors and others involved in the provision of goods and services adhere to established laws and mandatory standards. Consumer organizations should be encouraged to monitor adverse practices, such as the adulteration of foods, false or misleading claims in marketing and service frauds. Governments are also called upon to develop, strengthen or maintain, as the case may be, measures relating to the control of restrictive and other abusive business practices which may be harmful to consumers, including means for the enforcement of such measures. In this connection, Governments should be guided by their commitment to the Set of Multilaterally Agreed Equitable Principles and Rules for the Control of Restrictive Business Practices adopted by the General Assembly in resolution 35/63 of 5 December 1980.

In addition, Governments are called upon to adopt or maintain policies that make clear the responsibility of the producer to ensure that goods meet reasonable demands of durability, utility and reliability, and are suited to the purpose for which they are intended, and that the seller should see that these requirements are met. Similar policies should apply to the provision of services. Governments should also encourage fair and effective competition in order to provide consumers with the greatest range of choice among products and services at the lowest cost.

These principles should be entrenched and enforced in a regime dedicated to the consumer of air the transport product.

Finally, the question arises as to whether consumer rights can be considered as human rights. One commentator is of the view that:

> [A]ccording to some substantive tests of human rights, consumer rights can be recognized as human rights. These tests, although not conclusive, supply sufficient guidelines in delineating the main features of human rights... One of the requirements for acknowledging rights as human rights is the integration of the rights into formal international documents.[5]

It is submitted that consumer rights in aviation, although some may consider to be soft human rights in the context of the *Universal Declaration of Human Rights*, could mature into fully fledged human rights in terms of the passenger's health, safety and right to life in an aircraft that does not afford him or her the basic conveniences of life, such as assistance in the instance of disability; age or obesity and security against attacks on the passenger's life.

A compelling factor in considering consumer rights as fundamental human rights is the fact that some countries have incorporated consumer rights as human rights in their legislation. For example, the incorporation of consumer rights as constitutional provisions in Spain, Portugal, Brazil, and Switzerland well reflect the recognition that consumer rights are a pre requisite to the implementation of social justice. These facts of life more than justify the need to wake up moribund regulators to the reality that aviation, human rights and consumer protection are well meshed through public policy.

Reference

Deutch S (1994) Are consumer rights human rights? Osgood Hall Law J 32(3)

[5] Deutch (1994), Article (4) at 54 and 57.

Index

R. Abeyratne, *Aviation and International Cooperation*,
DOI 10.1007/978-3-319-17022-0

MIX
Papier aus verantwortungsvollen Quellen
Paper from responsible sources
FSC® C105338

If you have any concerns about our products,
you can contact us on
ProductSafety@springernature.com

In case Publisher is established outside the EU,
the EU authorized representative is:
Springer Nature Customer Service Center GmbH
Europaplatz 3, 69115 Heidelberg, Germany

Printed by Libri Plureos GmbH
in Hamburg, Germany